Mathematics Education in the Digital Era

Volume 21

The Mathematics Education in the Digital Era (MEDE) series explores ways in which digital technologies support mathematics teaching and the learning of Net Gen'ers, paying attention also to educational debates. Each volume will address one specific issue in mathematics education (e.g., visual mathematics and cyber-learning; inclusive and community based e-learning; teaching in the digital era), in an attempt to explore fundamental assumptions about teaching and learning mathematics in the presence of digital technologies. This series aims to attract diverse readers including: researchers in mathematics education, mathematicians, cognitive scientists and computer scientists, graduate students in education, policy-makers, educational software developers, administrators and teachers-practitioners. Among other things, the high quality scientific work published in this series will address questions related to the suitability of pedagogies and digital technologies for new generations of mathematics students. The series will also provide readers with deeper insight into how innovative teaching and assessment practices emerge, make their way into the classroom, and shape the learning of young students who have grown up with technology. The series will also look at how to bridge theory and practice to enhance the different learning styles of today's students and turn their motivation and natural interest in technology into an additional support for meaningful mathematics learning. The series provides the opportunity for the dissemination of findings that address the effects of digital technologies on learning outcomes and their integration into effective teaching practices; the potential of mathematics educational software for the transformation of instruction and curricula; and the power of the e-learning of mathematics, as inclusive and community-based, yet personalized and hands-on.

Submit your proposal:

Please contact the publishing editor at Springer: natalie.rieborn@springer.com

Gülseren Karagöz Akar · İsmail Özgür Zembat ·
Selahattin Arslan · Patrick W. Thompson
Editors

Quantitative Reasoning in Mathematics and Science Education

Springer

Editors
Gülseren Karagöz Akar
Boğaziçi University
Istanbul, Turkey

Selahattin Arslan
Trabzon University
Trabzon, Turkey

İsmail Özgür Zembat
University of Glasgow
Glasgow, UK

Patrick W. Thompson
Arizona State University
Tempe, AZ, USA

ISSN 2211-8136 ISSN 2211-8144 (electronic)
Mathematics Education in the Digital Era
ISBN 978-3-031-14555-1 ISBN 978-3-031-14553-7 (eBook)
https://doi.org/10.1007/978-3-031-14553-7

This Springer imprint is published by the registered company Springer Nature Switzerland AG
The registered company address is: Gewerbestrasse 11, 6330 Cham, Switzerland

Introduction

Quantitative Reasoning in Mathematics and Science Education in the Digital Era

The digital era is a period beginning in the mid-20th century and leading its way into the 21st century. Technology characterizes this era as it provides access to widespread information in various electronic forms; therefore, it "increases the speed and breadth of knowledge turnover within the economy and society" (Shepherd, 2004, p. 1). The digital era demands specific 21st-century skills and abilities such as critical thinking, creativity, collaboration, communication, and flexibility. These skills are central to STEM disciplines (Beswick & Fraser, 2019), with which the teachers and learners need to be equipped. Even though technology is a crucial driver for such a "skills agenda, simply assisting students to develop up-to-date technology skills is not sufficient" (Beswick & Fraser, 2019, p. 958) to promote such an agenda. This is where we believe quantitative reasoning comes to the fore as it lays the foundation for developing these skills within STEM subjects. This book focuses on quantitative reasoning as an orienting framework to analyze learning, teaching, and curriculum. Different chapters of the book delve into quantitative reasoning related to the learning and teaching diverse mathematics and science concepts, conceptual analysis of mathematical and scientific ideas, and analysis of school mathematics (K-16) curricula in different contexts.

Quantitative reasoning is "an individual's analysis of a situation into a quantitative structure" (Thompson, 1990, p. 13) such that it entails "the mental actions of an individual conceiving a situation, constructing quantities of his or her conceived situation, and both developing and reasoning about relationships between these constructed quantities" (Moore et al., 2009, p. 3). Thompson and Carlson (2017) point out that envisioning a situation in terms of a quantitative structure is advantageous for students' positioning "to propagate information about how to calculate values of quantities in the structure in terms of arithmetic or algebraic expressions that are implied by the structure" (p. 440). Particularly, quantitative reasoning provides "content and meaning for numerical and symbolic expression and computation"

(Smith III & Thompson, 2008, p. 41). Envisioning a situation in terms of quantities and relationships among quantities is important to establish a foundation for reasoning about covariation, which plays a crucial role in learners' development of more complex mathematical and scientific ideas in critical ways (Thompson & Carlson, 2017). These suggest that quantitative reasoning is a key in education, and the proposed book unveils its particulars. In this regard, Johnson's chapter focuses on the "relationships" as an intellectual need and uses mathematizing to describe a category of a way of thinking emerging from that need. These relationships are essential in both mathematics and science. Gonzales' chapter uses quantitative reasoning to develop an understanding of the energy budget as a system of interrelated quantities and utilizes covariational reasoning to investigate climate change with a critical lens. In addition, Brahmia and Olshon's chapter discusses physics quantitative literacy as the blending of conceptual and procedural mathematics to generate and apply models relating physics quantities to each other.

The relevant literature suggests that quantitative reasoning supports the learning of arithmetic and algebra and plays a vital role in learning concepts foundational to calculus, geometry, trigonometry, physics, and so on. The literature studies provided detailed accounts of how quantitative reasoning can play an essential role in learning and teaching different mathematical and scientific concepts. In this book, Moore et al. chapter provides an analysis of concept construction from a quantitative reasoning perspective. In addition, Paoletti et al. chapter further describes a task sequence to construct covariational relationships among quantities and distinguish nonlinear and linear relationships. Moreover, based on a 15-year research program, Carlson et al. chapter explores how to support instructors in making their precalculus teaching more engaging, meaningful, and coherent using quantitative relationships symbolically and graphically.

Quantitative reasoning also provides a propitious arena for the conceptual analysis of mathematical and scientific ideas. Thompson (2008) defined conceptual analysis of mathematical ideas as a method "to describe ways of understanding ideas that have the potential of becoming goals of instruction or of being guides for curricular development" (p. 58). Conceptual analyses are "extremely powerful" because they offer concrete examples of learning trajectories (Thompson, 2008). The book gives examples of such analyses from different areas of mathematics. For instance, Akar, Zembat, Arslan, and Belin's chapter provides such analysis of isometries and their conceptualization. Nunes and Bryant's chapter considers numbers and number systems as models of quantitative relations and investigates how action schemas used in different situations support students' understanding of quantities and numbers. Ellis et al. chapter provides examples from linear and quadratic functions by identifying a sequence of conceptual activities and examples of associated student reasoning and task design principles to guide curricular decisions.

The use of quantitative reasoning in the development of ideas in curricula has also been given prominence since 2010 in Common Core State Standards for Mathematics (CCSSM) (Johnson, 2016). However, Johnson argued that despite greater inclusion of quantity and quantitative reasoning in CCSSM, a lack of emphasis on forming and interpreting relationships between quantities that change together remains a challenge. Thompson and Carlson (2017) proposed researching the systematic analysis of different curricular approaches that support students in developing quantitative and covariational reasoning. Akar, Watanabe, and Turan's chapter exemplifies such systematic analysis based on quantitative reasoning for a Japanese textbook series and curricular resources.

Quantitative reasoning is also crucial for other disciplines, including science. Duschl and Bismack (2013) stated, "quantitative reasoning is represented as a component of model-based reasoning that bridges the divide between mathematics and science" (p. 122). Similarly, further elaborating on quantitative reasoning, Thompson (2011) offered a detailed definition of quantification as "the process of conceptualizing an object and an attribute of it so that the attribute has a unit of measure, and the attribute's measure entails a proportional relationship (linear, bi-linear, or multi-linear) with its unit" (p. 37). Thompson considered this definition as a link between mathematics and science education. One can undoubtedly establish such connections between mathematics and other disciplines, and this book contributes to such an initiative. For example, Jin et al. chapter uses the mathematization of science dwelling on quantitative reasoning to quantify phenomena and construct knowledge and as a cross-cutting theme to build curricular coherence in physical and life sciences.

Although not exhausting all quantitative reasoning work, we point to the importance of quantitative reasoning and its crucial role in mathematics and science education with this book. Thompson's introductory chapter highlights that many scholars have based their work on quantitative reasoning as a framework to investigate and think about learning and teaching, conceptual analyses, curricular efforts, and links to other disciplines for decades. However, there seems to be a void in collecting this work together and pondering quantitative reasoning from different angles. This book provides ways to cluster the work established so far and can be considered as a reference book to be used by researchers, teacher educators, curriculum developers, and pre- and in-service teachers. We hope that it finds its place in the mathematics and science education literature within the digital era.

References

Beswick, K., & Fraser, S. (2019). Developing mathematics teachers' 21st century competence for teaching in STEM contexts. *ZDM*, *51*, 955–965.

Duschl R. A., & Bismack, A. S. (2013). Standards for science education: Quantitative reasoning and modeling concepts. In R. A. Duschl & A. S. Bismack (Eds.), *Reconceptualizing STEM education: The central role of practices* (pp. 117–146). University of Wyoming.

Johnson, H. L. (2016). Quantitative reasoning in mathematics education: Directions in research and practice. In R. A. Duschl & A. S. Bismack (Eds.), *Reconceptualizing STEM education: The central role of practices* (pp. 117–146). Routledge, Taylor & Francis.

Moore, K. C., Carlson, M. P., & Oehrtman, M. (2009). The role of quantitative reasoning in solving applied precalculus problems. In *Twelfth Annual Special Interest Group of the Mathematical Association of America on Research in Undergraduate Mathematics Education (SIGMAA on RUME) Conference.* North Carolina State University.

Shepherd, J. (2004). What is the digital era? In G. Doukidis, N. Mylonopoulos, & N. Pouloudi (Eds.), *Social and economic transformation in the digital era* (pp. 1–18). Idea Group Publishing.
Smith, J. III., & Thompson, P. W. (2007). Quantitative reasoning and the development of algebraic reasoning. In J. J. Kaput, D. W. Carraher, & M. L. Blanton (Eds.), *Algebra in the early grades* (pp. 95–132). Lawrence Erlbaum Associates.
Thompson, P. W. (1990). *A theoretical model of quantity-based reasoning in arithmetic and algebra.* Center for Research in Mathematics & Science Education: San Diego State University.
Thompson, P. W. (2008). Conceptual analysis of mathematical ideas: Some spadework at the foundation of mathematics education. In O. Figueras, J. L. Cortina, S. Alatorre, T. Rojano, & A. Sepulveda (Eds.), *Plenary Paper Delivered at the 32nd Annual Meeting of the International Group for the Psychology of Mathematics Education* (Vol. 1, pp. 45–64). PME.
Thompson, P. W. (2011). Quantitative reasoning and mathematical modeling. In S. Chamberlin, L. L. Hatfield, & S. Belbase (Eds.), *New Perspectives and Directions for Collaborative Research in Mathematics Education: Papers from a Planning Conference for WISDOMe* (pp. 33–57). University of Wyoming.
Thompson, P. W., & Carlson, M. P. (2017). Variation, covariation and functions: Foundational ways of mathematical thinking. In J. Cai (Ed.), *Third handbook of research in mathematics education.* National Council of Teachers of Mathematics.

Contents

Quantitative Reasoning as an Educational Lens

Patrick W. Thompson

I must begin by thanking Gülseren Karagöz Akar, Ismail Özgür Zembat, and Selahattin Arslan for including me in their effort to produce this book. While I am listed as an editor, they did the heavy lifting of conceptualizing the book and working with authors. My role was more as a consultant than an editor. I am nevertheless grateful they thought to include me.

1 Origins of a Theory of Quantitative Reasoning and Its Applicability

Humans have been reasoning quantitatively for thousands of years. I did not invent quantitative reasoning. I developed a *theory* of quantitative reasoning—a theory with the aim of explaining how individuals might come to reason about the world as they see it through a measurement lens (including not seeing it through a measurement lens) and implications for students' mathematical learning. My early work was motivated by wanting to understand students' difficulties with story problems—descriptions of settings designed by textbook authors that included a question about the setting. This interest was sparked in the spring of 1985 by James Greeno in his presentation of Valerie Shalin's work (Shalin, 1987; Shalin & Bee, 1985) to the mathematics education faculty at San Diego State University. Shalin designed a computer interface of notecards to represent quantities and arrows among notecards to show relationships. I realized Shalin had devised a way to represent relationships among quantities without having to rely on formulas or expressions. Shalin had not, however, explicated what she meant by quantity or quantitative relationship, nor did

P. W. Thompson (✉)
Arizona State University, Tempe, AZ, USA
e-mail: pat@pat-thompson.net

G. Karagöz Akar et al. (eds.), *Quantitative Reasoning in Mathematics and Science Education*, Mathematics Education in the Digital Era 21,
https://doi.org/10.1007/978-3-031-14553-7_1

she include a theory of how relationships among quantities imply methods for evaluating them. However, I immediately saw the theoretical power of having a way to represent quantities and relationships without formulas or expressions.

In 1986 I was invited to contribute a chapter on artificial intelligence (AI) in mathematics education to an NCTM publication on learning and teaching algebra (Thompson, 1989). I wanted to include a discussion of Shalin's and Greeno's computer program, but was unsuccessful in obtaining more information about it. I therefore decided to write an AI program, *Word Problem Analyst* (WPA), inspired by Shalin's interface and discuss the aspects of quantitative reasoning as I conceived it embodied in the program. I will not recap all the insights I gained from writing WPA (and revising it over the next four years) except to say writing it, with support from the US National Science Foundation, provided a testbed for creating a scheme theory for ideas of quantity and the development of mathematical reasoning from quantitative reasoning (Thompson, 1990, 2011).

The following problem and Figs. 1, 2, 3, 4, 5, 6 and 7 illustrate the use of WPA to model someone conceptualizing a problem in terms of quantities and relationships among quantities and the algebra that can be inferred from this structure.

> MEA Export is to deliver an oil valve to Costa Rica. The valve's price is $5000. Freight charges to Costa Rica are $100. Insurance is 1.25% of Costa Rica's total cost. Costa Rica's total cost includes the costs of the valve, insurance, and freight. What is Costa Rica's total cost? (Thompson, 1990, p. 39)

Figure 1 shows a person's (say, José's) conception that there are six quantities involved in this situation: Total Cost to Costa Rica, the costs of Freight, Valve, and Insurance, the Insurance Rate, and the cost of Insurance and Freight together. At

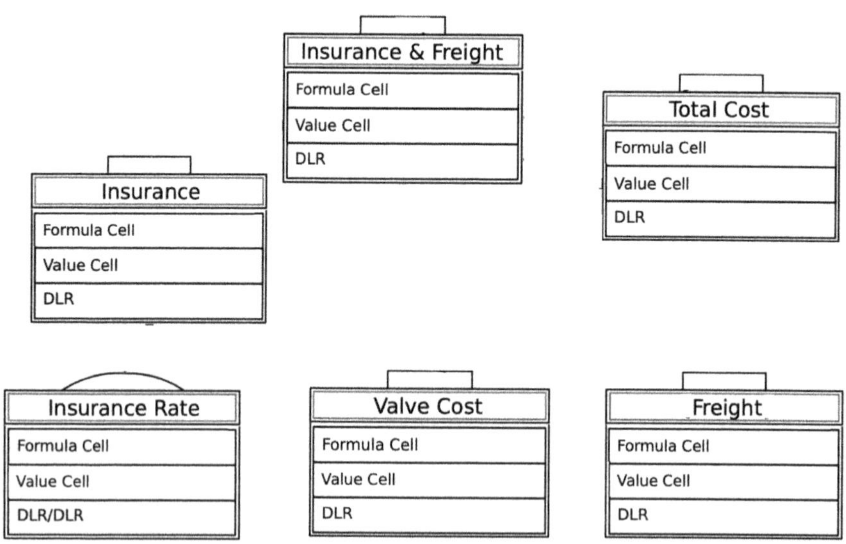

Fig. 1 José's understanding of quantities involved in the situation

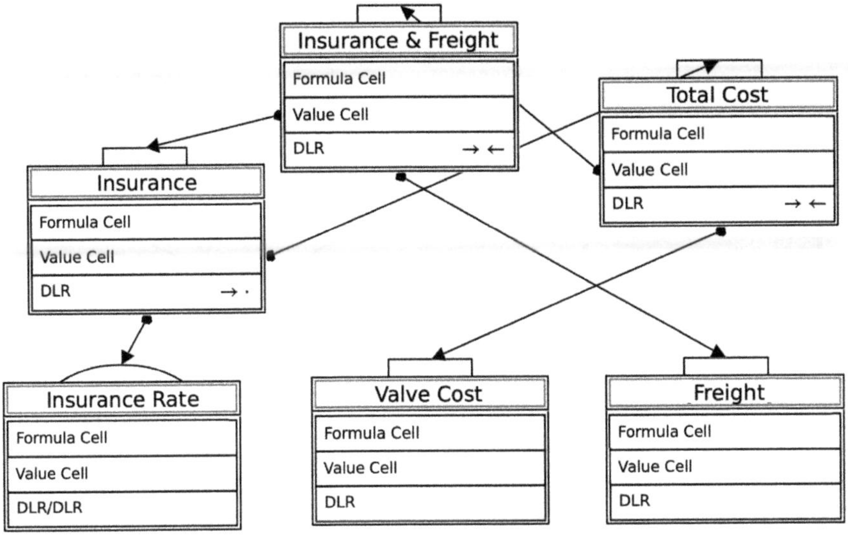

Fig. 2 José's conception of relationships among quantities in the situation

this moment José has not conceptualized any relationships among quantities. Each notecard reflects the schematic nature of a conceived quantity—a natural language description of an object's attribute, a unit in which the attribute is measured, and a potential value for the quantity's measure. Each notecard also has a "Formula Cell". This represents José's anticipation that a quantity's value might be calculated from relationships with other quantities.

Figure 2 shows the relationships José envisioned among quantities: *Total Cost* is made by an additive combination of *Insurance and Freight* and the cost of the *Valve*. *Insurance and Freight* is made by an additive combination of the cost of *Insurance* and the cost of *Freight*. The cost of *Insurance* is made by instantiating the *Insurance Rate* with the *Total Cost* to Costa Rica. Notice that at this moment, José has not thought about any calculations.

Figure 3 shows that José has now attended to the information given in the problem statement. *Freight* has a value of \$100, *Valve Cost* has a value of \$5000, and *Insurance Rate* has a value of \$1.25/100 of insurance per dollar of cost. Notice that, at this moment, José cannot make any inferences about values of other quantities.

Figure 4 shows José's decision to let C stand for the value of *Total Cost* to Costa Rica. Figure 5 shows an immediate consequence of letting C stand for the value of *Total Cost*—since *Total Cost* is made by an additive combination of *Insurance & Freight* and *Valve Cost*, and *Valve Cost* has a value of 5000, the value of *Insurance and Freight* must be C—5000\$.

Figure 6 shows the next propagation. Since *Insurance* is made by instantiating *Insurance Rate* with the value of *Total Cost*, the value of *Insurance* will be $C * 0.0125$ dollars. Figure 7 reflects José's openness to deriving a formula for a quantity for which he already knows a value. *Insurance & Freight* is made by an additive combination

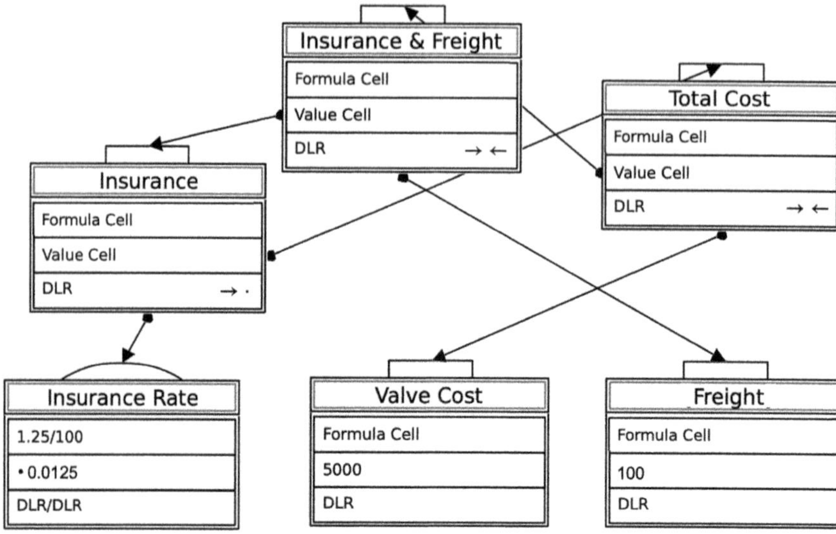

Fig. 3 Adding information given in the problem to José's conception of the situation

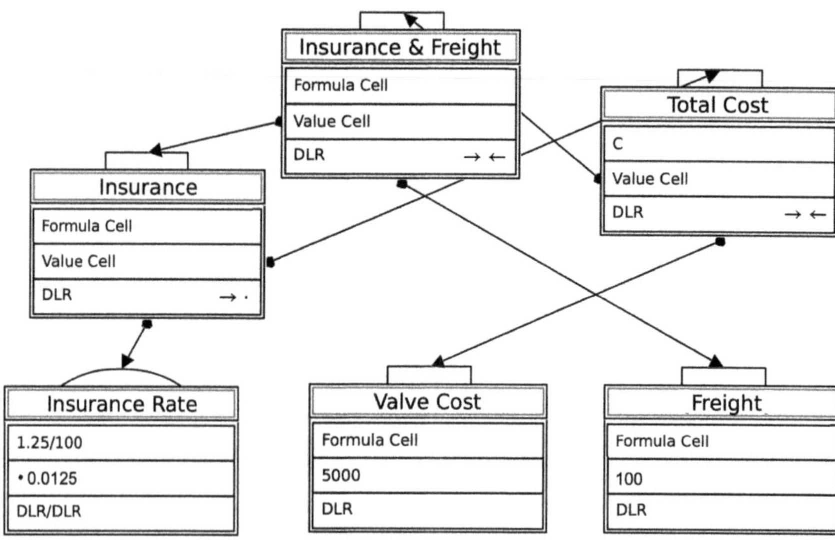

Fig. 4 Using "C" to stand for the value of *total cost*

of *Insurance* and *Freight*, and since its value is $C - 5000$ and *Freight's* value is 100, José infers that a formula to compute *Freight's* value is $C - 5000 - 0.0125C$. But the

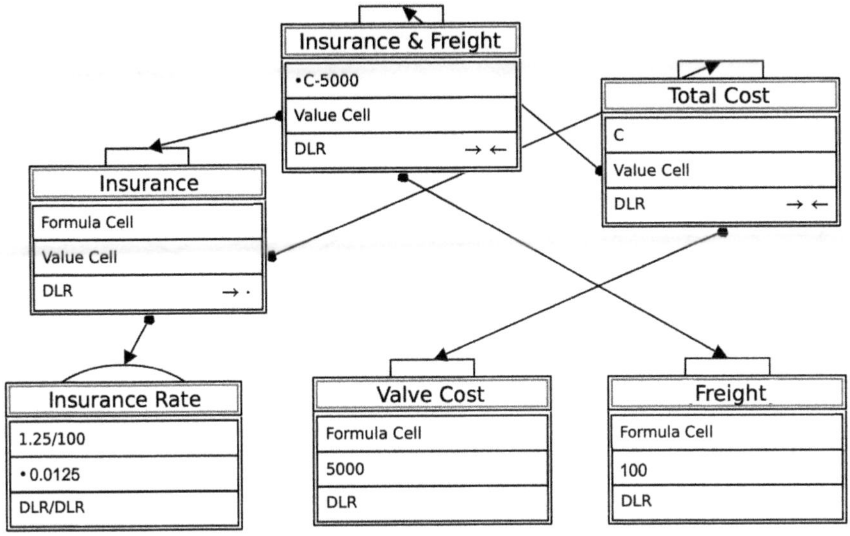

Fig. 5 Inferring a formula to compute the value of *insurance and freight*

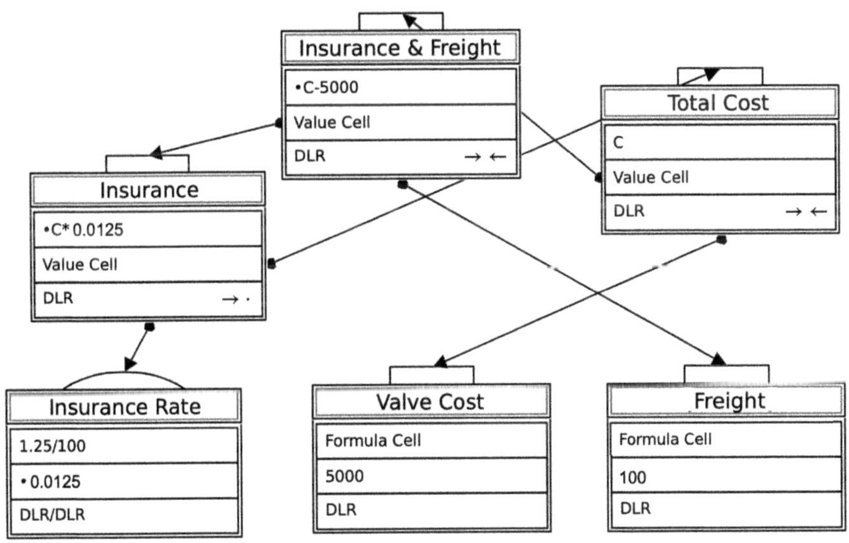

Fig. 6 Inferring a formula to calculate the value of *insurance*

value this formula must yield is the value of *Freight*, which is 100.[1] In other words, by reasoning quantitatively, José ended with the equation $C - 5000 - 0.125C = 100$.

[1] The brackets in the *Freight* notecard indicate that José ignored the fact he already knows a value of *Freight* in order to infer a formula to compute *Freight's* value.

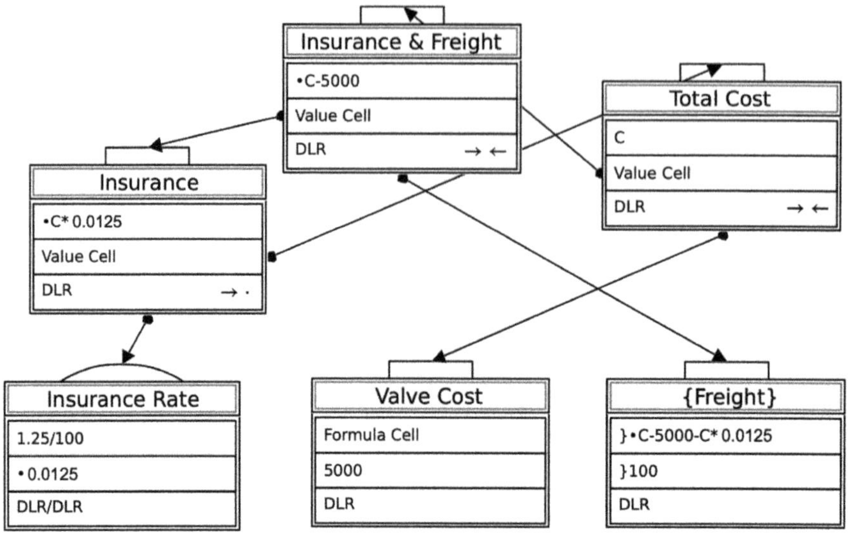

Fig. 7 Inferring a formula for the value of *freight* even though it has a known value of 100

José's conceptualization of the Costa Rica situation is not unique. It can be conceptualized in many ways. Indeed, in Thompson (1990) I illustrate how even simple problems can have very different underlying conceptualizations in terms of quantities and relationships composing it yet yield the same arithmetic or algebra.

There are three significant differences between Shalin's model and the theory I developed. First, Shalin's model did not have an underlying theory of quantity or quantification, except for the arithmetic or units developed by Schwartz (1988). An arithmetic of units, such as $\text{cm} \cdot \text{cm} = \text{cm}^2$, or $(\text{ft/s})/\text{s} = \text{ft/s}^2$, conflates arithmetic operations and quantitative operations. It is not a theory of quantitative reasoning. Rather, units are treated as if they are numbers or variables. An arithmetic of units is implied by quantitative reasoning, but it is not a theory of it. Second, the theory addressed how one propagates information throughout a quantitative structure when knowing only partial information about the context. The theory of propagation is the foundation of the model's hypotheses about students' transitions from quantity-based arithmetic to quantity-based algebra (and beyond). Third, Shalin did not make a distinction between quantitative operations and arithmetic operations, which resulted in confounding type of quantity with an arithmetic operation to calculate its value, such as describing a quantity as a difference simply because, in a particular situation, subtraction is used to calculate its value (see Greeno, 1987, p. 77).

Finally, the WPA model of José's conception of the Costa Rica situation presumed he had mature schemes for the quantities and quantitative operations depicted therein. WPA was meant to model implications of reasoning quantitatively for algebraic reasoning. It did not address ways learners *construct* quantities and quantitative operations. The theory I expressed in Thompson (1990) provided a foundation for later studies that brought coherence to understanding the development of students' schemes for quantitative comparisons, variation and covariation, ratio and rate, geometric and exponential growth, uses of notation, function, probability and statistics, and many ideas specific to calculus.

2 Chapters in This Book

I am surprised and gratified that many people found this early work, and later expansions of it, useful in their research. The chapters in this book show creative uses of quantitative reasoning as a lens for making sense of students' reasoning, for design of instruction, for curriculum design and evaluation, for teacher professional development, and for design of assessments. Johnson's use of Harel's notion of intellectual need as a motive for why students might *seek* relationships between quantities whose values vary is novel and powerful. Moore et al.'s focus on students' creation of abstract quantitative structures addresses the question of how students might generalize their quantitative reasoning in specific contexts to broader areas of application. Karagöz Akar, Watanabe and Turan created a novel way of examining mathematics textbooks by the criterion of ways they support or inhibit students' quantitative reasoning. Paoletti extends a framework for thinking about students' variational and covariational reasoning by filling a gap in it, while Ellis et al. build a learning progression based in variational and covariational reasoning to address students' development over early grades of schemes for function. Karagöz Akar, Zembat, Arslan and Belin leverage quantitative reasoning to address the issue of students' difficulties in conceiving motions in the plane as functions mapping $\mathbb{R}^2$ to $\mathbb{R}^2$. Carlson et al. leverage quantitative reasoning to address the question of how to support teachers in transitioning from speaking to students as if to themselves to engaging students in reflective discourse aimed at students' construction of coherent systems of mathematical meanings. I am especially gratified to see three chapters by science educators leveraging a theory originally aimed to support learning and teaching mathematics to address issues within science education. Jin et al. apply quantitative reasoning as a theme to enhance curricular coherence across grade levels and across a broad array of scientific concepts. González uses quantitative reasoning, especially distinctions between ratio as a quantity and rate as a quantity, to examine students' meanings for ideas central to understanding climate change. White Brahmia and Olsho turn the lens around. Instead of using quantitative reasoning as a lens on students' reasoning in physics, they use physics as a context to assess students' quantitative reasoning. Nunes and Bryant take an approach to quantitative reasoning more in line with Schwartz (1988), in which numbers represent quantities and arithmetic operations imply operations on quantities.

I suspect one reason quantitative reasoning has found such broad applicability is its fundamental stance that quantities are in a mind, not in the world. This stance forces anyone adopting it to examine ways *learners* understand situations presented to them. It forces us to ask, "What is this situation *to the learner*?" As Carlson et al. (this volume) document, adopting this stance is nontrivial for instructors who are accustomed to apply criteria of coherence only to their own understandings, not to ways their students might understand the situations presented to them or might understand their instructor's actions and utterances regarding a situation.

Another possible reason quantitative reasoning has been found broad applicability is that using it forces one to employ a level of qualitative precision that is uncommon in mathematics instruction, yet beneficial for students' learning. Distinctions among object, attribute, and measure are often unaddressed by mathematics teachers—as witnessed by the common proclivity among teachers and students to write statements like "D = distance". Carlson et al. (this volume) document difficulties precalculus instructors create for themselves and their students by their lack of precision about contextual meanings of numbers, variables, and expressions.

3 Conceptualizing Units and Conceptualizing Quantification: Aspects of Quantitative Reasoning Needing Greater Attention

Early on in developing this theory of quantitative reasoning I proposed that a quantity is a scheme—someone's conception of an object and an attribute of it the person has conceived as measurable in an appropriate unit. I also spoke repeatedly of the synergy among a person's conceptions of object, attribute, and measurability—they each mature as the person gains clarity on the others. In Thompson (2011) I gave a brief recount of 8th-graders' construction of "explosiveness of a grain silo" as a quantity. They engaged in extended discussions of just what was it that was explosive: The silo? The grain in the silo? Dust in the silo? Dust in the air within the silo? They also had to settle on a mechanism for explosions, eventually settling on oxidation at the surface of grain dust particles. This led them eventually to a unit of grain silo explosiveness: cm^2 of "dust surface area" per cm^3 of "dust volume" per ft^3 of "silo volume" in which the dust is dispersed.

I offered the example of grain silo explosiveness to illustrate the messiness of quantitative reasoning that often is unaddressed in studies employing a quantitative reasoning lens. But we need not go to uncommon quantities like "grain silo explosiveness" to see the interdependence among conceptualizations of object, attribute, and unit. In Thompson (2000) I spoke of ways students often understand area and volume as one-dimensional quantities. Area is one-dimensional when one conceives the unit as having one dimension—a square region of a particular size. Then all areas are just counts of that one-dimensional unit. Similarly, volume is one-dimensional when one conceives the unit as having one dimension—a cubic object of a particular size. Then all volumes are just counts of that one-dimensional unit. Brady and Lehrer (2020) clarified that a unit of area is conceived as two-dimensional when one conceives it as generated by two segments, one being swept along the other. This is the imagistic equivalent of understanding the interior of a rectangle being formed by the cross product of two perpendicular lines viewed as sets of points. Karagöz

Akar et al.'s chapter on isometries makes a similar point with respect to conceptualizing the Cartesian plane as $\mathbb{R}^2 = \mathbb{R} \times \mathbb{R}$ You obtain a two-dimensional object by the quantitative operation of multiplicative combination of two one-dimensional objects. Area and volume are just two instances of quantities teachers and researchers take as unproblematic in conceiving their unit when in fact students often conceive them in ways that are problematic for their comprehension of situations involving them.

In the following paragraphs, I offer two additional examples to illustrate the messiness of quantification and how attention to units can be helpful to students in understanding mathematical or scientific ideas. The first is conceptualizing interest rate as actually being a rate of change of one quantity with respect to another. The second is the quantification of kinetic energy.

3.1 Quantification of Interest Rate as a Rate of Change

To specify a quantity as a rate of change, we must state two quantities whose values covary. They vary with respect to each other. The "rate of change" attribute of two quantities covarying is captured by a statement of the amount one varies in relation to variations in the other.

Here are three definitions of interest rate by commonly accepted authorities:

1. "The cost of borrowing money from a lender is represented as a percentage of the principal loan amount, called the interest rate." U.S Federal Housing Administration https://www.fha.com/define/interest-rate
2. "The amount earned on a savings, checking, or money market account, or on an investment, as a certificate of deposit or bond, typically expressed as an annual percentage of the account balance or investment sum." Dictionary.com https://www.dictionary.com/browse/interest-rate
3. "The percentage usually on an annual basis that is paid by the borrower to the lender for a loan of money." Meriam-Webster.com https://www.merriam-webster.com/dictionary/rateofinterest.

I find it peculiar that, despite purporting to define interest rate, none of these statements actually defines a rate of change of one quantity with respect to another.

Imagine a bank advertisement as follows:

> We pay 3% interest per year on your deposit.

What quantities are involved in this practice of charging or paying interest? What are their units? What is the rate of change of one quantity with respect to another that is the "rate"?

The quantities are interest paid (dollars of interest), dollars on balance (basis of the percentage), and an amount of time (number of years balance is on deposit). Regarding the rate—what is it? Is it a rate of change of balance with respect to time? The rate of change of interest earned with respect to time?

The crux of the matter is to understand that "3%" has a unit: dollars of interest per dollar on balance. The unit of "3% interest per year" is ($interest/$balance)/year. The bank will pay interest at the rate of 0.03 dollars interest per dollar of balance per year. There is yet one open question: What constitutes the balance upon which interest is computed? Is it the current balance at the time of computing interest, or is it the initial balance at the time of opening the account?

The difference between simple interest and compound interest is much easier to understand when we answer these questions explicitly. "We pay 3% interest per year on your deposit, compounded quarterly" means that at the end of each quarter they will add to your balance the amount earned at the rate of (($0.03 interest per $1.00 balance at beginning of compounding period) per year) earned in 1/4 year. You earn interest over a quarter year at 1/4 the rate you would earn over a year. This is like speeding up at a rate of 10 (km/h)/h for 1/4 h. Your speed increases at a rate per 1/4 hour that is 1/4 the rate for an hour, or at a rate of (2.5 km/h) per 1/4 h.

The idea of the unit of an interest rate is related to students' difficulties distinguishing between linear and geometric growth. Graphs given in Fig. 8 show two ways to understand the phrase "… increases at a rate of 20% per month." Fig. 8a shows 20% of the original amount (e.g., $2) added to the current value (e.g., $10) to get the next value (e.g., $12). The same amount is added at the end of each month. Figure 8b shows 20% of the *current* month's value added to get the next month's value. Since the current value increases each month, the amount added at the end of each month increases.

The phrase "… increases at a rate of 20% per month" is ambiguous regarding which interpretation the speaker intends a listener to make. Being clear about the quantities and their units is clarifying. The first would be "… increases at a rate of

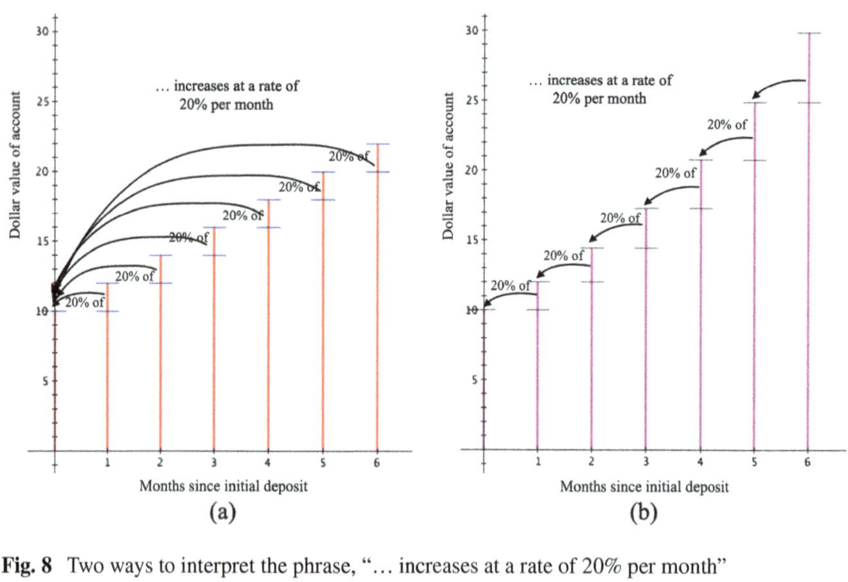

Fig. 8 Two ways to interpret the phrase, "… increases at a rate of 20% per month"

(\$0.20 interest per dollar of initial balance) per month", whereas the second would be, "... increases at a rate of (\$0.20 interest per dollar of current balance) per month".

3.2 *Quantification of Kinetic Energy*

A characteristic of physical quantities is how deeply their conceptualizations are interconnected. Energy is commonly defined as "the capacity to do work" (*Encyclopedia* Britannica, 2022). The idea of work is tied to the idea of applying a force to move an object some distance, while force is the idea of accelerating an object (having mass) from one velocity to another velocity. The meaning of *kinetic* energy is the work required to bring an object having mass m from velocity v to velocity 0.[2]

Jin et al. (this volume) speak of students' understanding of kinetic energy in terms of implications they draw from a formula for quantifying its measure, namely $E = \frac{1}{2}mv^2$, for how an object's kinetic energy changes when its velocity changes. Some students think doubling an object's velocity doubles its kinetic energy. Other students think doubling its velocity quadruples its kinetic energy. The issue Jin et al. addressed is students' abilities to reason about the implications of a quantification expressed in a formula. I address a more foundational issue—the quantitative reasoning involved in *quantifying* kinetic energy to *end* with the formula $E = \frac{1}{2}mv^2$. My aim here is to illustrate how conceptualizations of object, attribute, and quantifications are intertwined.

To quantify kinetic energy, we must identify an object and its attribute as a starting point of quantification—to determine a method by which to measure it and the unit in which it will be measured. In the case of kinetic energy, the "object" is anything having mass. One attribute is its motion—it is moving (at least momentarily) at a constant velocity. Another attribute is the effort (work) required to stop its motion. Work, as a quantity, is a force applied over a distance. The object's velocity, however, is not constant. Its velocity decreases as work is applied to it.

A slight twist which makes envisioning kinetic energy easier is to realize the energy required to bring an object from velocity v to velocity 0 is the same as the energy required to bring it from velocity 0 to velocity v.

Breaking down these components, and envisioning the object's velocity changing in little bits as it accelerates from 0 to v, we get

- a force of measure F is created by accelerating a mass of measure m at a rate of measure a,
- a small bit of acceleration is created by changing an object's velocity by a variation of measure dv during a variation of time of measure dt,
- a small variation in distance ds is made by going at velocity v for a small variation in time dt,

[2] I have limited these descriptions to mechanical quantities to avoid dealing with the complexities of their electro and thermal equivalents.

- a small variation of work is created by applying a force of measure F over a small variation in distance of measure *ds*, and
- a small variation in an object's kinetic energy of measure *dE* is created by a small variation in work of measure Fds that varies its velocity.

Symbolically, taking F as a measure of force, E as a measure of kinetic energy, and *dE*, dv, *dt*, and *ds* as infinitesimal variations in kinetic energy, velocity, time, and distance, respectively:

$$F = ma, a = \frac{dv}{dt}, ds = vdt, dE = Fds$$

$$\begin{aligned} dE &= Fds \\ &= mads \\ &= m\left(\frac{dv}{dt}\right)vdt \\ &= mvdv \end{aligned}$$

So, a small variation in an object's kinetic energy is its momentum times a small variation in its velocity. This says an object's momentum at any velocity is its *rate of change of kinetic energy with respect to velocity.*

Recalling that the work required to decelerate an object from v to 0 is the same as the work required to accelerate it from 0 to v, an object's kinetic energy is the (hyper) sum of all infinitesimal variations in its kinetic energy as velocity varies from 0 to v. Symbolically[3]:

$$\begin{aligned} E(v) &= \int_0^v mudu \\ &= \frac{1}{2}mu^2\Big|_{u=0}^{u=v} \\ &= \frac{1}{2}mv^2 \end{aligned}$$

As I said earlier, a full, robust understanding of this quantification of kinetic energy requires understanding constituent quantities' units (units of mass, time, distance) and the units of quantities created from them (acceleration, force, momentum, work, kinetic energy)—but not in the sense of an arithmetic of units. Rather, I mean one

---

[3] I acknowledge that this derivation relies on students' understanding of integrals as a (hyper) sum of infinitesimal variations and on their understanding of the relationship between a rate of change function and its accumulation functions. However, they could approximate any object's kinetic energy to an acceptable accuracy with Desmos using the finite sum $E_{\text{approx}}(v) = \sum_{i=1}^{v/\Delta v} m(i\Delta v)\Delta v$, where Δv is a small increment in velocity. See Thompson et al. (2019, Ch 5) for a full development of these ideas.

must understand units in the sense of Bridgman's (1922) dimensional analysis, which attends to the *creation* of quantities from other quantities while attending to the nature of their attributes. Bridgman wrote, for example, $[F] = [m][a]$ to convey that the quantity force is formed by the quantitative operation of multiplicative combination—by accelerating an object having mass. He wrote $F = ma$ to represent how you calculate a measure of force, ending with a number with a unit that is consistent with the quantity's dimension. If you measure a mass in kg and acceleration in ((meters per second) per second), the unit of force is kg ((m/s)/s), meaning a mass measured in kg is accelerated at a rate measured in ((meters per second) per second).

How might students *know* to multiply m and a to calculate a measure of force? Hopefully, from schemes they constructed through experimentation,[4] that force is proportional to both mass and acceleration. If we increase by a factor of j the mass being accelerated at a rate a, the force of accelerating it increases by a factor of j; if we increase the acceleration of an object by a factor of k, meaning its velocity increases k times as rapidly with respect to time, the force of accelerating it increases by a factor of k. Let $F(j, k)$ represent a measure of the force of accelerating an object of j mass units at a rate of k acceleration units. Then $F(j, k) = F(j \cdot 1, k \cdot 1) = j \cdot kF(1, 1)$. This says the measure of force that accelerates a mass of measure j mass units at a rate of measure k acceleration units is $j \cdot k$ times as large as the force of accelerating a mass of measure 1 mass unit at a rate of change of velocity with respect to time of 1 acceleration unit.

Lastly, there is another question we should hope students ask with respect to quantification of kinetic energy. Since kinetic energy is equivalent to an amount of work, they hopefully ask whether $\frac{1}{2}mv^2$, our quantification of kinetic energy, actually quantifies an amount of work. If it does, then the derived unit of $\frac{1}{2}mv^2$ must, in line with Bridgman, accord with a force applied over a distance. Its unit must be of dimension $[F][d]$. Here is where arithmetic of units is useful.

The standard unit of force in the kg-meter-second system is the Newton (N), or 1 kg accelerated at 1 (m/s)/s. Keeping track of units, and using m as a measure of mass and v as a measure of velocity in the kg-meter-second system, we get

$$\begin{aligned}\frac{1}{2}mv^2 &\to \mathrm{kg\,m^2/s^2} \\ &\to \left(\mathrm{kg}\left(\mathrm{m/s^2}\right)\mathrm{m}\right. \\ &\to \mathrm{(kg\ (m/s)/s)\ m} \\ &\to N\,\mathrm{m} \\ &\to [F][d]\end{aligned}$$

The unit of $\frac{1}{2}mv^2$ in the kg-meter-second system is the Newton-meter, which is of dimension $[F][d]$, so it is a unit of work.

[4] Of course, the experimentation that affords students an opportunity to construct such schemes must be crafted carefully so their abstractions are from their own activity.

4 Connections with Chapters in This Book

The examples of conceptualizing and quantifying force and kinetic energy tie together themes developed in several chapters of this volume: Brahmia and Oshlo's focus on quantification as a central aspect of scientific reasoning, Johnson's focus on mathematizing *a la* Freudenthal via an intellectual need for relationships, Jin et al.'s focus on mathematizing as a bridge between mathematics and science, Paoletti et al.'s and Ellis et al.'s focus on variation and covariation as foundational ways of thinking for students to develop understandings of functions, Moore et al.'s focus on abstracted quantitative structures as a target for students' quantitative reasoning, Gonzalez' proposal of quantitative reasoning and quantification as a central theme in climate science.

Moreover, if we consider these quantifications of force and kinetic energy as conceptual analyses of understandings we hope students construct—as a teacher's key developmental understandings of force and kinetic energy—then Carlson et al.'s analysis comes into play. As they explain, teachers must reflect upon their own quantitative understandings to become conscious of the intricacies entailed in their goals of instruction and must decenter to consider how one might support students in developing these understandings via conventions of speaking with meaning and emergent symbolization.

The example of work as a quantity relates to Moore et al.'s construct of abstract quantitative structure in a profound way. Understanding work dimensionally, as $[F][d]$, is to understand the quantitative structure of work and to understand that units will be involved, but the exact units need not be specified—they just need to be coherent with the quantities of force and distance. The example of kinetic energy also is related to Karagöz Akar, Watanabe and Turan's use of quantitative reasoning as a lens to examine mathematics textbooks' coherence. Does a textbook support teachers to engage students in reflective discourse aimed at their conceptualization of quantities, their quantification, and situations involving them that textbook authors purport to address?

The representation of kinetic energy as a function of velocity, $E(v) = \frac{1}{2}mv^2$, relates to Johnson's stance regarding intellectual need for relationships, Ellis et al.'s conceptual analysis of functions, and Paoletti's analysis of covariational reasoning. For a student (or instructor) to even consider writing "$E(v)$" requires they (1) seek a relationship between velocity and kinetic energy that remains invariant as velocity varies, (2) envision velocity varying smoothly from 0 to v regardless of the amount of time this acceleration takes, and (3) understand the notation "$E(v)$" through a scheme that entails an image of velocity and kinetic energy varying simultaneously and varying in a way that each value of velocity determines a value of kinetic energy (see Yoon & Thompson, 2020).

I can imagine mathematics educators questioning the examples of quantifying force and kinetic energy as being largely relevant to science education and less relevant to mathematics education. I disagree. Anyone who has taught arithmetic, algebra, precalculus or calculus in the United States has seen their students arrive at solutions to applied problems with little meaning or inappropriate meanings for numbers or variables in their answer. This is a serious problem. The solution to the problem of meaning, however, must be systemic. To take quantitative reasoning seriously in mathematics and science education requires attention to having students conceptualize quantities *and methods and meanings of their measures* throughout their schooling. This can range from asking students what quantity their arithmetic has evaluated, to asking them what an appropriate unit for the area of a rectangle of height 3 jibs and width of 4 jabs would be, to how one might convert measures of fuel efficiency from miles per gallon to kilometers per liter, to asking them for a useful unit of effort to complete a job (e.g., person-hour), and so on.

Moore et al.'s construct of *abstract quantitative structure* might be behind experts' utterances like "speed times time equals distance". They of course do not mean speed in *any* unit times time in *any* unit equals distance in *any* unit. Rather, they presume, without saying, this is true for a coherent system of units for speed, time, and distance. This brings to mind Carlson et al.'s explanation of the necessity for instructors to examine their own understandings and presumptions in order to consider how their expressions of them might be interpreted by students who will interpret teacher's utterances and actions through schemes quite unlike the teacher's.

5 Conclusion

I once again praise the authors' work expressed in this volume and my colleagues' who brought this collective work to our attention. I hope my call to give greater attention to the details of students' and teachers' conceptualizations of object, attribute, and measure is useful for those employing quantitative reasoning as a lens in mathematics and science education. I suspect doing this will give greater insight into difficulties students experience in learning mathematics and science and difficulties teachers experience in promoting such learning.

References

Brady, C., & Lehrer, R. (2020). Sweeping area across physical and virtual environments. *Digital Experiences in Mathematics Education, Online*. https://doi.org/10.1007/s40751-020-00076-2

Bridgman, P. W. (1922). *Dimensional analysis (ABS5571)*. Yale University Press.

Encyclopedia Britannica. (2022). Kinetic energy. https://www.britannica.com/science/energy

Greeno, J. G. (1987). Instructional representations based on research about understanding. In A. H. Schoenfeld (Ed.), *Cognitive science and mathematics education* (pp. 61–88). Erlbaum.

Schwartz, J. (1988). Intensive quantity and referent transforming arithmetic operations. In J. Hiebert & M. Behr (Eds.), *Number concepts and operations in the middle grades* (pp. 41–52). National Council of Teachers of Mathematics.

Shalin, V. L. (1987). *Knowledge of problem structure in mathematical problem solving* [Dissertation]. University of Pittsburgh.

Shalin, V. L., & Bee, N. L. (1985). *Analysis of the semantic structure of a domain of word problems*. University of Pittsburgh.

Thompson, P. W. (1989). Artificial intelligence, advanced technology, and learning and teaching algebra. In C. Kieran & S. Wagner (Eds.), *Research issues in the learning and teaching of algebra* (pp. 135–161). Erlbaum.

Thompson, P. W. (1990). *A theoretical model of quantity-based reasoning in arithmetic and algebra*. https://doi.org/10.13140/RG.2.2.23908.45447

Thompson, P. W. (2000). What is required to understand fractal dimension? *The Mathematics Educator, 10*(2), 33–35.

Thompson, P. W. (2011). Quantitative reasoning and mathematical modeling. In L. L. Hatfield, S. Chamberlain, & S. Belbase (Eds.), *New perspectives and directions for collaborative research in mathematics education* (Vol. 1, pp. 33–57). University of Wyoming. http://bit.ly/15II27f

Thompson, P. W., Ashbrook, M., & Milner, F. A. (2019). *Calculus: Newton, Leibniz, and Robinson meet technology*. Arizona State University. http://patthompson.net/ThompsonCalc

Yoon, H., & Thompson, P. W. (2020). Secondary teachers' meanings for function notation in the United States and South Korea. *Journal of Mathematical Behavior, 60*, 100804. https://doi.org/10.1016/j.jmathb.2020.100804

An Intellectual Need for Relationships: Engendering Students' Quantitative and Covariational Reasoning

Heather Lynn Johnson

People encounter situations involving change and variation as citizens of the world. For instance, sea levels are rising as the oceans continue to absorb heat from the atmosphere. One may read about this phenomenon in newspaper articles or encounter graphs representing rising sea levels over time. By engaging in quantitative and covariational reasoning (Carlson et al., 2002; Thompson, 1994, 2011; Thompson & Carlson, 2017), people can interpret and make meaning of such situations (e.g., González, 2021). Not only are these forms of mathematical reasoning productive for being informed citizens, but they also underlie key mathematical concepts such as rate and function (Thompson & Carlson, 2017). Hence, it is crucial for students to develop and engage in such reasoning, and for opportunities to occur throughout their schooling, across K-12 and university mathematics courses. Yet, from a student's point of view, what may serve as a catalyst, so students can actualize potential opportunities? Drawing on Harel's construct of "intellectual need" (1998, 2008b, 2013), I offer an intellectual need for relationships, which is a need to explain how elements work together, as in a system. I argue that this need can engender students' quantitative and covariational reasoning.

To illustrate, consider a situation involving Sam, who is walking from home to the corner store. There are a number of attributes that students may separate from the situation; two include Sam's distance from home and Sam's distance from the store. Engaging in quantitative reasoning (Thompson, 1994, 2011), a student can conceive of the possibility of measuring those attributes, even if they do not find particular amounts of measure. For instance, a student may have a sense of a length of a stretchable cord extending from Sam's current location to home or the store. As Sam is walking, each distance changes, increasing or decreasing depending on Sam's route. Engaging in covariational reasoning (Carlson et al., 2002; Thompson & Carlson,

---

H. L. Johnson (✉)
University of Colorado Denver, Denver, CO, USA
e-mail: heather.johnson@ucdenver.edu

G. Karagöz Akar et al. (eds.), *Quantitative Reasoning in Mathematics and Science Education*, Mathematics Education in the Digital Era 21,
https://doi.org/10.1007/978-3-031-14553-7_2

2017), a student can conceive of relationships between the changing distances. For instance, with a direct route, Sam's distance from home increases while the distance from the store decreases. By forming and interpreting relationships between attributes, students can mathematize (Freudenthal, 1973) such situations in terms of quantities and covariation.

Results of researchers' investigations of students' quantitative and covariational reasoning represent both challenge and promise. Even accomplished university students have demonstrated difficulty (e.g., Carlson et al., 2002; Moore et al., 2019a, 2019b), while middle and secondary students have shown promising evidence (e.g., Ellis et al., 2020; Johnson, 2012). I argue that students' intellectual need for such reasoning may account, in part, for differences in these findings. For example, consider a task in which students are to sketch a Cartesian graph relating Sam's distance from home and Sam's distance from the store. Some students may find such a task problematic; they may wonder how to measure and relate the different distances as they sketch their graph. In contrast, other students may think the task is an exercise in finding a resulting graph that is an instance of some familiar graph. If students are focused on getting end results, they may miss opportunities to engage in quantitative and covariational reasoning.

Harel (1998, 2008b, 2013) put forth the construct of intellectual need, rooted in Piaget's constructivist theory. To illustrate, say a student encounters a situation that is problematic for them, and as a result of engaging with that situation, they develop some new mathematical knowledge. The "problematic-ness" of that situation, from the student's point of view, is the student's intellectual need. For example, one student may intend for Sam's graph to represent a relationship between distances. Another student may intend to represent Sam's physical motion on the walk. While both students find the situation problematic, the first student's goal is more compatible with quantitative and covariational reasoning.

Harel (2008a) has posited two different forms of mathematical knowledge that can emerge from students' intellectual needs: ways of understanding (products of mental action) and ways of thinking (characteristics of mental action). For example, a conception of function can be a product of mental action, and a correspondence approach can be a characteristic of mental action. Through broad categories, Harel has illuminated three ways of thinking (2008a) and five forms of intellectual need (2013), leaving room for the possibility that more categories can emerge. I argue for an expansion of the ways of thinking and forms of intellectual need put forward by Harel.

I organize this chapter into six sections. First, I discuss theoretical underpinnings of quantitative and covariational reasoning. Second, I offer Freudenthal's term, "mathematizing" (Freudenthal, 1973), to represent an additional category of a way of thinking that can emerge from students' intellectual need. Third, I explain what I mean by an intellectual need for relationships, and how that need may engender students' quantitative and covariational reasoning. Fourth, I put forward four facets of such a need. Fifth, I address task design considerations for each facet, using a digital Ferris wheel task to illustrate. Sixth, I discuss implications for theory and practice.

1 Theorizing Quantitative and Covariational Reasoning

Thompson rooted the theory of quantitative reasoning (1994, 2011) in Piaget's constructivist theory, which assumes that individuals develop new understandings by reorganizing their existing conceptions. From this lens, the distances I identified in the situation of Sam walking from home to the store would not be "out there" for a student to observe. Rather, they would be a person's conception of the situation. In the theory of quantitative reasoning, Thompson explains how individuals may conceive of situations in terms of attributes that are possible to measure, such as the distances in Sam's situation. Engaging in quantitative reasoning involves conceptions of quantities, a quantification process, and quantitative operations. A student's quantitative reasoning can entail some or all of these elements.

Quantities are a foundational element of the theory. Per Thompson (1994), a quantity is an individual's conception of an attribute in a situation as being possible to measure. This means that quantities are human creations; through their conceptions, individuals transform attributes into quantities. For example, in Sam's situation, a student can transform attributes into quantities by separating those attributes (e.g., distance) from the physical motion described in the situation (e.g., Sam's walking). Essential to Thompson's theory is a distinction between conceiving of the possibility of measurement and the act of determining particular amounts of measure. This means that students can think of measuring Sam's distance from the store without finding certain amounts of distance.

With quantification, Thompson (2011) explained a three-part process by which an individual can formalize this "possible to measure-ness." First, they would conceive of an attribute that could be measured, such as Sam's distance. Second, they would conceive of a unit of measure for the attribute. This might be a standard unit, such as a meter or foot, or a nonstandard unit, such as one of Sam's steps. Third, they would conceive of a proportional relationship between the unit and the attribute's measure. That is, they could iterate one of the units, such as a step length, to measure Sam's distance from the store. As with quantity, an essential aspect of quantification was that an individual did not need to actually measure Sam's distance from the store with the indicated unit, just think of the possibility of doing so.

Thompson (1994, 2011) put forward quantitative operations to describe mental activity in which individuals could employ a quantitative lens on situations and conceive of new kinds of quantities. Thompson identified a "difference" as one such quantity that students could create via additive comparison. For example, at any instant in Sam's walk, a student might compare Sam's distance to the store and Sam's distance from home to create a new quantity, the difference between the distances. As with quantity and quantification, students could engage in quantitative operations without determining particular amounts of difference.

With Fig. 1, I express interconnections between quantity, quantification, and quantitative operations. Because both quantification and quantitative operations extend from quantities, I have placed unidirectional arrows between quantity and those elements. Conceiving of an attribute as being possible to measure is the first part

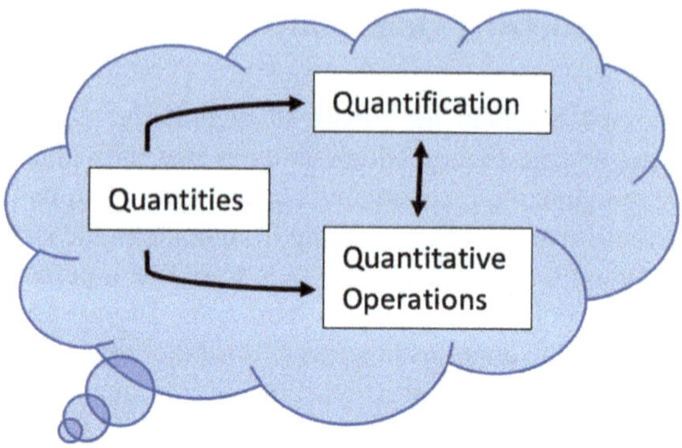

Fig. 1 Key elements of Thompson's theory of quantitative reasoning

of the quantification process. By engaging in quantitative operations, students can create new quantities in relationship to quantities they already know. I have placed bidirectional arrows between quantification and quantitative operations to indicate a reflexive relationship between them. Students can engage in quantification of some quantities, create new quantities, and then engage in quantification yet again.

Covariational reasoning (Carlson et al., 2002; Thompson & Carlson, 2017) entails conceptualizing relationships between attributes, which individuals perceive to be capable of varying and possible to measure. For example, one student may conceive that Sam's distances change in harmony with each other, their values continually changing together: "Sam's distance from home increases while Sam's distance from the store decreases." In contrast, another student may think about snapshots of the distances at particular instances in Sam's walk: "Now Sam is 2 blocks from home and 18 blocks from the store; now Sam is 5 blocks from home and 15 blocks from the store." As suggested by these examples, when individuals conceive of variation, or covariation in attributes, they have mental images of how those attributes have changed or are changing. Castillo-Garsow (2012) proposed the terms "chunky" and "smooth" to distinguish these images of change. One way to conceive of a distinction between these images is in their "countable-ness." Chunky images entail countable units, whereas smooth images entail a continual flow of change (Castillo-Garsow et al., 2013). The first example suggests smooth thinking because Sam's distances are continually changing together. The second example suggests chunky thinking because the focus is on particular instances in Sam's walk. While both images of change have utility, there is something special about smooth images of change, which comprise conceptions of continual change in attributes.

Not only did Thompson and Carlson (2017) position smooth thinking at the highest levels of variational and covariational reasoning, they contend that opportunities for students to engage in such thinking should happen early and often. Given the centrality of quantitative and covariational reasoning for students' mathematical development (Thompson & Carlson, 2017), I argue that they are more than just a means to promote students' learning of new mathematical concepts, such as rate or function. They are worthy ways of reasoning in and of themselves.

2 Mathematizing as a Way of Thinking Emerging from Students' Intellectual Need

To sketch out a landscape of ways of thinking, Harel (2008a) has provided three different, yet interrelated categories: proof schemes, problem-solving approaches, and beliefs about mathematics. Broadly, these ways of thinking involve how people determine the viability of an assertion, think while solving problems, and view mathematics itself. I posit that students' quantitative and covariational reasoning point to a categorically different way of thinking from those put forward by Harel. By engaging in quantitative and covariational reasoning, students can conceive of the possibility of measuring different attributes in a situation and form and interpret relationships between those attributes. This may or may not involve assessing the truth of an assertion, solving a given problem, or considering the nature of mathematics itself. I offer Freudenthal's (1973) term, "mathematizing," to characterize this fourth category of a way of thinking.

When mathematizing a situation, people conceive of some "thing" from a mathematical lens (Freudenthal, 1973). For example, I have provided descriptions of different ways students might mathematize Sam's situation, from quantitative and covariational lenses. These lenses are not "out there" for people to see, rather they are ways of thinking that people bring to a situation. By positioning mathematizing as a complementary, yet distinct way of thinking, from those put forward by Harel (2008a), I foreground mental actions involved in this human activity.

To provide a rationale for this fourth category, I appeal to the construct of goals (Simon & Tzur, 2004). By goal, I mean some achievable outcome that a person has set in an educational setting, rooted in their current conceptions and tasks they encounter (Simon & Tzur, 2004). A person's goal is a goal from their perspective; it can be different from a teacher or researcher's goal. For example, a teacher may intend for a student to sketch a graph of Sam's situation. One student may have a goal of sketching a graph that shows Sam's movement from home to the store. Another student may have a goal of exploring how Sam's distances are changing together. While the first student has a goal of solving the problem, the second student's goal involves investigating relationships between attributes in the situation. Mental actions compatible with the second student's goal are crucial for mathematizing Sam's situation in terms of quantity and covariation.

Like a student's goal, with the construct of intellectual need, a researcher employs their perspective of a student perspective, because an intellectual need is from the perspective of the person engaging in the thinking, rather than an outside observer. I conceive of a person's intellectual need as akin to a goal, with the caveat that an intellectual need emerges when a person finds a situation to be problematic for them. Whereas, a student may have a goal without problematizing anything. For example, one student may sketch a graph with the goal of showing Sam's literal movement, get feedback that a correct graph looks different, and experience nothing problematic about the situation. In contrast, another student with the same goal may wonder what could account for a graph's different features, and adjust their goal based on their wondering. The adjustments may entail separating the attributes from the situation and conceiving of how those attributes might be measured. As suggested by these examples, students may have the same "task" presented to them, yet they can conceive of that task in very different ways.

Students' covariational reasoning has potential to serve as a catalyst for their intellectual need. In a study of two university students, Paoletti and Moore (2017) have shown how aspects of students' covariational reasoning can engender an intellectual need for conceiving of a quantity, such as time, which may only be implicit in a problem situation. In particular, they have found that students conceived of time in a conceptual way (Thompson & Carlson, 2017), not just as elapsing, but as something possible to measure, in terms of duration. As Paoletti and Moore (2017) argued, such a conception can promote students' understanding of parametric function and represents something beyond mental actions in covariational reasoning. Describing this finding in terms of students' intellectual need can go like this: A way of reasoning (covariational reasoning) can create an intellectual need for a way of understanding (a new understanding of a quantity implicit in a situation), which can promote students' development of mathematical concepts (parametric function).

In light of the foundational nature of quantitative and covariational reasoning, I posit they are not only catalysts for, but also products of students' intellectual need. With Freudenthal's "mathematizing," I have described such ways of reasoning in broad terms, to illuminate a new category beyond the three offered by Harel (2008a). Broadening categories of ways of thinking can, in turn, make room for new categories of intellectual need. To this end, I propose an "intellectual need for relationships," which can engender students' quantitative and covariational reasoning.

3 An Intellectual Need for Relationships

Leaving room for the possibility of expansion, Harel (2013) put forward five categories of intellectual need: certainty, causality, computation, communication, and structure. Harel (2013) defined certainty as a need to determine the truth of some conjecture, causality as a need to explain why some assertion is true, computation as a need to determine values of measurable attributes in a situation, communication as a need to formalize and formulate mathematical ideas, and structure as a need to reorganize what is known in a logical way. Together, these intellectual needs provided a landscape to explain how students may reconcile situations they find to be problematic for them.

Students' quantitative and covariational reasoning point to a new category of intellectual need, beyond those put forward by Harel. Broadly, this new category involves a desire to explain, which shares some similarities with a need for causality. A key difference lies in the object of the explanation. This new category reflects a need to explain a situation from a mathematical lens, which does not necessitate explaining why something is true or determining the truth of a proposition. Furthermore, this new category of need is different from computation. As Thompson (1994,

2002, 2011) has asserted, students can conceive of the possibility of measuring attributes without finding particular amounts of measure. Forming and interpreting relationships between attributes does not necessitate formalizing ideas into symbolic expressions or formulating those expressions into the spoken word. Although a need to reorganize existing structures may follow from students' quantitative or covariational reasoning, it addresses a different kind of problem. Thus, I offer a sixth category of intellectual need: relationships.

Harel (2013) has posited that intellectual needs have three main characteristics, which I summarize here. First, they are from the perspective of a person, not an outside observer such as a researcher or teacher. Second, they are something people learn, not something innate. Third, they are linked to a person's desire to resolve some "problematic-to-them" situation. I view this third characteristic to be a key distinction between goals and intellectual needs. Goals may or may not result from a desire to resolve a problematic situation; they may just be part of engaging in some task. Intellectual needs resolve something problematic for a learner. Laying out each category of intellectual need, Harel (2013) has provided a three-part discussion: definition of the need, description of inchoate conceptions underlying the need, and historical evidence to account for the need. I follow this approach.

An intellectual need for relationships is a need to explain how elements work together, as in a system. This may apply to scientific phenomena, such as global warming, or to everyday situations, such as a filling bottle. While a need for causality is a need for directionality (e.g., A leads to B), a need for relationships is a need to understand how A and B work together. For instance, in the classic filling bottle problem (Shell Centre for Mathematical Education (University of Nottingham), 1985), students are to sketch graphs representing the height and volume of liquid in differently shaped bottles. I view an intellectual need for relationships to stem from people's desire to form connections between attributes, so they may mathematize the world around them. Across history, humans have connected measures of attributes, such as the length of an object's shadow, with a duration of time (Barnett, 1999). In the history of mathematics, a need for relationships has played a role in mathematicians' conceptualization of what is now called function.

Appealing to historical accounts of Boyer, which were compatible with those of Kleiner, Thompson and Carlson (2017) identified four broad eras in the evolution of the idea of function: proportion, equation, function (I), and function (II). In their discussion, Thompson and Carlson (2017) threaded the representation of relationships throughout the eras. In the proportion era, "people represented relationships between quantities geometrically" (p. 421). The equation era was "characterized by the use of equations to represent constrained variation in related quantities' values" (p. 422). The first function era was "characterized by explicit representations of a relationship between values of two quantities so that values of one determined values of another" (p. 422). The second function era, which is still continuing, was "characterized by values of one variable being determined uniquely by values of another" (p. 422). Thompson and Carlson (2017) emphasized how ideas of variation and covariation were central to people's development of the function concept, even though the evolution of people's meaning for function relegated those ideas to lesser, or even seemingly absent, roles.

In reflecting on the discussion of Thompson and Carlson (2017), I note a shift in the foreground and background, coinciding with the invention of algebraic representations. As algebraic representations have become more formal, causality has come to the foreground (e.g., the possibility of determining one variable's value given another). In turn, an explanation of how elements in a system work together has faded to the background (e.g., relationships between quantities given constraints in their variation). By proposing a need for relationships, I mean to foreground ways of reasoning, including quantitative and covariational reasoning, which are crucial for students' mathematical development (Thompson & Carlson, 2017).

4 Four Facets of an Intellectual Need for Relationships

I put forward four facets of an intellectual need for relationships: attributes in a situation (What are the things?), measurability of attributes (How can things be measured?), variation in attributes (How do things change?), and relationships between attributes (How do things change together?). I think of an intellectual need like a mental gemstone; a sparkling, multifaceted conception that can illuminate things once puzzling or mysterious. I include a question with each facet to emphasize a person's point of view, what they may be wondering in a situation. The first two facets address quantitative reasoning and the mental action of quantification. The last two address variational and covariational reasoning, respectively. I view the first facet, attributes in a situation, to ground the other facets, because it focuses on the "things" which people can separate from a situation, then conceive of as possible to measure or capable of varying.

While I present four facets, I leave room for the possibility for more to be included. I propose these facets based on theoretical underpinnings of quantitative and covariational reasoning, and on empirical results of fine-grained studies that I have led to investigate middle and secondary students' reasoning. My colleagues and I have found these facets to illuminate students' progressions in (or challenges with) their engagement in covariational reasoning (Johnson & McClintock, 2018; Johnson et al., 2017a, 2017b, 2020). I describe conceptions related to each facet, then highlight results to demonstrate how those conceptions were (or could have been) productive for students' reasoning.

5 Attributes in a Situation: What Are the Things?

To conceive of relationships between elements in a system, people need to separate those elements, or attributes, from the system itself. A conception of attributes themselves is foundational to quantitative (and covariational) reasoning. For example, to begin quantifying Sam's situation, students would separate attributes, such as Sam's distances from home and the store, from the physical situation itself. This conception may sound too obvious to highlight (e.g., of course students will separate distances from a situation); however, students' long-standing challenges with sketching and interpreting graphs suggest otherwise. Two enduring challenges involve conceptions of graphs as providing a static picture of a situation (Leinhardt et al., 1990), such as a physical map, or as showing the physical motion in a situation (Kerslake, 1977).

If students approach a graphing task with a goal of representing the physical motion they perceive in a situation, they likely will sketch a graph inconsistent with constraints of a Cartesian coordinate system. For example, a student may expect that a graph of Sam's walk from home to the store should share some physical characteristics with Sam's journey, and in turn, that student may sketch a graph that resembles the literal path that Sam took, regardless of the distances labeled on the axes. Even in the face of such inconsistencies, this goal can remain persistent for secondary students (Johnson et al., 2020).

6 Measurability of Attributes: How Can Things Be Measured?

To explain how elements work together in a system, people can mathematize different elements, or attributes, in a situation. I focus on a person's conception of the possibility of measuring some attribute they have separated from a situation, or their conception of a quantity, per Thompson's theory. Such a conception may or may not entail all three aspects of Thompson's (2011) process of quantification. For instance, a student may think of Sam's distance from home as a path represented by a line drawn on a map or a trail of breadcrumbs Sam may have left while walking. This student is doing more than just thinking of Sam engaging in the physical activity of walking to the store. They are separating an attribute from the situation and conceiving of how they might measure it. Students can extend from this conception to engage in all aspects of the quantification process by conceiving of a unit of measure and a multiplicative relationship between the unit of measure and the attribute.

When students conceive of how attributes may be measured, they are in a ripe position to mathematize variation in attributes. Evan McClintock and I have found that when middle school students conceived of an attribute as being possible to measure, it impacted their conceptions of variation in that attribute (Johnson & McClintock, 2018) when interacting with dynamic computer tasks involving "filling" polygons. For example, one task was a "filling triangle," in which students were to watch an animation of a right triangle "fill" with color, moving vertically from its horizontal base to the opposite vertex. All students who discerned variation in unidirectional change in that attribute (e.g., The "fill" increases, but the increases are slowing.) were those who conceived of the triangle's "fill" as an attribute possible to measure (e.g., the area of a polygon).

7 Variation in Attributes: How Do Things Change?

When exploring how elements work together in a system, students can conceive of how those elements, or attributes, vary. I liken this to a conception of a variable as something whose values can vary, rather than just a placeholder for some unknown value. When students engage in smooth thinking, they can conceive of continual variation in an attribute. Yet, at some point, there is reason to stop a continuation of ongoing values, at which point a person can conceive of some accumulated amount (Castillo-Garsow et al., 2013). Johnson (2012) used the term "smooth chunks" to describe products of this way of thinking, to distinguish them from "chunks" emerging from chunky images of change. It is beneficial for students to have space to wonder, "How do things change?," before determining, "By how much have things changed?," because they can conceive of values in an interval, and not just find beginning and ending amounts.

When students conceive of continuing variation in individual attributes, it is a productive time for teacher/researchers to engage them in tasks to promote their covariational reasoning. In two different studies that I have led, with secondary students from different school settings, when students conceived of continuing variation in individual attributes in a situation, they were able to shift to covariational reasoning via their work on digital task sequences (Johnson et al., 2017a, 2017b, 2020). Not only did students shift their reasoning, but they also were aware of a change in their thinking and found the new way of thinking to be powerful for them.

8 Relationships Between Attributes: How Do Things Change Together?

To explain how elements work together in a system, students conceive of how those elements change together, forming and interpreting relationships between those attributes. Put another way, they engage in covariational reasoning. Such reasoning can promote students' conception of nuances in relationships between attributes. For example, a student may wonder why a graph has a particular kind of curvature, or whether a linear or nonlinear graph may best represent a relationship. This can allow students to fine-tune their interpretations of graphs, and they can discern "new-to-them" distinctions.

Secondary students' engagement in covariational reasoning can foster their attention to and accounting for distinctions and nuances in graphs that represent relationships between attributes in linked animations (Johnson et al., 2017a, 2017b, 2020). I discuss two instances, in which students had a spontaneous question or noticing, during an individual task-based interview. These instances illustrate how a student's intellectual need for relationships can intertwine with their graphing.

One student, Alan, spontaneously questioned how it could be possible for a graph to be piecewise linear, when he noticed the linked animation was slowing down

(Johnson et al., 2020). The graph represented a relationship between distance and height, with time as an implicit variable, because each of the attributes were varying with elapsing time. The researcher invited Alan to explore the situation further. Relating different amounts of distance and height, Alan convinced himself that a piecewise linear graph best represented the relationship (Johnson et al., 2020). Another student, Ana, spontaneously noticed differences in the curvature between a graph that she had drawn and a computer graph. (Ana's graph looked more like a parabolic arch and the computer graph was a sine curve.) By conceiving of how two different attributes were varying together in the situation (e.g., The Ferris wheel cart is gaining a lot of distance, but only a little bit of height), she decided it made more sense for her graph to curve in a way that would account for that kind of covariation (Johnson et al., 2017a, 2017b).

9 Task Design Considerations: A Ferris Wheel "Techtivity"

By a task, I mean something more than an artifact, such as a problem written on paper or a computer activity. Tasks include the intentions and activities of those designing the task, implementing the task, and engaging with the task (Johnson et al., 2017a, 2017b). To illustrate task design considerations, I provide an example of a task, a dynamic computer activity, that my colleague, Gary Olson, termed a "Techtivity." My purpose is to illuminate how task designers may work to nurture students' intellectual need for relationships, and in turn engender students' quantitative and covariational reasoning.

The Techtivity that I share is part of a set of seven freely available digital tasks (Desmos, n.d.), usable by a broad range of instructors. Each Techtivity consists of a series of screens which students can move through at their own pace. There are four main components. First, students are to view an animation of a dynamic situation involving change in progress, a move common for researchers designing tasks to investigate students' conceptions of change and variation. Second, students are to manipulate dynamic segments representing measures of two attributes in the situation, a move that operationalizes Thompson's (2002) recommendation that students use their fingers as tools to represent change in individual attributes. Third, students are to sketch a Cartesian graph to represent a relationship between variables in the situation. Fourth, students are to repeat the second and third components for the same situation, with attributes represented on different axes, a move that shares similarities with tasks designed by Moore and colleagues (e.g., Moore et al., 2014). In addition, at the end of each Techtivity there are questions designed to promote students' reflection on relationships between attributes in the situation.

Figure 2 depicts the four components of a Ferris wheel Techtivity. The animation shows a green cart that begins in the middle left of the Ferris wheel (Fig. 2, bottom left), and moves clockwise around the Ferris wheel for one rotation. Throughout my description of this Techtivity, I use parentheticals to highlight how the design components link back to the four facets of an intellectual need for relationships.

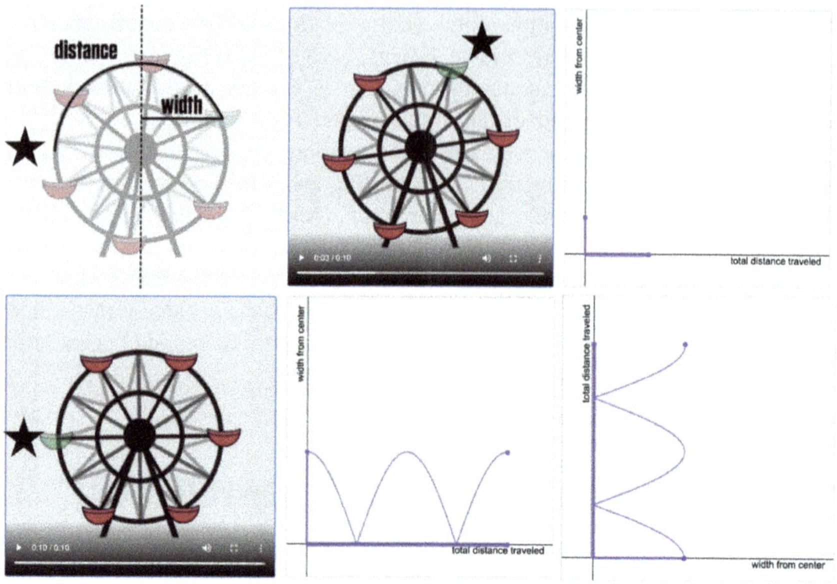

Fig. 2 Components of a Ferris wheel techtivity

There are many attributes which students may conceive of in the Ferris wheel situation (What are the things?). For this Techtivity, students are to consider two attributes: the cart's width from the center and the cart's distance traveled around the wheel. The width is measured by the cart's horizontal distance from a vertical line extending through the center of the wheel (How can things be measured?). While the width might seem like an arbitrary attribute, people riding on a Ferris wheel may feel this "width" as a sensation of moving out and back while the cart goes around the wheel. Figure 2 (top left) shows a trace of each attribute for a partial turn of the wheel. When students represent change in the width and distance (How does each thing change?), at first the attributes are represented on the vertical and horizontal axes, respectively (Fig. 2, top right). Both Cartesian graphs are shown at the bottom of Fig. 2. In the second graph (bottom right), the width and distance are represented on the horizontal and vertical axes, respectively (How do things change together?).

At the end of this Ferris wheel Techtivity, there are two different types of reflection questions. One is an interpretation of a single point on the graph, and the other is a comparison of two different graphs. First, students are to predict the green cart's location on the wheel given a point on the graph (Fig. 3). Second, they are to determine whether they agree or disagree with a student's claim that the two different looking graphs generated by the computer can represent the same relationship between attributes (Fig. 4). In addition, Figs. 3 and 4 include selected responses from undergraduate students enrolled in a College Algebra course.

The selected responses to each reflection question provide examples of how students may conceive of the four different facets in their work on the task. Responses

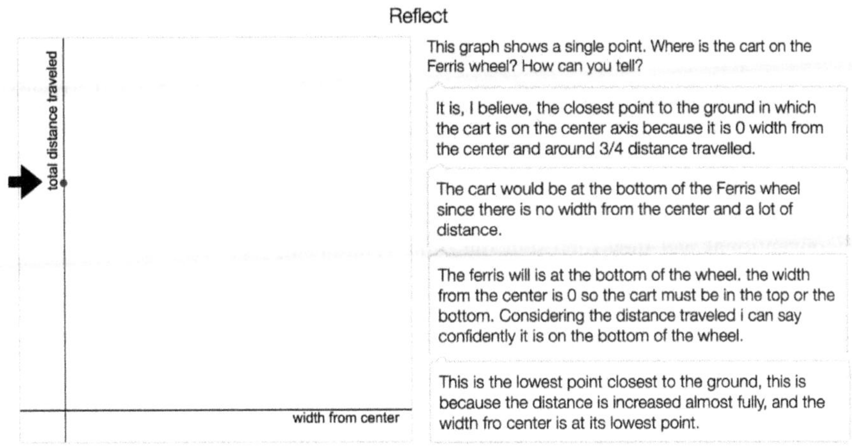

Fig. 3 Reflection question: graph shows a single point

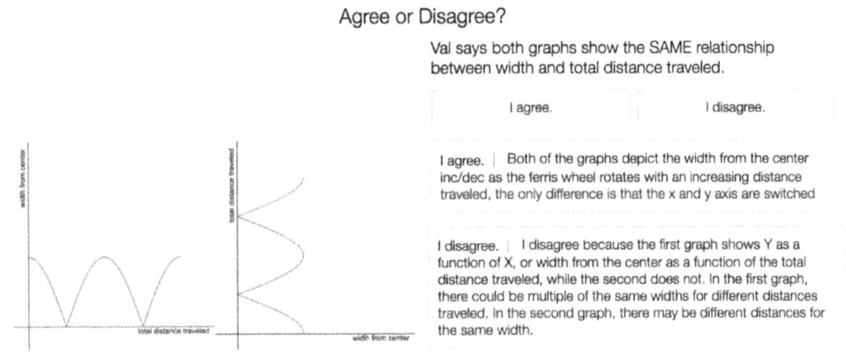

Fig. 4 Reflection question: agree or disagree?

to the first reflection question (Fig. 3) represent a range of reasons given by students who predicted the cart to be at the bottom of the wheel, and mentioned both width and distance in their response (What are the things? How can the things be measured?). Some students have provided specific amounts of width and distance to support their responses (e.g., "0 width from the center," "around ¾ distance travelled"), while others have discussed in more general terms (e.g., "no width," "a lot of distance"). The two responses to the second reflection question (Fig. 4) give evidence of students' conceptualization of change in, and relationships between attributes, even when they provide differing views of Val's claim (How is each thing changing? How do things change together?).

With the reflection questions, I have intended to invite students to relate, or even mentally fuse different attributes to make sense of the situation (How do things change together?). Put another way, I have worked to create space to engender students' mental forming of multiplicative objects (Thompson & Carlson, 2017). In the first question (Fig. 3), the point students are to interpret is on the vertical axis, representing a location when the green cart's width from center is equal to zero. There are two possibilities, the top and bottom of the wheel. By taking into account both the width from the center and the distance traveled, students can determine the point to represent when the cart is located at the bottom of the wheel. In the second question (Fig. 4), the two graphs look different, with the second graph being unconventional, yet they represent the same relationship between attributes. Even advanced university students can have challenges interpreting function relationships represented by unconventional graphs (Moore et al., 2014). With this in mind, I have designed the reflection question in terms of whether students agree or disagree with Val, rather than whether Val is right or wrong. With this move, I intend to make room for students to consider Val's statement as a sensible claim made by a human, rather than rushing to a judgment of the validity of Val's claim. For instance, students may think Val's claim is reasonable, yet state Val is wrong, because they do not think such a claim is viable to make in a mathematics class, given the unconventional looking graph.

10 Discussion

I have posited an expansion in Harel's categories of students' intellectual need, to include a need for relationships; a need to explain how elements work together, as in a system. While interconnected, this need is distinct from the needs that Harel (2013) offered (certainty, causality, computation, communication, and structure). Thompson and Carlson (2017) have discussed how relationships are woven throughout scholars' development of the function concept. Yet, the relationships are something more than just a stepping stone in students' development of the concept of function. Variational and covariational reasoning are theoretical constructs, ways of thinking that can explain students' conceptualizations of situations in ways that are both quantitative and dynamic (Thompson & Carlson, 2017).

When engaging in quantitative reasoning, students mathematize attributes, by conceiving of them as being possible to measure. Both Thompson and Harel discuss mental actions of quantifying and quantification, drawing on Piaget's theory. Harel (2013) explains quantifying in broad terms, such that a person could transform some perceptible "thing," for example a feeling of movement, into a measurable attribute. Thompson's (2011) definition of quantification illuminates three mental actions in such a transformation: a conception of an attribute as possible to measure, a unit with which to measure the attribute, and a multiplicative relationship between the unit and attribute. Harel (2013) has located quantifying within an intellectual need for computation, yet quantifying is not limited to a need for computation. Quantifying entails relationships, which Thompson's definition addresses. By positioning an intellectual need for relationships as a need in and of itself, I aim to raise the status of quantifying, and related forms of reasoning, to position them as something more than a means to compute a result.

When engaging in covariational reasoning, students form and interpret relationships between attributes they can conceive to be capable of varying and possible to measure. Images of change are part of such mental action, and those images make a difference. Chunky images of change involve only beginning and ending amounts, while smooth images of change allow for all values in an interval. Accordingly, Thompson and Carlson (2017) position smooth images of change at the highest levels of variational and covariational reasoning. Furthermore, they assert that students' opportunities to engage in smooth thinking come early and often. With the Ferris wheel Techtivity, I provide an example of a task designed to engender such opportunities.

Nurturing students' intellectual need for relationships may help them to develop further abstractions. One possibility is an "abstracted quantitative structure" (Moore et al., this volume; Moore et al., 2019a, 2019b). Moore et al., (2019a, 2019b, p. 1879) have characterized such a structure as "a system of quantitative relationships a person has interiorized to the extent they can operate as if it is independent of specific figurative material (i.e., representation free)." To illustrate, they report on a preservice mathematics teacher who conceived of the inverse sine function as a relationship that was irrespective of a particular representation type. Questions, such as the second reflection question in the Ferris wheel Techtivity (Fig. 4), can afford opportunities for students to conceive of relationships that remain invariant, even if physical characteristics of graphs vary. Both students' responses (Fig. 4, right), provide evidence of their conception of relationships between attributes in the situation. Physical artifacts, such as Cartesian graphs, are products of a representation system. I conjecture that students who conceive of relationships between quantities in ways that are independent of such artifacts, can discern aspects of the representation system itself (see also Johnson, 2020). Integrating multiple theoretical lenses can be productive for researchers investigating students' quantitative and covariational reasoning while engaging in tasks involving socially shared artifacts, such as Cartesian graphs.

An intellectual need for relationships can serve as a starting point to reimagine curricular materials focused on function. One consequence can be opportunities to conceive of quantities as covarying, for which Thompson and Carlson (2017) advocate. A second consequence can be the ways in which students encounter different types of functions. In U.S. secondary math classrooms, students typically see, in a particular order, different types of functions and graphs representing those function types (e.g., linear, then quadratic, then exponential). With such an approach, students may miss out on the relationships themselves. Our field has spent much time arguing about the order in which to present different function types (e.g., Should linear functions come first? Should exponentials come before quadratics?). I recommend reframing the argument. Rather than organizing materials around function types, center relationships between attributes, then introduce different function types as a way to explain different kinds of relationships.

11 Conclusion

I proposed an expansion in Harel's categories of intellectual needs and ways of thinking. With my choice of Freudenthal's term, mathematizing, I intended to communicate that conceiving of some "thing" from a mathematical lens was a viable way of thinking in its own right. "Mathematizing" is more than a part of a problem-solving approach or a proof scheme. It is a way of thinking that can entail conceptions of variables as actually varying together, a fundamental mathematical idea not to be backgrounded in service of a formal definition. For students to develop quantitative and covariational reasoning across K-12 mathematics, for which Thompson and Carlson (2017) advocate, it is important that ways of thinking be positioned as just as valuable as ways of understanding. While worthwhile, such a goal is at odds with high-stakes testing pressure rampant in K-12 education in the U.S. in which students and teachers can hear messages that test results, and consequently answer finding and computation, are the only things that matter. Hence, stakeholders have work to do at multiple levels so that each and every student can have opportunities to engage in mathematical reasoning in spaces where they feel safe and valued.

There is a "tension of intention" (Johnson et al., 2017a, 2017b) with task design to engender students' quantitative and covariational reasoning, taking into account students' intellectual need for relationships. While some students may have goals consistent with satisfying an intellectual need for relationships, other students may have different goals. As task designers, it is important to wrestle with the tension of anticipated versus actual intellectual needs in students' engagement with a task situation. As researchers, it is crucial to critique one's own task and research design, to guard against deficit approaches in investigations of student cognition (Johnson et al., 2020).

Promoting students' reasoning and sense making in mathematics classrooms is not a neutral activity. Despite the utility of an intellectual need for relationships, students may not perceive mathematics classrooms to be places where they could exert such a need. They may have internalized that to "play the game" (Gutiérrez, 2009) of mathematics, answers are what matter. Furthermore, even if instructors make space for students' reasoning, existing classroom power dynamics can become more apparent, for example, which student voices get amplified (or marginalized). This can be compounded if the intended reasoning is something that their instructors still need to develop. Thompson and Carlson (2017) have called for research investigating teachers' experiences fostering students' covariational reasoning, highlighting how teachers may need to develop such reasoning themselves. Such a problem is complex, and I argue that its investigation could benefit from teams of researchers coordinating different theoretical lenses to explain multiple phenomena at play.

Broadly, I view an intellectual need for relationships to be compatible with broader needs for play and exploration in mathematics. When students are mathematizing a situation via their quantitative and covariational reasoning, they conceptualize that situation in terms of attributes which they conceive to be measurable and the ways in which those attributes can change together. They can play with different possibilities and explore results of their efforts (e.g., What happens if?). Students' quantitative and covariational reasoning are more than just a means to develop their function understanding. When their intellectual need for relationships is nurtured in mathematical spaces, students can feel that their ways of reasoning, as well as the results of their reasoning, are welcomed and valued.

Acknowledgements U.S. National Science Foundation Grants (DUE-1709903, DUE-2013186) have supported the development and implementation of the Ferris wheel Techtivity. Opinions and conclusions are those of the author. I thank Evan McClintock for his feedback on this manuscript.

References

Barnett, J. E. (1999). *Time's pendulum: From sundials to atomic clocks, the fascinating history of timekeeping and how our discoveries changed the world*. Houghton Mifflin Harcourt.

Carlson, M., Jacobs, S., Coe, E., Larsen, S., & Hsu, E. (2002). Applying covariational reasoning while modeling dynamic events: A framework and a study. *Journal for Research in Mathematics Education, 33*(5), 352–378.

Castillo-Garsow, C. (2012). Continuous quantitative reasoning. In R. Mayes & Hatfield (Ed.), *Quantitative reasoning and mathematical modeling: A driver for STEM integrated education and teaching in context* (pp. 55–73). University of Wyoming.

Castillo-Garsow, C., Johnson, H. L., & Moore, K. C. (2013). Chunky and smooth images of change. *For the Learning of Mathematics, 33*(3), 31–37.

Desmos. (n.d.). *How graphs work*. bit.ly/HowGraphsWork.

Ellis, A., Ely, R., Singleton, B., & Tasova, H. (2020). Scaling-continuous variation: Supporting students' algebraic reasoning. *Educational Studies in Mathematics, 104*, 87–103.

Freudenthal, H. (1973). *Mathematics as an educational task*. D. Reidel Publishing.

González, D. A. (2021). The progression of preservice teachers' covariational reasoning as they model global warming. *The Journal of Mathematical Behavior, 62*, Article 100859.

Gutiérrez, R. (2009). Framing equity: Helping students "play the game" and "change the game." *Teaching for Excellence and Equity in Mathematics, 1*(1), 4–8.

Harel, G. (1998). Two dual assertions: The first on learning and the second on teaching (or vice versa). *The American Mathematical Monthly: The Official Journal of the Mathematical Association of America, 105*(6), 497–507.

Harel, G. (2008a). DNR perspective on mathematics curriculum and instruction, part I: Focus on proving. *ZDM: The International Journal on Mathematics Education, 40*(3), 487–500.

Harel, G. (2008b). A DNR perspective on mathematics curriculum and instruction. Part II: With reference to teacher's knowledge base. *ZDM: The International Journal on Mathematics Education, 40*(5), 893–907.

Harel, G. (2013). Intellectual need. In K. R. Leatham (Ed.), *Vital directions for mathematics education research* (pp. 119–151). Springer.

Johnson, H. L. (2012). Reasoning about variation in the intensity of change in covarying quantities involved in rate of change. *The Journal of Mathematical Behavior, 31*(3), 313–330.

Johnson, H. L. (2020). Task design for graphs: Rethink multiple representations with variation theory. *Mathematical Thinking and Learning*. https://doi.org/10.1080/10986065.2020.1824056

Johnson, H. L., Coles, A., & Clarke, D. (2017a). Mathematical tasks and the student: navigating "tensions of intentions" between designers, teachers, and students. *ZDM: The International Journal on Mathematics Education*. https://link.springer.com/article/https://doi.org/10.1007/s11858-017-0894-0

Johnson, H. L., & McClintock, E. (2018). A link between students' discernment of variation in unidirectional change and their use of quantitative variational reasoning. *Educational Studies in Mathematics, 97*(3), 299–316.
Johnson, H. L., McClintock, E., & Gardner, A. (2020). Opportunities for reasoning: Digital task design to promote students' conceptions of graphs as representing relationships between quantities. *Digital Experiences in Mathematics Education, 6*(3), 340–366.
Johnson, H. L., McClintock, E., & Hornbein, P. (2017b). Ferris wheels and filling bottles: A case of a student's transfer of covariational reasoning across tasks with different backgrounds and features. *ZDM: The International Journal on Mathematics Education, 49*(6), 851–864.
Kerslake, D. (1977). The understanding of graphs. *Mathematics in School, 6*(2), 22–25.
Leinhardt, G., Zaslavsky, O., & Stein, M. K. (1990). Functions, graphs, and graphing: Tasks, learning, and teaching. *Review of Educational Research, 60*(1), 1–64.
Moore, K. C., Liang, B., Tasova, H. I., & Stevens, I. E. (2019a). Abstracted quantitative structures. In S. Otten, A. Candela, Z. de Araujo, C. Haines, & C. Munter (Eds.), *Proceedings of the forty-first annual meeting of the north American chapter of the international group for the psychology of mathematics education* (pp. 1879–1883). University of Missouri.
Moore, K. C., Silverman, J., Paoletti, T., & LaForest, K. (2014). Breaking conventions to support quantitative reasoning. *Mathematics Teacher Educator, 2*(2), 141–157.
Moore, K. C., Stevens, I. E., Paoletti, T., Hobson, N. L. F., & Liang, B. (2019b). Pre-service teachers' figurative and operative graphing actions. *The Journal of Mathematical Behavior, 56*, Article 100692.
Paoletti, T., & Moore, K. C. (2017). The parametric nature of two students' covariational reasoning. *The Journal of Mathematical Behavior, 48*, 137–151.
Shell Centre for Mathematical Education (University of Nottingham). (1985). *The language of functions and graphs: An examination module for secondary schools.* JMB/Shell Centre for Mathematical Education.
Simon, M. A., & Tzur, R. (2004). Explicating the role of mathematical tasks in conceptual learning: An elaboration of the hypothetical learning trajectory. *Mathematical Thinking and Learning, 6*(2), 91–104.
Thompson, P. W. (1994). The development of the concept of speed and its relationship to concepts of rate. In G. Harel & J. Confrey (Eds.), *The development of multiplicative reasoning in the learning of mathematics* (pp. 179–234). State University of New York Press.
Thompson, P. W. (2002). Didactic objects and didactic models in radical constructivism. In K. Gravemeijer, R. Lehrer, B. Van Oers, & L. Verschaffel (Eds.), *Symbolizing, modeling and tool use in mathematics education* (pp. 197–220). Springer.
Thompson, P. W. (2011). Quantitative reasoning and mathematical modeling. In Chamberlain, S. A., & Hatfield, L. (Ed.), *New perspectives and directions for collaborative research in mathematics education: Papers from a planning conference for wisdom^e* (Vol. 1, pp. 33–56). University of Wyoming College of Education.
Thompson, P. W., & Carlson, M. P. (2017). Variation, covariation, and functions: Foundational ways of thinking mathematically. In J. Cai (Ed.), *Compendium for research in mathematics education* (pp. 421–456). National Council of Teachers of Mathematics.

Abstracted Quantitative Structures: Using Quantitative Reasoning to Define Concept Construction

Kevin C. Moore, Biyao Liang, Irma E. Stevens, Halil I. Tasova, and Teo Paoletti

Steffe and Thompson enacted and sustained research programs that have characterized students' (and teachers') mathematical development in terms of their conceiving and reasoning about measurable or countable attributes (see Steffe & Olive, 2010; Thompson & Carlson, 2017). Thompson (1990, 2011) formalized such reasoning into a system of mental operations he termed *quantitative reasoning*, and researchers have since adopted quantitative reasoning to characterize individuals' meanings within topical and related reasoning areas. For instance, researchers have adopted a quantitative reasoning perspective to explore how individuals construct or understand exponential relationships (Castillo-Garsow, 2010; Ellis et al., 2015), graphs or coordinate systems (Frank, 2017; Lee, 2017; Lee et al., 2019), and trigonometric functions (Moore, 2014; Thompson et al., 2007). Relatedly, researchers have adopted quantitative reasoning to explore individuals' meanings for concepts including rate of change (Byerley & Thompson, 2017; Johnson, 2015a, 2015b), function (Oehrtman et al.,

K. C. Moore (✉)
University of Georgia, Athens, Georgia
e-mail: kvcmoore@uga.edu

B. Liang
The University of Hong Kong, Pok Fu Lam, Hong Kong SAR
e-mail: biyao@hku.hk

I. E. Stevens
University of Rhode Island, Kingston, RI, USA
e-mail: irma.stevens@uri.edu

H. I. Tasova
CSU San Bernardino, San Bernardino, CA, USA
e-mail: halil.tasova@csusb.edu

T. Paoletti
University of Delaware, Newark, DE, USA
e-mail: teop@udel.edu

G. Karagöz Akar et al. (eds.), *Quantitative Reasoning in Mathematics and Science Education*, Mathematics Education in the Digital Era 21,
https://doi.org/10.1007/978-3-031-14553-7_3

2008; Paoletti & Moore, 2018), and accumulation (Thompson & Silverman, 2007), to name a few. More generally, researchers have related quantitative reasoning to other types of reasoning processes including multiplicative reasoning (Hackenberg, 2010; Tzur, 2004), algebraic reasoning (Ellis et al., 2020; Smith III & Thompson, 2007), generalization (Ellis, 2007), problem solving (Carlson et al., 2003), and transfer (Lobato & Siebert, 2002). An additional example that we leverage significantly in this chapter is reasoning about quantities changing in tandem, or *covarying* (e.g., Carlson et al., 2002; Johnson, 2012, 2015b; Saldanha & Thompson, 1998; Stalvey & Vidakovic, 2015).

The aforementioned studies and their authors' research agendas vary in the extent they focus on local or longitudinal development. Carlson et al. (2002) conducted a set of clinical interviews with calculus students in order to provide a localized description of their covariational reasoning. In contrast, Steffe and Olive (2010) provide a longitudinal model for children's construction and development of fractional schemes and concepts. Both grain sizes are critical to mathematics education research, with localized studies forming the foundation for longitudinal and generalized descriptions of individuals' cognition. But, each grain size has an associated cost. Localized studies can make it difficult for a researcher to characterize an individual's abstraction of a concept due to the extensive work necessary to make claims about an individual's in-the-moment reasoning. Longitudinal studies can make it difficult for the researcher to incorporate nuanced discussions of individual's in-the-moment construction of a concept due to the focus on described operations and their development in ways independent of particular contexts.

In this chapter, we introduce the construct *abstracted quantitative structure* (AQS) to marry the two aforementioned grain sizes and enable sensitivity to both localized and longitudinal development. Defined generally, an AQS is a system of quantitative operations a person has interiorized to the extent they can operate *as if* it is independent of specific figurative material.[1] That person can thus re-present this structure in several ways, and an AQS enables an individual to accommodate to novel experiences permitting the associated quantitative operations. Importantly, the AQS construct provides researchers (and teachers) criteria for characterizing individuals' construction and abstraction of concepts, and the AQS criteria are generalizable across concepts rather than specific to a particular concept. Researchers conducting work in the area of quantitative reasoning have made significant progress in articulating its role in the learning of particular concepts, yet generalized descriptions of how concept construction can be framed in terms of quantitative reasoning are less detailed or prevalent. This lack of specificity likely limits researchers' abilities to apply quantitative reasoning to their work (Drimalla et al., 2020; Thompson, 2008). This is especially problematic due to the extent researchers referenced above have shown quantitative reasoning to be a key component to students constructing a mathematics that is generative, coherent, generalizable, and sophisticated. It is thus important that researchers have the tools necessary to clarify meanings associated with these reasoning processes. In addition to providing generalizable criteria

[1] As we elaborate below, no conceptual structure is truly representation free.

by which to describe concept construction, the AQS construct helps make more explicit the role of quantitative reasoning in students' learning and development of key mathematical concepts.

In this chapter, we first introduce the AQS construct and its criteria. Due to its criteria being informed by numerous perspectives and extant constructs, we then introduce background information that underpins the AQS construct. With the background information in place, we provide a more detailed discussion of the AQS construct and its criteria. As part of this discussion, we draw from data to illustrate both indications—those actions that are consistent with particular AQS criteria—and contraindications—those actions that are inconsistent with particular AQS criteria—of individuals having constructed an AQS.[2] Following our empirical examples, we discuss the AQS in terms of how it can inform both research and teaching.

1 Introducing the Abstracted Quantitative Structure Construct

von Glaserfeld (1982) defined a *concept* as, "any structure that has been abstracted from the process of experiential construction as recurrently usable…must be stable enough to be re-presented in the absence of perceptual 'input'" (p. 194). Our notion of an AQS applies and extends von Glasersfeld's definition of concept to the area of quantitative and covariational reasoning. In the introduction, we defined an AQS as a system of quantitative (including covariational) operations a person has interiorized to the extent he or she can operate as if it is independent of specific figurative material. Adapting and extending von Glaserfeld's definition, an AQS is a system of quantitative operations (or quantitative structure) that an individual has interiorized so that it:

(1) is recurrently usable beyond its initial experiential construction;
(2) can be re-presented in the absence of available figurative material including that in which it was initially constructed;
(3) can be transformed to accommodate to novel contexts permitting the associated quantitative operations;
(4) is anticipated as re-presentable in any figurative material that permits the associated quantitative operations.

Our development of the Criteria 1–4 (C1–C4) defining an AQS is informed by perspectives on quantitative reasoning and covariational reasoning, distinctions between figurative and operative thought, and different forms of re-presentation. To provide the proper foundation for further defining and illustrating the criteria

[2] We underscore that we do not consider an AQS to be an exhaustive description of the meanings that can be associated with some concept. An AQS is a construct that can be used to describe a meaning for a concept that is rooted in quantitative and covariational reasoning, and we limit our discussion to such meanings except when contrasting them with alternative meanings to clarify AQS defining criteria.

associated with an AQS, we introduce critical definitions and perspectives in this section.

1.1 Quantitative Reasoning

Thompson (2011) defined quantitative reasoning as the mental operations involved in conceiving a context as entailing measurable attributes (i.e., quantities) and relationships between those attributes (i.e., quantitative relationships). A premise of quantitative reasoning is that quantities and their relationships are idiosyncratic constructions that occur and develop over time (e.g., hours, weeks, or even years). A researcher or a teacher cannot take quantities or their relationships as a given when working with students or teachers (Izsák, 2003; Moore, 2013; Thompson, 2011). Furthermore, and reflecting the criteria of an AQS presented below, a researcher or teacher should not assume a student has constructed a system of quantities and their relationships based on actions within only one context (e.g., situation, graph, or formula).

An important distinction is Thompson's (Smith III & Thompson, 2007; Thompson, 1990) use of *quantitative operation/magnitude* and *arithmetic operation/measure*. The former refers to the mental actions involved in constructing a quantity via a quantitative relationship. The latter are actions used to determine a quantity's numerical measure. Following Thompson (1990), we illustrate these distinctions using a comparison between two heights. Thompson (1990) described that an additive comparison requires one to construct an image of the measurable attribute that indicates by how much one height exceeds the other height (Fig. 1). Constructing such a quantity (i.e., a difference in heights) through the quantitative operation of comparing two other quantities additively does not depend on having specified measures. Returning to Fig. 1, no arithmetic operations are needed to conceive of a difference in heights as a measurable attribute, nor are they needed to conceive that difference as the black segment. *Arithmetic operations*, on the other hand, are those operations between specified or generalized measures such as addition, subtraction, or multiplication, that one uses to evaluate a quantity's measure. Such operations often occur in the context of symbols like inscriptions or glyphs. The arithmetic operations used to evaluate a quantity may reflect those quantitative operations that form the quantity (Fig. 2a, in which subtraction is used to calculate the difference quantity) or they may reflect other contextual relationships and properties (Fig. 2b, in which multiplication is used to calculate the difference quantity) (Thompson, 1990).

We provide Thompson's distinction between quantitative operations and arithmetic operations to emphasize that the *enactment* of quantitative operations occurs in the context of figurative material that permit those operations. As Steffe described in the context of the use of formulas versus graphs, "operations have to operate on something and that something is the figurative material contained in the operations, figurative material that has its origin in the construction of the operations" (L. P. Steffe, personal communication, July 24, 2019). To illustrate, consider Fig. 3. For

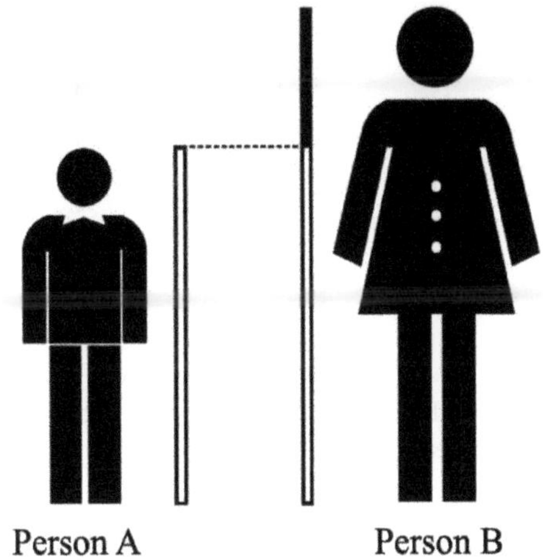

Fig. 1 An image of an additive comparison based in magnitudes

(a)

Height A: 45 inches
Height B: 60 inches

$$60 - 45 = 15$$

Height B exceeds Height A by 15 inches.

(b)

Height B is 4/3 times as large as Height A. Height A is 45 inches.

$$\frac{1}{3} \cdot 45 = 15$$

Height B exceeds Height A by 15 inches, because it exceeds Height A by 1/3 of Height A.

Fig. 2 The arithmetic operations that might be used to evaluate the measure of a quantity

example, the symbol "3" was socially negotiated as a way to signify those operations involved in measuring some magnitude as three of some unit. The symbol "3" is not designed to afford the enactment of quantitative operations.[3] On the other hand, a segment (more naturally) provides figurative material to assimilate via quantitative operations as having a measure of "3". Figure 3 illustrates a sequence that involves operations associated with units coordination including creating a unit and iterating that unit to determine the segment is some number of times as large as the unit (Steffe & Olive, 2010).

Addressing this distinction's implications for the AQS criteria, because symbols including glyphs and inscriptions are typically not used for the purpose of providing the figurative material to operate on quantitatively, students operating with symbols as such provides limited evidence of them enacting quantitative operations (Liang & Moore, 2021; Moore, Stevens, et al., 2019a, 2019b; Van Engen, 1949). On the other

[3] We acknowledge that we can identify creative ways to partition the symbol, but it is not used for such purposes. A stronger example of the distinction between a symbol and figurative material that permits quantitative operations is provided in the following section.

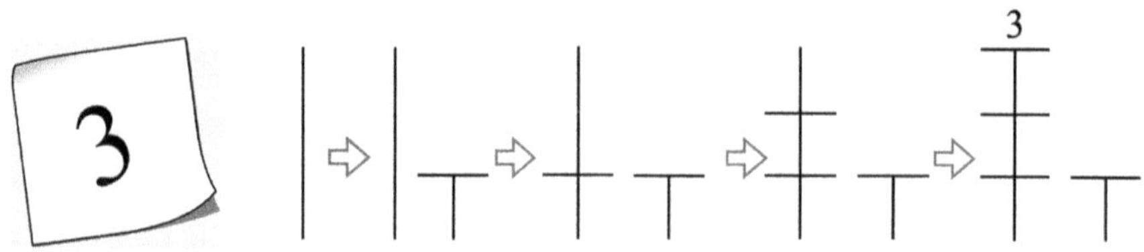

Fig. 3 The symbol "3" and a segment that affords enacting the mental operations signified by "3"

hand, coordinate systems, displayed graphs, phenomena, physical objects (e.g., a Ferris wheel or fraction blocks), and the like provide figurative material in which quantities can to be conceived and quantitative operations can be constructed and enacted. Thus, a student's activity with them can provide evidence of that student's engagement in reasoning quantitatively at that moment. It is for that reason that our defining the AQS criteria is with reference to contexts like coordinate systems, phenomena, and physical objects. We again illustrate the relevance of our focus in the following section and in the context of covarying quantities (Fig. 5) rather than quantities in a fixed state (Fig. 1 or Fig. 3).

1.2 Covariational Reasoning

A form of quantitative reasoning is covariational reasoning, which is defined as the actions involved in constructing relationships between two quantities that vary in tandem (Carlson et al., 2002; Saldanha & Thompson, 1998; Thompson & Carlson, 2017). Researchers have identified that covariational reasoning is critical for key concepts of K–16 mathematics including function (Carlson, 1998; Oehrtman et al., 2008), modeling dynamic situations (Carlson et al., 2002; Johnson, 2012, 2015b; Paoletti & Moore, 2017), and calculus (Johnson, 2015a; Thompson & Silverman, 2007; Thompson, 1994b). Researchers have also illustrated that covariational reasoning is critical to constructing function classes (Ellis, 2007; Hohensee, 2014; Lobato & Siebert, 2002; Moore, 2014).

Carlson et al. (2002), Confrey and Smith (1995), Ellis and colleagues (Ellis, 2011; Ellis et al., 2020), Castillo-Garsow and colleagues (Castillo-Garsow, 2012; Castillo-Garsow et al., 2013), Johnson (2015a, 2015b), and Thompson and Carlson (2017) have each detailed covariation frameworks and mental actions. Reflecting the emphasis of the empirical examples we use below, we narrow the present chapter's focus to *Mental Action 3* (Fig. 4, MA3) identified by Carlson et al. (2002). MA3 refers to coordinating and comparing quantities' amounts of change, which is a critical mental action to differentiating between nonlinear and linear growth (Paoletti & Vishnubhotla, this volume) and various function classes (Ellis et al., 2015; Moore, 2014). MA3 is also important for understanding and justifying that a graph and its curvature

Mental Action	Descriptions of Mental Actions
MA1	Coordinating the value of one variable with changes in the other
MA2	Coordinating direction of change of one variable with changes in the other variable
MA3	Coordinating amount of change of one variable with changes in the other variable
MA4	Coordinating the average rate-of-change of the function with uniform increments of change in the input variable
MA5	Coordinating the instantaneous rate of change of the function with continuous changes in the independent variable for the entire domain of the function

Fig. 4 Carlson et al., (2002, p. 357) covariational reasoning mental actions

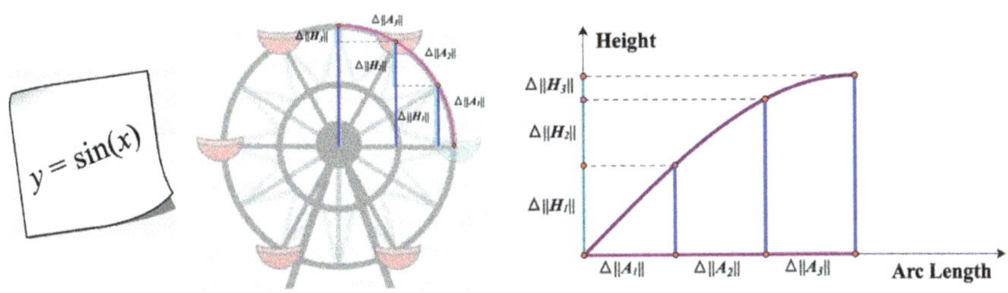

Fig. 5 For equal increases in arc length (colored in pink) from the 3 o'clock position, height (colored dark blue) increases by decreasing amounts (colored in light blue)

appropriately model covarying quantities of a situation (Fig. 5) (Stevens & Moore, 2016), and MA3 provides a foundation for rate of change reasoning (Johnson, 2015b; Thompson, 1994a). Furthermore, such reasoning enables understanding invariance among different representations of quantities' covariation (Liang & Moore, 2021; Moore et al., 2013; Norton, 2019), which is an AQS criterion we detail below.

Returning to the previous section's discussion of figurative material and the enactment of quantitative operations, Fig. 5 illustrates how a coordinate system and phenomenon provide material to assimilate via quantitative operations associated with MA3 and the sine relationship (Liang & Moore, 2021; Moore, 2014). The written "$y = \sin(x)$" was socially negotiated as a way to signify the operations involved in conceiving the sine relationship including the quantities' covariation. The symbol "$y = \sin(x)$" is not designed to afford the enactment of those quantitative operations, whereas the coordinate system and phenomenon provide contexts more organic to investigations of individuals' quantitative reasoning and, hence, defining the AQS criteria.

1.3 Figurative and Operative Thought

Although contexts like coordinate systems and phenomenon afford the enactment of quantitative operations, they also afford numerous other ways of reasoning, including those that might be incompatible with quantitative operations. We thus draw on Piagetian notions of *figurative* and *operative* thought (Piaget, 2001; Steffe, 1991), and particularly Thompson's (1985) extension of Piaget's distinction (see Moore, Stevens and et al., (2019b) for more), in our defining an AQS. The distinction between the two forms of thought enable us to differentiate between foregrounded aspects of thought including how those foregrounded aspects of thought may or may not support conceiving invariance among different actions and contexts. Defined generally, figurative thought is often based in and constrained to carrying out activity including physical actions, mental actions, motion, and imitations so that such activity produces a particular state. Operative thought is dominated by logico-mathematical operations, their re-presentation, and possibly their transformations. Operative thought foregrounds "intrinsic necessity, as opposed to successful solutions by chance or successful solutions that have simply been observed" (Piaget, 2001, p. 272), and available figurative material including that of the physical and perceptual kind is subordinate to the associated mental operations. Furthermore, meanings rooted in operative thought enable the reproduction or imagining of unavailable figurative material such that said material is a consequence of the construction of those mental actions. Quantitative and covariational reasoning are examples of operative thought due to their basis in logico-mathematical operations (Steffe & Olive, 2010; Thompson, 1994b).

To illustrate the figurative and operative distinction, Steffe (1991) characterized a child's counting scheme as figurative if his counting *required* re-presenting particular sensorimotor actions and operative if it entailed unitized records of counting that did not require the child to re-present particular perceptual material or sensorimotor experience. Relevant to the present chapter, Moore, Stevens and et al., (2019b) illustrated figurative graphing meanings in which prospective secondary teachers' graphing actions were constrained to particular figurative features (e.g., drawing a graph solely left-to-right) even when they perceived those features as constraining their ability to graph a relationship. In contrast, Moore, Stevens and et al., (2019b) described that a prospective secondary teacher's graphing meaning is operative when mental operations associated with quantitative and covariational operations persistently dominate perceptual and sensorimotor features of their graphing actions. Such a meaning enables an individual to graph a relationship across different coordinate orientations and coordinate systems and come to understand those graphs as quantitatively equivalent despite their perceptual differences or differences that occur in the sensorimotor experience of drawing a graph (Fig. 6). A student's construction of such a meaning illustrates Thompson's (1985) emphasis on the distinction of figurative and operative thought as an issue of "figure to ground" (p. 195), in which that which is operative on one level can become figurative on another level as it becomes the source material for subsequent operations and transformations.

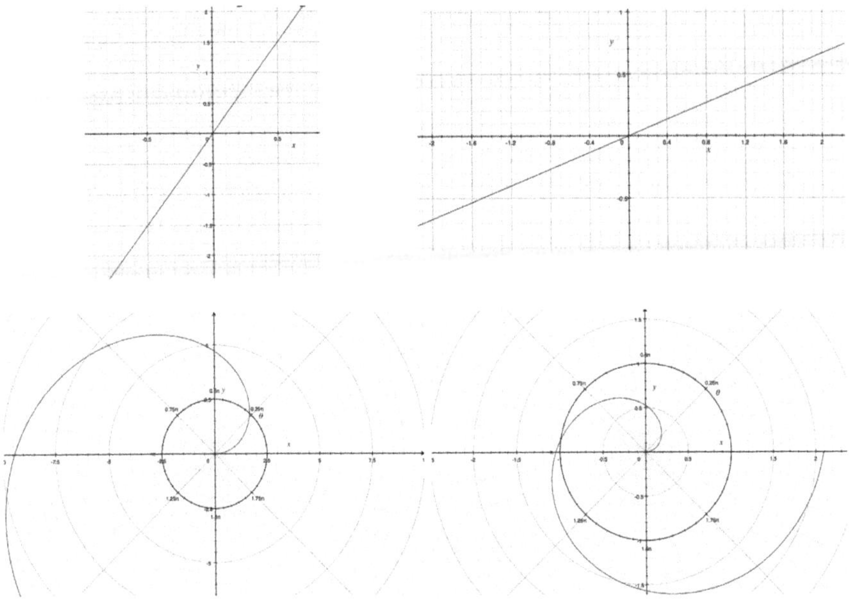

Fig. 6 Four graphs that differ perceptually and each represent one varying quantity being three times as large as another varying quantity (clockwise from top-left: $y = 3x, x = 3y, r = 3\theta, \theta = 3r$)

1.4 Three Forms of Re-presentation

Building on these aforementioned scholars' work on students' figurative and operative thought, Liang and Moore (2021) further operationalized these constructs in terms of a critical feature of an individual's cognition—an individual's ability to *re-present* their thought. According to Piaget (2001), von Glasersfeld (1995), and other constructivist scholars (e.g., Hackenberg (2010) and Steffe and Olive (2010)), re-presentation refers to an individual's ability to bring forth an image of schemes and operations that were enacted previously. Building on these scholars' collective works, Liang and Moore conceptualized three manifestations of an individual's re-presentational activities. The first form of re-presentation requires an individual to mentally generate some substitute for *all* sensory material that was present in prior experience but is absent currently (von Glasersfeld, 1995). For example, an individual can re-present MA3 in the context of a blank sheet of paper by recalling a Ferris wheel (or circle) and partitions (drawn or imagined) that correspond to amounts of change in two quantities.

The second form of re-presentation is similar to the first form, but it allows for the presence or supply of minimal figurative material or stimuli whose reconstructions are trivial to an individual. Using the same example as above, we can offer the individual a drawn circle after they've experienced a Ferris wheel animation and ask them to reconstruct MA3 in such context. This form of re-presentation is less strict than the first form, because it allows for some figurative material, and particularly that of the perceptual kind (e.g., the drawn circle), to be made available to an individual by a researcher.

The third form of re-presentation involves an individual *transforming* and regenerating operations enacted from a prior experience to accommodate a novel context. For example, after an individual constructs MA3 in the Ferris wheel (or circle) context, they can recall those operations and then transform and regenerate those operations under the constraints of the Cartesian coordinate system's quantitative organization to produce a graph representing the equivalent covariational relationship (see the graph given in Fig. 5). We use the word transform because this latter construction requires the individual to modify the operations associated with figurative material from the circle (e.g., incremental arcs and vertical segments) into material that differs perceptually due to the orthogonal orientation on the coordinate system. We underscore that this third form of re-presentation is different from the previous two forms in that it requires a transformation to occur in order to accommodate a novel context in a way that preserves some form of mathematical equivalence, and it thus forms an apropos example of operative thought. Collectively, these three forms of re-representation enable us to draw distinctions between each of the AQS criteria.

2 Further Defining and Illustrating the Abstracted Quantitative Structure Criteria

Recall that the criteria for an AQS is a system of quantitative operations that an individual has interiorized so that it:

(C1) is recurrently usable beyond its initial experiential construction;
(C2) can be re-presented in the absence of available figurative material including that in which it was initially constructed;
(C3) can be transformed to accommodate to novel contexts permitting the associated quantitative operations;
(C4) is anticipated as re-presentable in any figurative material that permits the associated quantitative operations.

C1 and C2 are consistent with the first two forms of re-presentation and associated examples discussed in the prior section. With respect to C2, it involves an individual having constructed a quantitative structure that is re-presentable in thought, and the individual can regenerate the operations with respect to those contexts experienced previously. It does not require that the individual be able to transform those operations to accommodate or generate a novel context.

Clarifying C3, a feature of an AQS is that it can accommodate novel contexts through additional processes of experiential construction with figurative material of which such construction has not previously occurred. We use *accommodation* to refer to when an individual modifies or reorganizes their meanings in order to establish a state of equilibrium or understanding (Montangero & Maurice-Naville, 1997; von Glasersfeld, 1995). Some forms of accommodation can be quite significant and require a fundamental change in an individual's ways of operating (Steffe & Olive, 2010), while other forms of accommodation can be more subtle and involve a way of operating being used with respect to novel sensory material (i.e., generalizing assimilation as defined by Steffe and Thompson (2000)). To illustrate, and building off the example provided in the previous section, after an individual has constructed some quantitative structure in the Cartesian coordinate system, they might recall those operations and then transform and regenerate them under the constraints of the Polar coordinate system's quantitative organization to produce a graph representing the equivalent covariational relationship (see the graphs given in Fig. 6). Such an accommodation is consistent with the third form of re-presentation discussed in the prior section, and it is a hallmark of operative thought because it entails an individual transforming and using operations of their quantitative structure to accommodate to novel quantities and associated figurative material, as opposed to having fragments of figurative activity dominate their thought (Thompson, 1985).

Whereas C3 refers to the enactment (and transformation) of quantitative operations in specified contexts, C4 refers to an individual anticipating the mathematical properties (e.g., quantities' covariation) of the quantitative operations constituting an AQS independent of any particular instantiation of them. The individual understands the operations and associated properties as not constrained to any particular quantities and figurative material. It is in this way that the quantitative operations of an AQS are abstract; the individual not only understands that the operations are re-presentable in previous experiences, but they also anticipate that the operations and their properties *could be* relevant to novel but not yet had experiences. Or, similarly, the individual anticipates that the operations and their properties *could be* relevant to experiences so complex in their figurative material or specified quantities that the individual does not yet have the fluency to enact those operations.

C4 extends beyond the forms of re-presentation discussed in the previous section due to it involving the anticipation of a hypothetical experience that has not been previously experienced. As an example, after graphing some relationship in numerous coordinate systems and orientations (e.g., the Cartesian and polar coordinate systems), a student may anticipate that there likely exist coordinate systems not yet experienced such that those coordinate systems enable enacting and representing the AQS and its mathematical properties; the student understands that they will need to adjust their operations to the specific quantitative constraints of the yet-to-be-experienced coordinate system, while also anticipating that the properties of those operations will remain the same (i.e., a linear relationship entails a constant rate of change no matter the coordinate system it is graphed within). As an alternative example, an individual could experience a novel or complex pair of quantities in which they have difficulty or cannot enact particular quantitative operations. Despite that difficulty, C4 involves the individual being able to anticipate the mathematical properties of a particular AQS. For instance, a student might not yet have constructed the capacity to enact quantitative operations with a quantity like surface area, but the student could anticipate a linear relationship between the painted surface area and height of a sphere to mean the painted surface area of a spherical cap increases at a constant rate of change with respect to the painted height of a sphere (Fig. 7).

Fig. 7 For equal increases in height (marked on the right of the figure), the surface area of the spherical cap (colored in green) increases by equal amounts (each shaded band of surface area)

The student could then leverage that inference to explore the relationship between surface area and height without having to enact quantitative operations in the context of the sphere (Stevens, 2019).

2.1 *Empirical Illustrations*

The value of a construct aimed at explaining cognition rests in its ability to provide explanatory or descriptive accounts of individual activity. We use selected empirical examples in this section to illustrate the criteria of an AQS. Each example is drawn from a study that used clinical interview (Ginsburg, 1997) or teaching experiment (Steffe & Thompson, 2000) methodologies to build second-order models of student thinking (Ulrich et al., 2014). It was in our reflecting on second-order models that we identified themes in student reasoning, and the AQS construct provided a consistent way to frame data from those studies.

We acknowledge the way we have defined AQS presents inherent problems in attempts to characterize a student as having or having not constructed such. First, it is impossible to investigate a student's reasoning in every context in which an AQS could be relevant. This limits the strength of claims relative to C3. Second, to characterize a students' quantitative reasoning necessarily involves focusing on their enactment of operations in the context of particular figurative material. This limits the strength of claims relative to C3 and C4, and particularly attempts to characterize the extent an individual's reasoning is not constrained to specific figurative material or quantities. No conceptual structure is truly representation free, as "operations have to operate on something" (L. P. Steffe, personal communication, July 24, 2019), but a conceptual structure can be abstracted to the extent the individual can symbolize and project it as if it is representation free. For these above reasons, we find it productive to discuss a student's actions in terms of *indications* and *contraindications* of their having constructed an AQS per the defined criteria, as opposed to claiming whether a student *has* or *has not* constructed an AQS. Indications are those actions that are consistent with particular AQS criteria, and contraindications are those actions that are inconsistent with particular AQS criteria.

In what follows, we provide examples that illustrate indications and contraindications of C2 and C3. C1 is trivial in its illustrations and is indicated by an individual re-enacting the associated quantitative operations to assimilate some experience separate from but identical in form to the original enactment of those operations. A contraindication of C1 is an individual enacting associated quantitative operations only in-activity, with each enactment of them being effortful and somewhat anew even when presented with what an observer considers the same figurative material as considered in previous activity (e.g., an identical task and/or animation). As we illustrate below, C4 is also somewhat trivial in its illustration, as it typically involves an individual making a verbal statement acknowledging the possibility of re-presenting particular relationships in situations not yet experienced. The illustration in Fig. 7 was one example of a potential anticipated linear relationship in a novel context. Another example is after numerous experiences graphing some relationship in multiple coordinate orientations and systems, the individual might acknowledge the possibility that other coordinate systems exist that entail a quantitative organization that affords re-presenting the operations associated with that relationship. Yet another example, although more complex in its form, is an individual thinking of hypothetical experiences they may work to occasion with a learner in order to support the learner's developmental of an AQS.[4]

2.1.1 A Contraindication of Re-Presentation in the Absence of Figurative Material (C2)

A critical criterion of an AQS is the ability to re-present that structure in the absence of available figurative material (and, often, in the context of transforming its operations to accommodate to novel contexts, i.e., C3). As a contraindication of re-presentation, consider Lydia's actions during a teaching experiment focused on trigonometric relationships and re-presentation (Liang & Moore, 2021). Prior to the actions presented here, Lydia had constructed incremental changes compatible with those displayed with the Ferris wheel in Fig. 5 to conclude that the vertical segment (Fig. 8, in green) increases by decreasing amounts (circled in Fig. 8c) for equal changes of arc length (i.e., MA3). We took her actions to indicate her reasoning quantitatively, particularly as she was able to reproduce her actions repeatedly in the context of the Ferris wheel animation (i.e., C1). We subsequently presented her the *Which One?* task (Fig. 9, also see [https://youtu.be/2pVVGl8eEr0]).

This task included a simplified version of a Ferris wheel (the left side of Fig. 9) with the position of a rider indicated by a dynamic point. The topmost blue bar (the right side of Fig. 9) displayed the arc length the rider had traveled counterclockwise from the 3 o'clock position. Lydia could vary the length of this bar by dragging its endpoint or by clicking the "Vary" button, and the dynamic point on the circle moved correspondingly. We asked Lydia to determine which of the six red bars, if any,

[4] We thank a reviewer for pointing out the relationship between an educator seeking to engender learning and their having constructed an AQS.

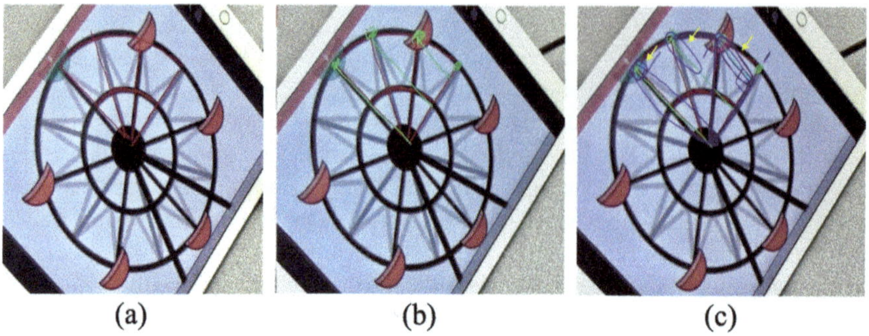

Fig. 8 Lydia's prior actions and their results (Liang & Moore, 2021, p. 303)

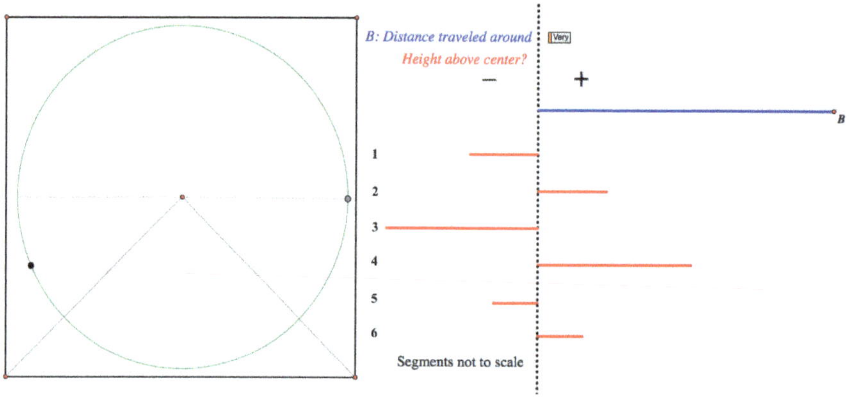

Fig. 9 The *Which One?* Task as presented to the student (Liang & Moore, 2021, p. 300)

accurately display the rider's height above the horizontal diameter as the rider's arc length varied (i.e., the sine relationship). The vertical dotted line provided a reference mark for the red bars, with a red bar emanating left being a negative magnitude and a red bar emanating right being a positive magnitude. The red bars were "free-moving" in that they could be repositioned and reoriented in the plane, including being reoriented and placed as a vertical segment emanating from the center of the circle. Our overall design was intended to determine the extent she could re-present her previous actions in a similar context with less figurative material immediately available than before (i.e., the Ferris wheel and its features, like the spokes in Fig. 8), but with novel material that might support her in enacting those operations (i.e., the red and blue segments oriented horizontally). For reference, the topmost bar is a normative solution, and the other bars vary with either different directional variation (e.g., positive or negative; decreasing or increasing) or different rates (e.g., constant, increasing, or decreasing rate) than the normative solution (see https://youtu.be/2pVVGl8eEr0).

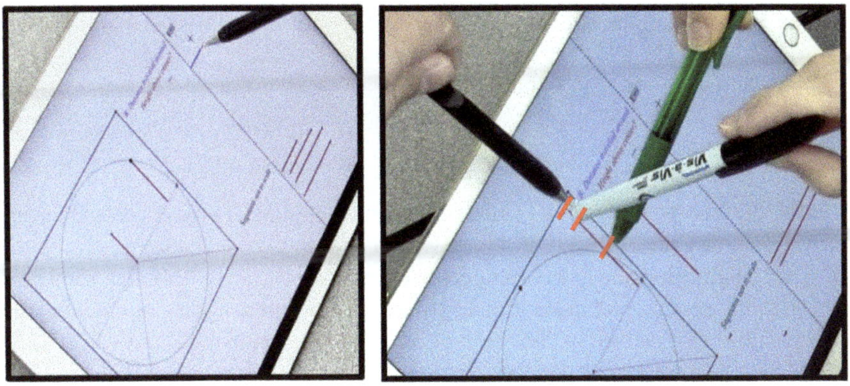

Fig. 10 Lydia re-orienting and checking the red segment that is the normative solution (left) and Lydia, with assistance, constructing partitioning activity (right, with the red partitions added to aid the reader) (Liang & Moore, 2021, p. 304)

As detailed in Liang and Moore (2021), Lydia became perturbed as to whether or not the horizontal red segment should vary at a changing rate with respect to the horizontal blue segment despite originally claiming that it should be based on her previous actions with the Ferris wheel (see identified changes circled in blue in Fig. 8). After much effort, she abandoned considering the segments in the horizontal orientation and re-oriented them vertically on the screen. Importantly, she persistently held in mind that her activity with the Ferris wheel led to the relationship of decreasing increases in height for successive equal changes in arc length (e.g., Fig. 8), but stated that she was unsure how to show such a relationship with the present segments. She eventually chose the correct segment solution by checking whether the heights matched at multiple static states within the displayed circle (Fig. 10, left).

At this point in the task, Lydia's activity had us question the extent she had constructed an AQS during her activity for the Ferris wheel. Specifically, she required re-orienting the red segments rather than either mentally rotating them vertically or leaving them horizontal and comparing their behavior with her image from the prior activity (i.e., a contraindication of C3). Furthermore, when re-oriented, she explicitly acknowledged having difficulty re-presenting the relationship from the Ferris wheel (i.e., a contraindication of C2). In an attempt to provide additional insights into her reasoning, after Lydia had chosen the normative solution, the teacher-researcher (TR) returned her to the question of whether the chosen red segment and blue segment entailed the same relationship as she identified in her previous activity (see Fig. 5, left):

Lydia: Not really...Um, I don't know. [*laughs*] Because that was just like something that I had seen for the first time, so I don't know if that will like show in every other case...Well, for a theory to hold true, it like – it needs to be true in other occasions, um, unless defined to one occasion.

TR.: So is what we're looking at right now different than what we were looking at with the Ferris wheel?

Lydia: No. It's – No…Because I saw what I saw, and I saw that difference in the Ferris wheel, but I don't see it here, and so –

TR.: And by you "don't see it here," you mean you don't see it in that red segment?

Lydia: Yes.

(Liang & Moore, 2021, p. 303).

In the present interaction, and as the interaction continued, Lydia expressed uncertainty as to how to determine if the blue segment and her chosen red segment entailed the same relationship she had illustrated in her previous activity, although she knew the segments were correct in static states. We underscore that Lydia held in mind the relationship she conceived in the Ferris wheel situation, and in the Ferris wheel illustration she could regenerate the operations as suggested by her work in Fig. 5 (i.e., an indication of C1). Consistent with a contraindication of C2, one possible explanation for her activity is that the Ferris wheel situation provided arms that supported her partitioning activity, and her activity at this time was reliant on the availability of that material (Liang & Moore, 2021). As a further contraindication of C2, it was only after the teacher-researcher introduced perceptual material using their pens (Fig. 10, right) that she conceived the red and blue segments' covariation as compatible with the MA3 relationship she had constructed in the Ferris wheel situation.

2.1.2 An Indication of Re-Presentation in the Absence of Figurative Material (C2)

As suggested by Lydia's activity, an indication of C2 would have been her regenerating amounts of change using the circle, the re-oriented red bars, and the blue bars, and without the assistance of the research team. As a more detailed indication of C2, consider Caleb's activity (Liang et al., 2018; Tasova et al., 2019) when engaging in the *Changing Bars Task* (Fig. 11). For the task, the red segment on the circle represents the magnitude of the point's height above the horizontal diameter and the blue segment represents the magnitude of the point's arc length from the 3 o'clock position (i.e., the sine relationship). The user was able to move the endpoint along the circle between the 3:00 position to the 12:00 position. On each displayed orthogonal pair, the user was able to drag the endpoint of the *red segment* in order to increase or decrease its magnitude. We asked Caleb to choose which, if any, of the orthogonal pairs—red representing height and blue representing arc length—accurately represents the relationship between the point's height and arc length as it moves a quarter of a rotation around the circle. Two of the pairs accurately represented the relationship.

Before Caleb engaged with the *Changing Bars Task,* he had engaged with numerous Ferris wheel and segment orientation tasks including *Which One?* His activity on those tasks indicated he had constructed a quantitative structure consistent with C1. We designed the *Changing Bars Task* to gain insights into the sophistication of his quantitative structuring. The *Changing Bars Task* thus includes subtle figurative differences including less figurative material than that of a Ferris wheel

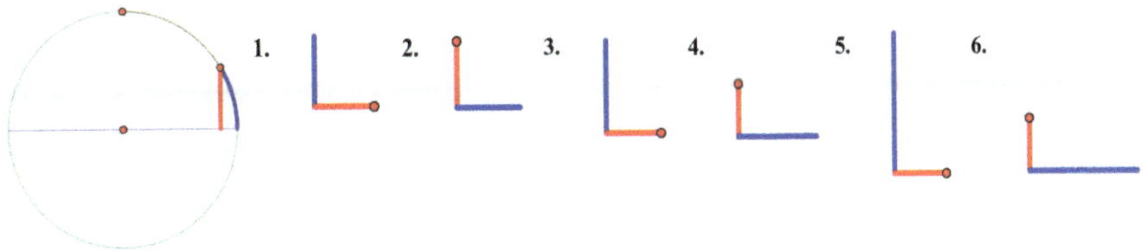

Fig. 11 *Changing Bars Task* (numbering of the six pairs provided for readers)

situation, pairs of orthogonally oriented bars (Fig. 11) rather than six height bars and one arc length bar (as in Fig. 9), and the ability to vary the red (height) segment rather than the blue (arc length) segment.

Summarizing Caleb's activity, he initially compared the amounts of change in arc length and amounts of change in height as the dynamic point traveled a small distance from the 3:00 position. He stated that, "…at the very beginning, … the height above the center and the distance traveled from 3:00 position should be similar." Caleb then repeated his actions near the 12:00 position, adding:

> *Caleb*: ...from this point [*pointing to the point denoted in orange in* Fig. 12b*]* ... to this point *[pointing to 12:00 position in* Fig. 12b], the height barely changes [*green segment in* Figs. 12b *and c (i.e.,* $\Delta||H_3||$*)*], but you're still traveling a fair distance around the circle [*blue annotation in* Fig. 12b *and blue segment (i.e.,* $\Delta||A_3||$*) in* Fig. 12c].

Caleb's actions suggest he constructed a gross additive comparison of $\Delta||H_1||$ with $\Delta||A_1||$ near the 3:00 position (i.e., $\Delta||H_1||$ is almost equal to $\Delta||A_1||$) and of $\Delta||H_3||$ with $\Delta||A_3||$ near the 12:00 position (i.e., $\Delta||H_3||$ is smaller than $\Delta||A_3||$). He also generalized this relationship across all cases in the first quarter of rotation. He stated, "the further you move away from the 3:00 position, the more variance there would be between the red (i.e., $\Delta||H||$) and the blue lines (i.e., $\Delta||A||$)." In this case, by "variance" he meant that $\Delta||A||$ became much bigger than $\Delta||H||$ as the dynamic point approached the 12:00 position, whereas an alternative meaning would be comparing $\Delta||H||$ magnitudes for successive changes in arc length. In other words, he was coordinating how the two quantities' changes compared to each other rather

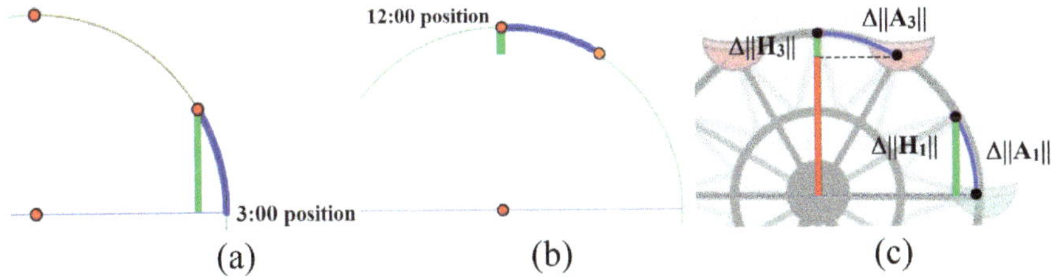

Fig. 12 Recreation of Caleb's activity in the *Changing Bars Task*. We introduce the Δ and magnitude notation to use in the narrative and highlight that Caleb's reasoning foregrounded magnitudes rather than (directed) measures

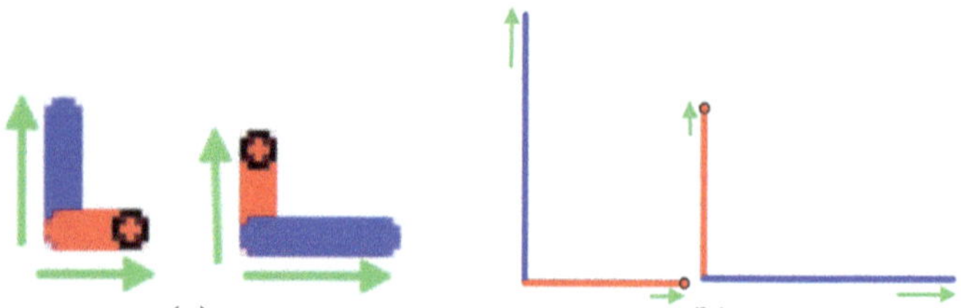

Fig. 13 Caleb's choice of two pairs of bars at **a** the beginning state and **b** the final state. The direction and length of each arrow indicates the direction and magnitude of change respectively

than fixing changes in one quantity and comparing the changes in the other quantity (Liang et al., 2018; Tasova et al., 2019).

Turning his attention to the orthogonal pairs, he dragged the end point of the red bar for a small amount from a start of near zero magnitude, and observed by how much the blue bar changed (Fig. 13a); he also dragged the end point of the red bar for a small amount towards the maximum length of both bars to observe by how much the blue bar changed (Fig. 13b). He moved aside all pairs which either the blue bar did not change by "almost equal" as the red bar near their minimal amounts or the blue bar did not change by noticeably larger amounts than the red bar near their maximum amounts. He selected two pairs (the normative solutions), the relevant behavior of which are described in the Fig. 13 caption.

In contrast to Lydia's activity above, Caleb's activity is an indication of C2 because the entirety of his solution suggests he did not need aspects of the Ferris wheel (e.g., the spokes) or denoted partitions on the segments to provide markers for his activity. Rather, he was able to mentally imagine particular actions and their results in the absence of available figurative material including that with which he had previously acted. Furthermore, he was able to conceive equivalence in his actions among both contexts without having to reflect upon and enact operations on produced figurative material.[5]

2.1.3 A Contraindication of Accommodation (C3)

As a contraindication of C3, we turn to Patty's activity when prompted to graph a covariational relationship in a different Cartesian orientation than she had previously graphed (Moore et al., 2019b). Patty was working the *Going Around Gainesville* (*GAG*) task (Fig. 14). Patty constructed a normative solution to *Part I* in ways that suggested her reasoning covariationally and re-presenting that relationship using a Cartesian graph (Fig. 15).

[5] We note that we do not consider Caleb's activity as an indication of C3 as he had previously experienced both circle and segment contexts repeatedly during the teaching experiment. His engagement suggested they were not novel relative to his perceived goal and activity.

Going Around Gainesville Part I

You've decided to road trip to Tampa Bay for Spring Break. Of course, this means traveling around Gainesville on your way down and back, because who would want to go through Gainesville? The animation represents a simplification of your trip there and back. Create a graph that relates your total distance traveled and your distance from Gainesville during your trip.

Going Around Gainesville Part II

Create a graph that relates your distance from Gainesville and your distance from Athens during your trip.

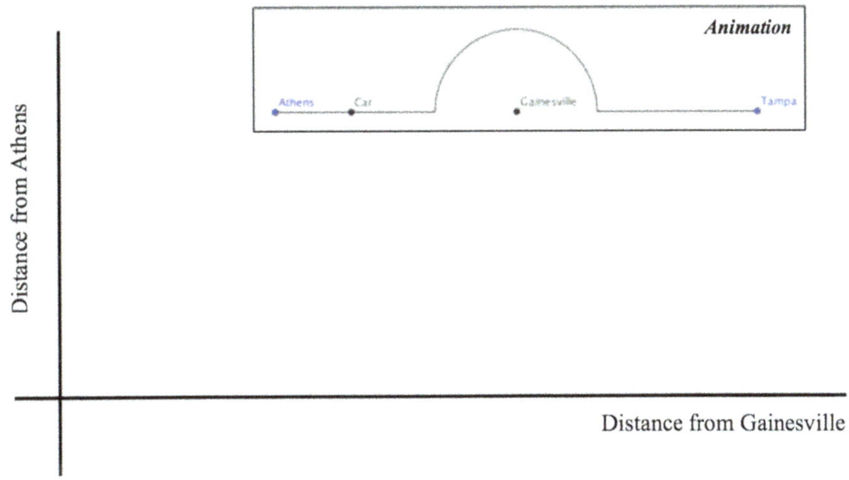

Fig. 14 The *Going around Gainesville* (*GAG*) task (Moore et al.; 2019b, p. 4)[6]

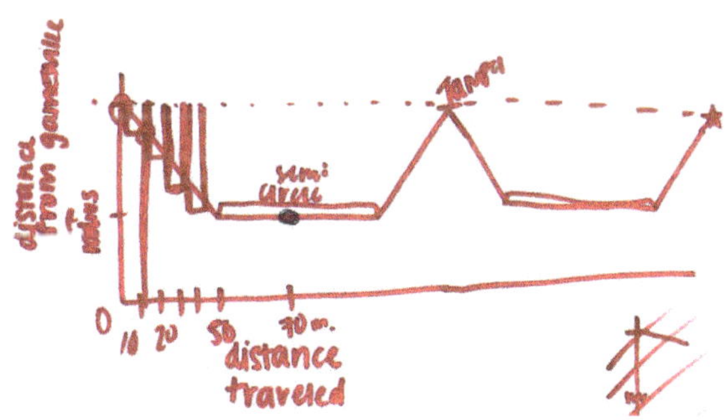

Fig. 15 Patty's work on *GAG* part I (Moore et al., 2019b, p. 13)

Moore et al. (2019b) provided a detailed account of Patty's solution to *GAG Part II*, which illustrated her experiencing a sustained, conscious perturbation that left her unable to complete the task to her satisfaction. We note that Patty anticipated graphing the same relationship she had previously constructed and graphed (i.e., a

[6] This task is a modification of the task Saldanha and Thompson (1998) presented.

quantity's magnitude decreasing as the other magnitude increased), but she encountered an irreconcilable perturbation when attempting to do so in the *Part II* coordinate orientation. Specifically, Patty had determined an initial point along the vertical axis and then motioned as if drawing a segment sloping downward left to right from this point. She later crossed out that point on vertical axis, as seen in the top-left of Fig. 16. She explained (see Fig. 16 for her work):

> *Patty*: I wanted to start here because I wanted to show that the distance was decreasing [motioning down and to the right from the point plotted on the vertical axis], but that means your distance from Athens is decreasing [tracing vertical axis from the initial point to the origin]…[turning her attention to the relationship from the animation] But your distance from Athens is growing. But your distance from Gainesville is decreasing. So, if that's growing and that's decreasing, so [draws an arrow pointing downward beside horizontal axis label and then an arrow pointing upwards beside the vertical axis label]
>
> [Patty then works for six additional minutes maintaining her 'starting' point on the vertical axis, without making progress, and explaining "this is so hard". She repeatedly identifies the distance from Gainesville as decreasing and the distance from Athens as increasing, including drawing a graph in an alternative axes orientation (i.e., Distance from Athens ("dA") being on the horizontal axis, see the bottom right of Fig. 16). She eventually has an insight.]
>
> *Patty*: Ohhhh, what if I started it like here [*plots point on the right end of the horizontal axis*]. Okay…but I don't want to start like, like I don't like starting graphs. You know I don't know work backwards that's weird…[*in the next minute and a half Patty draws in a normative initial segment of the graph, as seen in Fig. 16, hesitating throughout while explaining how the distances covary*] But it's backwards so I don't like it…My graph is from right-to-left, which is probably not right…[*describes the covariational relationship between the two distances*] I guess I just don't like this.
>
> *Int.*: And why don't you like it?
>
> *Patty*: Because it's backwards.
>
> *Int.*: And by backwards we mean?
>
> *Patty*: Backwards is traveling from right-to-left. But I think my graph is just, I think I'm just not clicking. I think I'm missing something.
>
> (Moore et al., 2019b, p. 14).

Recall that C3 involves transforming a system of quantitative or covariational operations in order to accommodate novel contexts permitting the associated quantitative operations. On one hand, Patty's activity is an indication of C3; she was able to transform MA3 operations enacted in the situation—a car traveling along a road—to construct a Cartesian graph representing the same amounts of change relationship (Fig. 15). On the other hand, Patty's activity is a contraindication of C3; she was unable to accommodate those operations enacted in the situation and previous graph to construct a Cartesian graph in a different quantitative orientation that she maintained as a correct representation of the relationship. Notably, figurative features of her graphing activity (e.g., "work[ing] backward," and "traveling right-to-left) constrained her ability to enact and sustain quantitative operations in the alternative Cartesian orientation. Patty's activity illustrates the complexity of C3, and we return to this complexity in the closing discussion.

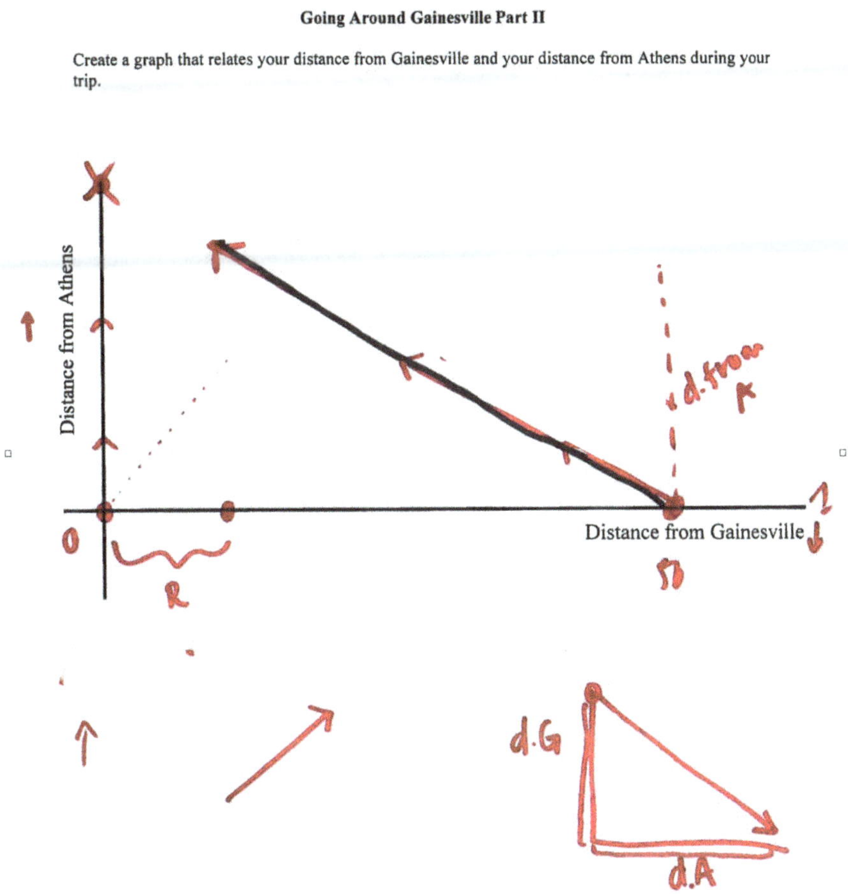

Fig. 16 Patty's graph for the first portion of the trip for *GAG* part II

2.1.4 An Indication of Re-presentation and Accommodation (C3)

As an indication of re-presentation and accommodation (i.e., C3), we turn to two prospective secondary teachers'—Kate and Jack—actions when asked to determine a formula for an unnamed polar coordinate system graph (Fig. 17, which is $r = \sin(\theta)$; see Moore et al. (2013) for the detailed study). After investigating a few points, Kate and Jack conjectured that $r = \sin(\theta)$ is the appropriate formula and drew from memory a Cartesian sine graph to compare to the polar graph. Important to note, Kate and Jack were not familiar with graphing the sine relationship in the polar coordinate system.

> *Kate*: This gets us from zero to right here is zero again [*tracing along Cartesian horizontal axis from 0 to* π]. So, we start here [*pointing to the pole in the polar coordinate system*].

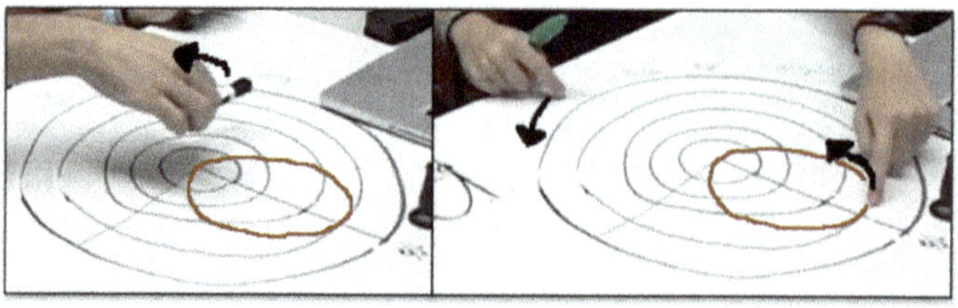

Fig. 17 Kate and Jack covary quantities with respect to the given graph (from Moore et al., 2013, p. 467)

Jack: Yeah, and you're sweeping around because [*making circular motion with pen*], theta's increasing, distance from the origin increases and then decreases again [*Jack traces along Cartesian graph from 0 to π as Kate traces along corresponding part of the polar graph*].

TR.: OK, so you're saying as theta increases the distance from the origin does what?

Jack: It increases until pi over 2 [*Kate traces along polar graph*] and then it starts decreasing [*Kate traces along polar graph as Jack traces along Cartesian graph*].

TR.: And then what happens from like pi to two pi.

Kate: It's the same.

Jack: Um, same idea except your, the radius is going to be negative, so it gets more in the negative direction of the angle we're sweeping out [*using marker to sweep out a ray from π to 3π/2 radians – see* Fig. 17] until three pi over two where it's negative one away and then it gets closer to zero [*continuing to rotate marker*].

TR.: OK, so from three pi over two to two pi, can you show me where on this graph [*pointing to polar graph*] we would start from and end at?

Kate: This is the biggest in magnitude, so it's the furthest away [*placing a finger on a ray defining 3π/2 and a finger at (1, π/2)*], and then [*the distance from the pole*] gets smaller in magnitude [*simultaneously tracing one index finger along an arc from 3π/2 to 2π and the other index finger along the graph – see* Fig. 17].

(Moore et al., 2013, p. 468).

Kate and Jack's actions indicate their having constructed (or constructing) a covariational relationship associated with the sine relationship such that they could take that relationship as a given in the Cartesian coordinate system. Furthermore, their actions indicate their transforming the associated operations to accommodate to a polar coordinate system displayed graph. Their activity enabled them to conceive two graphs as representing equivalent quantitative structures despite their perceptual differences, which is an indication of C3.

2.1.5 An Implication of Anticipation (C4) and Accommodation (C3)

In Kate and Jack's case, an indication of C4 would involve their identifying the *potential* of not yet experienced coordinate systems that enable re-presenting the same quantitative structure. In our experience it is difficult to gain evidence of C4, as it relies almost entirely on verbal descriptions of anticipation that stem from researcher prompting. Thus, rather than using this section to illustrate an indication of C4, we focus on an important implication of an individual's (whether they are a researcher, teacher, or student) actions that suggest their constructing a structure consistent with C4: their AQS is generative in their consideration of other individuals' work. By generative, we mean that their AQS is productive for assimilating a broad range of experiences with others in ways that are sensitive to the ways of operating of those others (Liang, 2021).

In our previous work, we documented prospective teachers' (PSTs') difficulties with attributing and valuing meanings rooted in quantitative and covariational reasoning to non-normative student work and coordinate system orientations, such as a student graphing x and y on the Cartesian vertical and horizontal axes, respectively (Lee et al., 2019; Moore et al., 2014, 2019a, 2019b). Attributing and valuing meanings rooted in quantitative and covariational reasoning to non-normative work (e.g., graphing under alternative axes orientations) necessitates the construction of an AQS. More specifically, an individual having constructed an AQS consistent with C1 through C4 positions the individual to anticipate some other person producing representations that, while novel to the individual, are viable (i.e., C4). Furthermore, the individual is positioned to accommodate to that person's produced representations via the transformation and regeneration of particular quantitative operations (i.e., C3).

As an illustration, we draw on Annika's activity as reported in Moore et al., (2019b). Annika was addressing a task (Fig. 18) presenting student work. Prior to this task, Annika's activity had indicated her anticipating rate of change as a coordination of quantities' variation that she could enact in any coordinate system or orientation so as long as she adjusted to the quantitative organization of that system and its orientation. Moore et al., (2019b) thus provided a task to see how she would attribute meaning to student work that was not clearly labeled but could be determined as viable in a number of ways.

As an indication of C4, Annika's immediate action was to consider the presented graph as a potentially viable graph of $y = 3x$, and she sought to determine coordinate orientations that enabled her to regenerate the quantitative operations she associated with a graph of $y = 3x$. Namely, she sought to determine labeling so that y varied by a magnitude 3 times as large as any corresponding variation in x. As potential labels, she identified positive x- and y-values oriented down and right of the origin, respectively, and she identified positive x- and y-values oriented up and left of the origin, respectively.

Following this interaction, and in an attempt to determine the extent Annika considered the graph as a viable representation of $y = 3x$, the teacher-researcher posed that a different student claimed that the line has "a negative slope" and thus is not a graph of $y = 3x$. She responded with the following, using positive x- and y-values oriented down and to the right of the origin (see Fig. 19), respectively, as her specific example:

> *Annika*: You'd have to notice that even though it looks like a negative slope [*making a hand motion down and to the right*] because we call it slope because it's visual and it's easy to visualize a negative and positive slope [*making hand motions to indicate different slopes*].

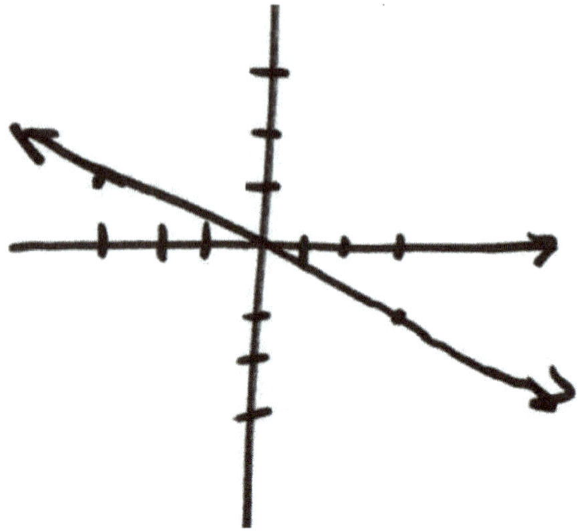

Fig. 18 The task presented to Annika, which was posed as a student solution to graphing $y = 3x$ (Moore et al., 2019b, p. 7)

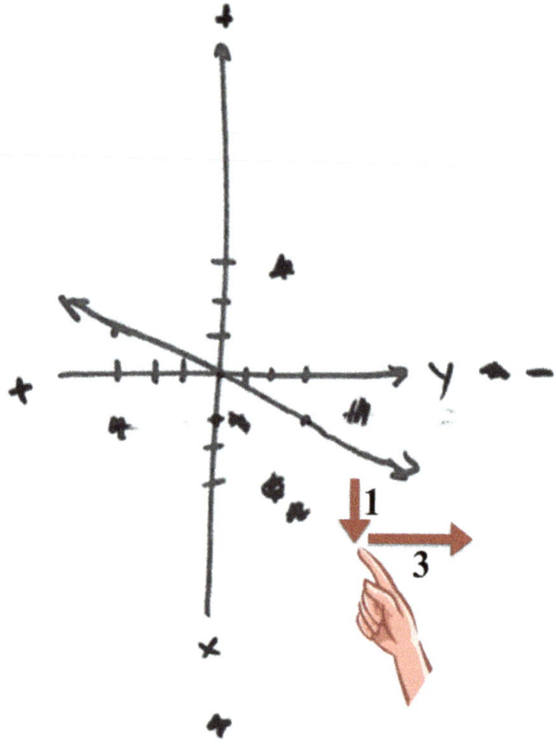

Fig. 19 Annika's annotated version of Fig. 18, each tick mark based on a unit change of 1 (Moore et al., 2019b, p. 13)

> But that's only visual on our conventions of how we set it up. Um, but like [*pointing to the graph*] if slope is rate of change, we can still see that for like equal increases of *x* [*making hand motions to indicate equal magnitude increases*] we have an equal increase of *y* [*making hand motions to indicate equal magnitude increases*] of three. And so for equal positive increase of one [*sweeping fingers vertically downward to indicate an increase of one*], we have an equal positive increase of three [*sweeping fingers horizontally rightward to indicate an increase of three*]. And so, it is a positive slope.
>
> (Moore et al., 2019b, p. 12).

Annika's response suggests her differentiating visual notions of slope from rate of change in a way that indicates her having constructed rate of change as a relationship between quantities' values as they vary. This relationship was not tied tightly to any particular coordinate system or orientation (i.e., C4). She then further enacted her scheme for constant rate of change to make sense of a graph in a particular orientation by re-presenting the quantitative operations she associated with a constant rate of change of 3 (i.e., changes in one value are a multiple of changes in the other value). Collectively, Annika's activity illustrates the powerful implications of an individual, and in this case a PST, having constructed an AQS in the context of their considering non-normative and unclear student work.

2.1.6 A Few Comments on Context

For consistency, we provided examples drawn from one population (prospective teachers) with a focus on particular mathematical ideas (e.g., covariation and graphing). However, our framing of an AQS is applicable to all age ranges and across numerous mathematical, scientific, and day-to-day domains (e.g. Steffe and colleagues work addressing students' counting and fraction schemes). Namely, we see the AQS criteria as relevant to topics and contexts including the construction of individual quantities (e.g., length), the combining of quantities to form other structures (e.g., coordinate systems or multiplicative objects), the construction of particular covariational relationships (e.g., the sine relationship), or the construction of a phenomenon itself (e.g., a Ferris wheel). For instance, an individual's construction of length as a quantity can start out tied to the experience of movement. Over time the individual may then construct height as a measurable attribute of a person, and then length as a more generalized measurable attribute of any number of concrete objects. Finally, that individual may conceive length as a measurable attribute they can impose on any span of space or object they come across in future experience. To illustrate further, we draw on the Faucet Task (Paoletti, 2019; Paoletti et al., accepted) to provide an example that spans students' mathematical understandings, understandings of systems, and lived experiences.

In the Faucet task, students interact with a dynamic applet that allows them to turn hot and cold knobs, with such turns resulting in changing amounts of water and temperature of water leaving the faucet (Fig. 20). Overall, the middle school students (aged 10–13) develop a quantitative structure that supports them in making additive (e.g., turning a knob on results in an increase in water) and ratio (e.g., turning the hot

Fig. 20 Several screenshots of the *faucet task* with the cold-water knob being turned on from an initial state

knob on results in increasing the hot water relative to the cold water, thereby making the water hotter) comparisons to determine how temperature and amount of water covary. As an initial stage of the task, and with the aid of the applet, students develop an understanding of how the faucet system operates. For instance, if only the cold knob is turned, all the water comes out at the constant temperature of groundwater. Or, if only the hot knob is turned, all the water comes out a constant temperature set by the hot water heater. Likely due to their numerous experiences with faucets, middle school students often quickly develop a quantitative structure for the faucet system that satisfies C1–C2, thus being able to imagine changes in amounts of water and temperature of water in thought, as well as creating the relevant states and variations of the system using the applet. In other words, they can enact quantitative operations in order to control the faucet system as they choose.

The Faucet tasks illustrates that a conversation on the extent an individual has constructed an AQS can exist with respect to different contexts. For instance, while middle school students are able to construct a sophisticated quantitative structure with respect to the faucet system (i.e., C1–C2), it is often non-trivial for them to use a graph to re-present relationships constructed within the faucet system environment (i.e., C3). Similarly, Patty's difficulty graphing a relationship she conceived highlights this complexity. A student having developed a sophisticated situational quantitative structure that provides evidence of C1–C2 does not immediately imply the student will be able to transform or regenerate this structure in a new context like a coordinate system. As Patty's actions illustrate, it could be the case that while their situational quantitative structure is sophisticated, their meaning for the representational system in which they are attempting to re-present that structure might not afford such actions. Or, as Lydia's actions indicate, it could be the case that an individual's situational understanding is not sophisticated enough to either be re-presented in the absence of specific figurative material or be transformed and regenerated in a different representational system. We return to this point in the next section when discussing research and teaching implications.

3 Discussion and Implications

Throughout the results, we used students' activity to provide indications and contraindications of the four criteria for AQSs (C1–C4). We leveraged these criteria to highlight the different forms of re-presentation and to distinguish between students' meanings in terms of their foregrounding figurative material and activity and their foregrounding logico-mathematical operations (i.e., quantitative operations). For example, Lydia's initial activity was illustrative of the importance of the first two manifestations of re-presentation, as she required some figurative material available to coordinate which segment length represented the rider's vertical distance. We underscore that Lydia did not encounter much difficulty once all necessary figurative material was available; she assimilated the segments and their variation to quantitative operations. Rather, Lydia struggled (and explicitly acknowledged said struggle) to accommodate the relationship she constructed in a way that she could re-present it with novel, and partially unavailable figurative material.

The complexities Lydia experienced further demonstrates the power of Kate, Jack, and Annika's reasoning. Kate and Jack's activity exemplifies students leveraging the third manifestation of re-presentation by transforming and regenerating operations enacted from a prior experience to accommodate a novel context, which is an indication of operative thought. For example, not only did they re-present a quantitative structure and regenerate that structure in a novel context, they also abstracted the associated operations such that they could identify the same relationship within a perceptually different representational system. In Annika's case, her activity underscores that the construction of an AQS better positions an individual to understand the reasoning of others as their meanings are more malleable in the presence of novel figurative material.

Because this chapter serves as an introduction of the AQS construct and criteria, we spend the remainder of this section discussing potential research and teaching implications. We envision these implications to provide avenues by which researchers and teachers can move the AQS construct forward. As we mentioned above, the value of a construct aimed at explaining cognition is measured by the extent it affords explanatory or descriptive accounts of individual activity, and such an affordance is best judged in the context of subsequent research and attempts to engender learning.

3.1 *Research Implications*

We find the criteria associated with an AQS to provide a grounding for researcher claims regarding students' (and teachers') quantitative and covariational reasoning in two primary ways. First, the AQS criteria provide a way to characterize the sophistication of a student's quantitative reasoning whether with respect to a phenomenon (e.g., a Ferris wheel or a faucet), a representational system (e.g., a coordinate system or number line), or a concept (e.g., rate of change or the sine relationship). Prior to developing the AQS criteria, our research team often found it difficult (and unproductive) to characterize a student as reasoning quantitatively or not. For instance, for students for which we only had gathered data indicating C1, we were unsure whether those students were or were not reasoning quantitatively. But, as we gathered indications (or contraindications) of C2 through C4 through the design of interactions that afforded such indications, we found that we were able to develop more nuanced models of the students' quantitative reasoning. Furthermore, as we gathered indications or contraindications of C2 through C4, we found that we had more evidence to make viable claims regarding students' quantitative reasoning across each context in which their reasoning occurred. Thus, we see the AQS criteria as providing guidance for researchers as it relates to building evidence and making claims regarding the affordances and constraints of a student's quantitative reasoning. The AQS criteria emphasize that a researcher's sensitivity to figurative and operative distinctions in an individual's thought should be based in the researcher's sustained interactions with the individual and iterative testing of hypotheses regarding the individual's re-presentation and regeneration capacities.

Second, and relatedly, the AQS construct and criteria emphasize the importance of situating models of students' quantitative reasoning, including the perturbations students experience when enacting such reasoning. By providing criteria that draw attention to recurrent usability, re-presentation of prior experience, regeneration within novel contexts, and the anticipation of regeneration in future experience, the AQS criteria enable researchers to situate their claims regarding students' quantitative reasoning by being explicit about both the quantitative operations under study, as well as the contexts and figurative material in which those operations can be enacted or anticipated. Actions like those of Patty and Lydia highlight that it is important for researchers to simultaneously attend to students' meanings for various phenomena (e.g., a faucet system or Ferris wheel ride), various representational systems (e.g., Cartesian coordinate system and polar coordinate system), and the quantitative relationships they construct within a phenomenon or representational system. For instance, Patty's experienced perturbation did not stem from the relationship she constructed within the phenomenon. It instead stemmed from her meaning for graphing in the Cartesian coordinate system. The AQS criteria draw attention to this distinction with a focus on quantitative operations, their enactment, and their regeneration, and this distinction can prove powerful when designing for other interactions with a student as we illustrate in the next section.

With respect to students' mathematical development, we acknowledge to this point we have not explicitly defined *abstraction* in this chapter. This is notable given Piaget's (2001) extensive use of different forms of abstraction. It is not an oversight that we have not explicitly defined abstraction to this point, but rather an indication that we have not yet operationalized the construct of an AQS in terms of its construction and development. We are in the process of conducting and designing additional studies to provide insights into the construction and development of such structures as it relates to particular relationships, topics like rate of change, and representational systems. We hypothesize that the construction of AQSs occurs through cyclical processes of pseudo-empirical, reflecting, and reflected abstraction, in which what becomes operative and conscious at one level becomes the figurative ground for further processes of abstraction. We also note that our studies to date suggest the developmental interdependence of particular structures. Particularly, a student having constructed an AQS for a particular relationship can support them in constructing an AQS for a representational system as they attempt to re-present the relationship in that system. One can imagine that Patty's perturbation could have led to a powerful accommodation to her Cartesian graphing meanings had the setting afforded particular instructional interventions. Similarly, a student is afforded the opportunity to construct an AQS associated with a topic like rate of change through repeated opportunities of constructing AQSs of particular relationships that entail different rates of change as those relationships can become a source of reflection and abstraction.

Lastly, providing a set of criteria for a construct invites the question whether said criteria have a hierarchy. There is a natural hierarchy with C2 through C4, as they move from re-presenting a previous experience, to accommodating to a novel context, and ultimately anticipating hypothetical future contexts. With that said, we have not conducted the empirical work necessary to articulate developmental stages or shifts, and thus hesitate to make claims relative to the hierarchy of the criteria. We do note that C2 and C3 are each a subset of C1, as both C2 and C3 require the enactment of the associated operations beyond their initial experiential construction. For instance, a student reenacting MA3 by reproducing a Ferris wheel (or circle) on a (provided or imagined) blank sheet of paper and producing (via drawing or imagining) partitions to construct amounts of change is an example of C2 and necessarily implies their constructed quantitative structure is recurrently usable beyond its initial experiential construction. C1 is broader than C2 and C3, and particularly C2, because it includes cases that do not require regenerating figurative/perceptual material, including that in which it was initially constructed. For example, a researcher might give a student their finished work from a previous experience, and the student could assimilate it with little effort in order to recall their previous actions and their results.

3.2 *Teaching Implications*

With respect to teaching, and mirroring several of the research implications, the AQS criteria provide a lens for instructional design, both with respect to curricula and a teacher's or student's interactions with students. With respect to curricula, our experience (at least in the US) leads us to believe a majority of mathematics curricula do not intentionally provide students opportunities to reason in ways consistent with C2 through C4 (Moore et al., 2013, 2014, 2019a, 2019b). As a notable example, within US 6–12 curricula function classes are almost presented and graphed exclusively in the Cartesian coordinate system with their independent variable along an axis oriented horizontally and with positive values oriented to the right of the origin. Furthermore, the vertical axis is almost exclusively oriented with positive values above the origin.

This practice constrains students' opportunities to construct an AQS because it does not afford repeated occasions to transform and regenerate the quantitative operations associated with a function class to accommodate to novel contexts (i.e., other coordinate orientations or systems). Instead, this practice affords the propagation of meanings that foreground figurative aspects of thought and lower forms of abstraction (Moore et al., 2019b; Paoletti et al., 2018a, 2018b; Thompson, 2013; Thompson et al., 2017). More generally, it likely restricts students in constructing a coordinate system as a form of an AQS, as students only experience one form of a coordinate system.

That the majority of curricula do not intentionally target C2 through C4 is also problematic for the type of classroom interactions that occur in the context of that curricula. Echoing our comments about building evidence for making claims relative to students' quantitative reasoning, the products students are likely to produce are constrained (cf. diSessa et al., 1991). Ultimately, this limits the variety and multitude of student products teachers and fellow students have to consider, compare, and leverage. For students, this limits their opportunities to construct and compare different forms of reasoning and representations. For a teacher, this limits their ability to assess the extent their students have constructed a sophisticated and flexible meanings for the topic under consideration. In short, the aforementioned limitations in curricula have the likely consequence of restricting or devaluing the variety of student actions necessary to support reasoning in ways consistent with C2 through C4.

Responding to this issue, we have found promising results in designing instructional experiences that incorporate the criteria set forth in this chapter (Moore et al., 2014). Specifically, when designing for instruction, we have found it productive to start with answering the question: What is critical to a concept and what are merely conventional practices for representing that concept? By answering that question and differentiating between the two, an educator can identify the mental operations they want to foster and then use the AQS criteria to inform their instruction and interactions with students. As a concise example, consider the sine relationship. With respect to covariation and MA3, we consider a critical aspect of the sine relationship to be that for successive equal increases from 0 in one quantity's value, the other quantity's value increases by decreasing amounts, then decreases by increasing amounts, then decreases by decreasing amounts, and then increases by increasing amounts before repeating that pattern.[7] With that aspect in mind, targeting C2 and C3 then involves providing students a variety of experiences to construct, coordinate, re-present, and regenerate the corresponding mental operations. Such experiences could include having students determine how the arc length traveled by various Ferris wheel riders varies in relation to the riders' heights above the center of the Ferris wheel, and then extending that to any object traveling along a circle (i.e., C2). From there, the students could be tasked with regenerating those operations in different Cartesian

[7] We note that there are several other key aspects to understanding the sine relationship, including measuring quantities in radius lengths, proportionality, and periodicity (Bressoud, 2010; Moore, 2014; Thompson, 2008).

and polar coordinate orientations (i.e., C3). Ideally, when reflecting on that collection of experiences and identifying the invariant properties of their operations, the students would be positioned to anticipate the corresponding MA3 relationship as potentially relevant for future experiences in not-yet experienced coordinate systems and phenomenon (i.e., C4).

In providing the concise example in the prior paragraph, we do not imply that such a process is simple or quick. Rather, each step and construction along the way is quite effortful on the part of the learner, and we view each AQS criterium as drawing educators' attention (whether teachers, researchers, or curricular designers) to an area deserving intense focus. As an example, consider a student's construction of a coordinate system, which is itself a quantitative structure. Lee (2017) illustrated, and underscoring the emphasis of C1 and C2, a student's construction of a coordinate system is a complex coordination of mental operations that develop over time. Students need repeated opportunities to construct and coordinate the operations associated with a coordinate system if they are to use coordinate systems productively. Instructional design should thus be built around student opportunities to construct coordinate systems that at least satisfy C1 through C3. Additionally, other mental operations are necessary when constructing a graph within a coordinate system, particularly when that graph involves regenerating and representing a relationship from a different context like a Ferris wheel (Lee et al., 2018; Moore, 2021; Moore & Thompson, 2015; Paoletti et al., 2018a). Here, the AQS criteria C1 through C3 draw attention to the importance of a student not only constructing a sophisticated quantitative structure in a context like the Ferris wheel, but also to their being able to transform and regenerate that structure in the context of the coordinate system. Student actions as such cannot be taken for granted and should instead be explicit targets of instructional design.

Acknowledgements This material is based upon work supported by the National Science Foundation under Grants No. DRL-1350342, No. DRL-1419973, and No. DUE-1920538. Any opinions, findings, and conclusions or recommendations expressed in this material are those of the author(s) and do not necessarily reflect the views of the National Science Foundation.

References

Bressoud, D. M. (2010). Historical reflections on teaching trigonometry. *Mathematics Teacher, 104*(2), 106–112.

Byerley, C., & Thompson, P. W. (2017). Secondary mathematics teachers' meanings for measure, slope, and rate of change. *The Journal of Mathematical Behavior, 48*, 168–193.

Carlson, M. P. (1998). A cross-sectional investigation of the development of the function concept. In E. Dubinsky, A. H. Schoenfeld, & J. J. Kaput (Eds.), *Research in collegiate mathematics education, III. Issues in mathematics education* (Vol. 7, pp. 114–162).

Carlson, M. P., Jacobs, S., Coe, E., Larsen, S., & Hsu, E. (2002). Applying covariational reasoning while modeling dynamic events: A framework and a study. *Journal for Research in Mathematics Education, 33*(5), 352–378. https://doi.org/10.2307/4149958

Carlson, M. P., Larsen, S., & Lesh, R. A. (2003). Integrating a models and modeling perspective with existing research and practice. In R. Lesh & H. M. Doerr (Eds.), *Beyond constructivism: Models and modeling perspectives on mathematics problem solving, learning, and teaching* (pp. 465–478). Lawrence Erlbaum Associates.
Castillo-Garsow, C. (2010). *Teaching the Verhulst model: A teaching experiment in covariational reasoning and exponential growth* [Ph.D. Dissertation]. Arizona State University: USA.
Castillo-Garsow, C. (2012). Continuous quantitative reasoning. In R. Mayes & L. L. Hatfield (Eds.), *Quantitative reasoning and mathematical modeling: A driver for STEM integrated education and teaching in context* (pp. 55–73). University of Wyoming.
Castillo-Garsow, C., Johnson, H. L., & Moore, K. C. (2013). Chunky and smooth images of change. *For the Learning of Mathematics, 33*(3), 31–37.
Confrey, J., & Smith, E. (1995). Splitting, covariation, and their role in the development of exponential functions. *Journal for Research in Mathematics Education, 26*(66–86). https://doi.org/10.2307/749228
diSessa, A. A., Hammer, D., Sherin, B. L., & Kolpakowski, T. (1991). Inventing graphing: Metarepresentational expertise in children. *The Journal of Mathematical Behavior, 10*, 117–160.
Drimalla, J., Tyburski, B., Byerley, C., Boyce, S., Grabhorn, J., Roman, C., & Moore, K. C. (2020). An invitation to conversation: Addressing the limitations of graphical tasks for assessing covariational reasoning. In A. I. Sacristán, J. C. Cortés-Zavala, & P. M. Ruiz-Arias (Eds.), *Mathematics education across cultures: Proceedings of the 42nd meeting of the North American chapter of the international group for the psychology of mathematics education* (pp. 2270–2278). Cinvestav/PME-NA.
Ellis, A. B. (2007). The influence of reasoning with emergent quantities on students' generalizations. *Cognition and Instruction, 25*(4), 439–478.
Ellis, A. B. (2011). Generalizing-promoting actions: How classroom collaborations can support students' mathematical generalizations. *Journal for Research in Mathematics Education, 42*(4), 308–345. http://www.eric.ed.gov/ERICWebPortal/detail?accno=EJ932248
Ellis, A. B., Ely, R., Singleton, B., & Tasova, H. (2020). Scaling-continuous variation: Supporting students' algebraic reasoning. *Educational Studies in Mathematics*. https://doi.org/10.1007/s10649-020-09951-6
Ellis, A. B., Özgür, Z., Kulow, T., Williams, C. C., & Amidon, J. (2015). Quantifying exponential growth: Three conceptual shifts in coordinating multiplicative and additive growth. *The Journal of Mathematical Behavior, 39*, 135–155. https://doi.org/10.1016/j.jmathb.2015.06.004
Frank, K. M. (2017). *Examining the development of students' covariational reasoning in the context of graphing* [Ph.D. Dissertation]. Arizona State University: USA.
Ginsburg, H. P. (1997). *Entering the child's mind: The clinical interview in psychological research and practice*. Cambridge University Press.
Hackenberg, A. J. (2010). Students' reasoning with reversible multiplicative relationships. *Cognition and Instruction, 28*(4), 383–432.
Hohensee, C. (2014). Backward transfer: An Investigation of the influence of quadratic functions instruction on students' prior ways of reasoning about linear functions. *Mathematical Thinking and Learning, 16*(2), 135–174. https://doi.org/10.1080/10986065.2014.889503
Izsák, A. (2003). "We want a statement that Is always true": Criteria for good algebraic representations and the development of modeling knowledge. *Journal for Research in Mathematics Education, 34*(3), 191–227. http://www.eric.ed.gov/ERICWebPortal/detail?accno=EJ667031
Johnson, H. L. (2012). Reasoning about variation in the intensity of change in covarying quantities involved in rate of change. *The Journal of Mathematical Behavior, 31*(3), 313–330. https://doi.org/10.1016/j.jmathb.2012.01.001
Johnson, H. L. (2015a). Secondary students' quantification of ratio and rate: A framework for reasoning about change in covarying quantities. *Mathematical Thinking and Learning, 17*(1), 64–90.

Johnson, H. L. (2015b). Together yet separate: Students' associating amounts of change in quantities involved in rate of change. *Educational Studies in Mathematics, 1–22*. https://doi.org/10.1007/s10649-014-9590-y

Lee, H. Y. (2017). *Students' construction of spatial coordinate systems* [Ph.D. Dissertation]. University of Georgia: USA.

Lee, H. Y., Hardison, H., & Paoletti, T. (2018). Uses of coordinate systems: A conceptual analysis with pedagogical implications. In T. E. Hodges, G. J. Roy, & A. M. Tyminski (Eds.), *Proceedings of the 40th annual meeting of the North American chapter of the international group for the psychology of mathematics education* (pp. 1307–1314). University of South Carolina & Clemson University.

Lee, H. Y., Moore, K. C., & Tasova, H. I. (2019). Reasoning within quantitative frames of reference: The case of Lydia. *The Journal of Mathematical Behavior, 53*, 81–95.

Liang, B. (2021). *Learning about and learning from students: Two teachers' constructions of students' mathematical meanings through student-teacher interactions* [Ph.D. Dissertation].

Liang, B., & Moore, K. C. (2021). Figurative and operative partitioning activity: A student's meanings for amounts of change in covarying quantities. *Mathematical Thinking & Learning, 23*(4), 291–317. https://doi.org/10.1080/10986065.2020.1789930

Liang, B., Stevens, I. E., Tasova, H. I., & Moore, K. C. (2018). Magnitude reasoning: A pre-calculus student's quantitative comparison between covarying magnitudes. In T. E. Hodges, G. J. Roy, & A. M. Tyminski (Eds.), *Proceedings of the 40th annual meeting of the North American chapter of the international group for the psychology of mathematics education* (pp. 608–611). University of South Carolina & Clemson University.

Lobato, J., & Siebert, D. (2002). Quantitative reasoning in a reconceived view of transfer. *The Journal of Mathematical Behavior, 21*, 87–116. https://doi.org/10.1016/s0732-3123(02)00105-0

Montangero, J., & Maurice-Naville, D. (1997). Piaget, or, the advance of knowledge. Mahwah, N.J.: L. Erlbaum Associates.

Moore, K. C. (2013). Making sense by measuring arcs: A teaching experiment in angle measure. *Educational Studies in Mathematics, 83*(2), 225–245. https://doi.org/10.1007/s10649-012-9450-6

Moore, K. C. (2014). Quantitative reasoning and the sine function: The case of Zac. *Journal for Research in Mathematics Education, 45*(1), 102–138.

Moore, K. C. (2021). Graphical shape thinking and transfer. In C. Hohensee & J. Lobato (Eds.), *Transfer of learning: Progressive perspectives for mathematics education and related fields* (pp. 145–171). Springer.

Moore, K. C., Paoletti, T., & Musgrave, S. (2013). Covariational reasoning and invariance among coordinate systems. *The Journal of Mathematical Behavior, 32*(3), 461–473. https://doi.org/10.1016/j.jmathb.2013.05.002

Moore, K. C., Silverman, J., Paoletti, T., & LaForest, K. R. (2014). Breaking conventions to support quantitative reasoning. *Mathematics Teacher Educator, 2*(2), 141–157. https://doi.org/10.5951/mathteaceduc.2.2.0141

Moore, K. C., Silverman, J., Paoletti, T., Liss, D., & Musgrave, S. (2019a). Conventions, habits, and U.S. teachers' meanings for graphs. *The Journal of Mathematical Behavior, 53*, 179–195. https://doi.org/10.1016/j.jmathb.2018.08.002

Moore, K. C., Stevens, I. E., Paoletti, T., Hobson, N. L. F., & Liang, B. (2019b). Pre-service teachers' figurative and operative graphing actions. *The Journal of Mathematical Behavior, 56*. https://doi.org/10.1016/j.jmathb.2019.01.008

Moore, K. C., & Thompson, P. W. (2015). Shape thinking and students' graphing activity. In T. Fukawa-Connelly, N. Infante, K. Keene, & M. Zandieh (Eds.), *Proceedings of the eighteenth annual conference on research in undergraduate mathematics education* (pp. 782–789).

Norton, A. (2019). An Erlangen program that empowers students' mathematics. *For the Learning of Mathematics, 39*(3), 22–27.

Oehrtman, M., Carlson, M. P., & Thompson, P. W. (2008). Foundational reasoning abilities that promote coherence in students' function understanding. In M. P. Carlson & C. L. Rasmussen

(Eds.), *Making the connection: Research and teaching in undergraduate mathematics education* (pp. 27–42). Mathematical Association of America.
Paoletti, T. (2019). Support students' understanding graphs as emergent traces: The faucet task. In M. Graven, H. Venkat, A. Essien, & P. Vale (Eds.), *Proceedings of the 43rd conference of the international group for the psychology of mathematics education* (Vol. 3, pp. 185–192). PME.
Paoletti, T., Lee, H. Y., & Hardison, H. (2018a). Static and emergent thinking in spatial and quantitative coordinate systems. In T. E. Hodges, G. J. Roy, & A. M. Tyminski (Eds.), *Proceedings of the 40th annual meeting of the North American chapter of the international group for the psychology of mathematics education* (pp. 1315–1322). University of South Carolina & Clemson University.
Paoletti, T., Stevens, I. E., Hobson, N. L. F., Moore, K. C., LaForest, K. R. (2018b). Inverse function: Pre-service teachers' techniques and meanings. *Educational Studies in Mathematics, 97*(1), 93–109. https://doi.org/10.1007/s10649-017-9787-y
Paoletti, T., & Moore, K. C. (2017). The parametric nature of two students' covariational reasoning. *The Journal of Mathematical Behavior, 48*, 137–151. https://doi.org/10.1016/j.jmathb.2017.08.003
Paoletti, T., & Moore, K. C. (2018). A covariational understanding of function: Putting a horse before the cart. *For the Learning of Mathematics, 38*(3), 37–43.
Paoletti, T., Gantt, A. L., & Corven, J. (Accepted). A local instructional theory for middle school students' emergent reasoning. *Journal for Research in Mathematics Education.*
Piaget, J. (2001). *Studies in reflecting abstraction.* Psychology Press Ltd.
Saldanha, L. A., & Thompson, P. W. (1998). Re-thinking co-variation from a quantitative perspective: Simultaneous continuous variation. In S. B. Berensen, K. R. Dawkings, M. Blanton, W. N. Coulombe, J. Kolb, K. Norwood, & L. Stiff (Eds.), *Proceedings of the 20th annual meeting of the North American chapter of the international group for the psychology of mathematics education* (Vol. 1, pp. 298–303). ERIC Clearinghouse for Science, Mathematics, and Environmental Education.
Smith, J. P., III., & Thompson, P. W. (2007). Quantitative reasoning and the development of algebraic reasoning. In J. J. Kaput, D. W. Carraher, & M. L. Blanton (Eds.), *Algebra in the early grades* (pp. 95–132). Lawrence Erlbaum Associates.
Stalvey, H. E., & Vidakovic, D. (2015). Students' reasoning about relationships between variables in a real-world problem. *The Journal of Mathematical Behavior, 40*, 192–210.
Steffe, L. P. (1991). The learning paradox: A plausible counterexample. In L. P. Steffe (Ed.), *Epistemological foundations of mathematical experience* (pp. 26–44). Springer. https://doi.org/10.1007/978-1-4612-3178-3_3
Steffe, L. P., & Olive, J. (2010). *Children's fractional knowledge.* Springer.
Steffe, L. P., & Thompson, P. W. (2000). Teaching experiment methodology: Underlying principles and essential elements. In R. A. Lesh & A. E. Kelly (Eds.), *Handbook of research design in mathematics and science education* (pp. 267–307). Erlbaum.
Stevens, I. E. (2019). *Pre-service teachers' constructions of formulas through covariational reasoning with dynamic objects* [Ph.D. Dissertation].
Stevens, I. E., & Moore, K. C. (2016). The Ferris wheel and justifications of curvature. In M. B. Wood, E. E. Turner, M. Civil, & J. A. Eli (Eds.), *Proceedings of the 38th annual meeting of the North American chapter of the international group for the psychology of mathematics education* (pp. 644–651). The University of Arizona.
Tasova, H. I., Liang, B., & Moore, K. C. (2019). Generalizing actions of forming: Identifying patterns and relationships between quantities. In A. Weinberg, D. Moore-Russo, H. Soto, & M. Wawro (Eds.), *Proceedings of the twenty-second annual conference on research in undergraduate mathematics education* (pp. 602–610).
Thompson, P. W. (1985). Experience, problem solving, and learning mathematics: Considerations in developing mathematics curricula. In E. A. Silver (Ed.), *Teaching and learning mathematical problem solving: Multiple research perspectives* (pp. 189–243). Erlbaum.

Thompson, P. W. (1990, March 27–31). *A cognitive model of quantity-based algebraic reasoning.* Annual Meeting of the American Educational Research Association, San Francisco, CA.
Thompson, P. W. (1994a). The development of the concept of speed and its relationship to concepts of rate. In G. Harel & J. Confrey (Eds.), *The development of multiplicative reasoning in the learning of mathematics.* SUNY Press.
Thompson, P. W. (1994b). Images of rate and operational understanding of the fundamental theorem of calculus. *Educational Studies in Mathematics, 26*(2–3), 229–274. https://doi.org/10.1007/BF01273664
Thompson, P. W. (2008). Conceptual analysis of mathematical ideas: Some spadework at the foundations of mathematics education. In O. Figueras, J. L. Cortina, S. Alatorre, T. Rojano, & A. Sépulveda (Eds.), *Proceedings of the annual meeting of the international group for the psychology of mathematics education* (Vol. 1, pp. 31–49). PME.
Thompson, P. W. (2011). Quantitative reasoning and mathematical modeling. In S. Chamberlin, L. L. Hatfield, & S. Belbase (Eds.), *New perspectives and directions for collaborative research in mathematics education: Papers from a planning conference for WISDOM^e* (pp. 33–57).
Thompson, P. W. (2013). In the absence of meaning. In K. Leatham (Ed.), *Vital directions for research in mathematics education* (pp. 57–93). Springer. https://doi.org/10.1007/978-1-4614-6977-3_4
Thompson, P. W., & Carlson, M. P. (2017). Variation, covariation, and functions: Foundational ways of thinking mathematically. In J. Cai (Ed.), *Compendium for research in mathematics education* (pp. 421–456). National Council of Teachers of Mathematics.
Thompson, P. W., Carlson, M. P., & Silverman, J. (2007). The design of tasks in support of teachers' development of coherent mathematical meanings. *Journal of Mathematics Teacher Education, 10*, 415–432.
Thompson, P. W., Hatfield, N., Yoon, H., Joshua, S., & Byerley, C. (2017). Covariational reasoning among U.S. and South Korean secondary mathematics teachers. *The Journal of Mathematical Behavior, 48*, 95–111. https://doi.org/10.1016/j.jmathb.2017.08.001
Thompson, P. W., & Silverman, J. (2007). The concept of accumulations in calculus. In M. P. Carlson & C. L. Rasmussen (Eds.), *Making the connection: Research and teaching in undergraduate mathematics* (pp. 117–131). Mathematical Association of America. https://doi.org/10.5948/upo9780883859759.005
Tzur, R. (2004). Teacher and students' joint production of a reversible fraction conception. *The Journal of Mathematical Behavior, 23*(1), 93–114. https://doi.org/10.1016/j.jmathb.2003.12.006
Ulrich, C., Tillema, E. S., Hackenberg, A. J., & Norton, A. (2014). Constructivist model building: Empirical examples from mathematics education. *Constructivist Foundations, 9*(3), 328–339.
Van Engen, H. (1949). An analysis of meaning in arithmetic. *Elementary School Journal, 49*(321–329), 395–400.
von Glasersfeld, E. (1982). Subitizing: The role of figural patterns in the development of numerical concepts. *Archives De Psychologie, 50*, 191–218.
von Glasersfeld, E. (1995). Radical constructivism: A way of knowing and learning. *Falmer Press.* https://doi.org/10.4324/9780203454220

Number Systems as Models of Quantitative Relations

Terezinha Nunes and Peter Bryant

This chapter starts with a theoretical and, in our view, crucial distinction between two types of meaning that number words have: a representational meaning and an analytical meaning. The representational meaning connects the number words to quantities and to relations between quantities. Quantities are constructs about objects and events in the world; "a person constitutes a quantity by conceiving of a quality of an object in such a way that he or she understands the possibility of measuring it." (Thompson, 1993, p. 165). The analytical meaning is defined by the rules of the number system. Because numbers are conventional systems of signs, there is nothing in the world nor in the organism that justifies that 2 plus 2 makes 4 (Piaget, 1952a); this meaning of the number 4 is based on the conventional rules of the system.

Mathematical thinking also has two distinct facets, quantitative reasoning, which relates to the representational meaning of number, and arithmetic, which relates to the analytical meaning. Quantitative reasoning is the ability to analyze "a situation into a quantitative structure a network of quantities and quantitative relationships. A prominent characteristic of reasoning quantitatively is that numbers and numeric relationships are of secondary importance, and do not enter into the primary analysis of a situation. What is important is relationships among quantities" (Thompson, 1993, p. 165).[1]

[1] The term *relationship(s)* is kept as in the original. Subsequently the term *relation(s)* is used with the same meaning to keep with the recommendation of the American Psychological Association that *relationship(s)* should be restricted to those between people.

T. Nunes (✉)
Terezinha Nunes, University of Oxford, Oxford, UK
e-mail: terezinha.nunes@education.ox.ac.uk

Department of Education, University of Oxford, Oxford, UK

P. Bryant
University of Oxford, Oxford, UK
e-mail: peter.bryant.edu@gmail.com

G. Karagöz Akar et al. (eds.), *Quantitative Reasoning in Mathematics and Science Education*, Mathematics Education in the Digital Era 21,
https://doi.org/10.1007/978-3-031-14553-7_4

In contrast, arithmetic has been defined as "the science of numbers, which analyzes the behavior of various numbers in four operations: addition, subtraction, multiplication, and division" (Guedj, 1998, p. 63); relations between quantities are not the point of arithmetic. When arithmetic is used to solve problems about quantities, numbers and operations work essentially as models of the world, and the model must be relevant to the problem in order for the solution to be a good one. From the psychological perspective, a fundamental question is how children come to understand these two meanings of number in such a way that they can connect quantitative reasoning and arithmetic in order to use mathematics to understand the world.

The aim of this chapter is to provide an account of how action schemas used in different situations support students' understanding of quantities and of numbers as ways of representing their thinking about quantities. In the first section of this chapter, we shall illustrate the distinction between the two types of meaning, representational and analytical, by considering words in natural language in order to draw on theories and research about the development of word meaning. In the second section we will contrast the schemas of action that give meaning to natural and rational numbers. In the third section we will use the pair situation/action-schema to analyze research about students' understanding of rational numbers. In the final section we shall draw implications from our analysis for the teaching of rational numbers.

1 Learning the Meanings of Words

Two eminent developmental psychologists, Piaget and Vygotsky, whose theories differ in many ways, argued that word meanings come from thought, not from associations between words and referents. Adults say words that children can repeat, but they cannot pass on the words' meanings directly to children. "Thought and speech have different genetic roots; the two functions develop along different lines and independently of each other" (Vygotsky, 1962, p. 41). Vygotsky suggested that thought develops from the child's activity whereas speech originates from social interactions. In Piaget's theory too, thought and language have different roots: the origin of thinking is in (sensorimotor) action schemas[2] and words are learned in social interactions. An action schema (or schema of action) is an organized action that can be applied to different objects and in different situations (Piaget, 1952a, 1953); for example, joining two collections is an action schema because collections can be composed of different objects (e.g., pencils, flowers, candies) and with different aims (e.g., to put them in a bag, to count them).

[2] In English translations, Piaget's term 'scheme' is often maintained, but we prefer 'schema' because it is included in English dictionaries with a meaning close to the one intended here. The noun 'scheme' in English means a large-scale plan to attain something (e.g., maternal leave scheme); the verb 'scheme' means to plot. When the full expression 'schema of action', is used, there is less room for confusion with other meanings.

Both psychologists theorized that, as children develop, their thinking processes change from relying on specific instances or images to relying on classes and relations. As children use their action schemas in interaction with the external world, they establish relations between objects and events, which give origin to classification and classes. As children interact with other people and learn words, adults direct the children's attention to specific relations and classes and use words, which can become signs to represent thinking about classes and relations. These interactions result in the socialization of thinking (Inhelder & Piaget, 1964) because children's thinking is now based on classes and relations similar to those used by adults.

This abstract description can be clarified by considering how children learn the meanings of words in natural language. When children start to learn the meaning of words, their first experiences are with the connection between words and specific exemplars of concepts because adults say a word and indicate a referent for that word: for example, the mother might say to a young child "this is your little brother" when she brings home a newborn baby. The child learns that this is his/her brother, but it is likely that the child will not give to the word "brother" the meaning that adults give. Children start to use the word "brother" at an early age to refer to specific persons (e.g., my brother, my friend's brother) and can use it correctly in this way; this functional similarity leads to the illusion that they know the meaning of the word (Vygotsky, 1962). However, the referential meaning of the word "brother" is not a particular person: it is based on classes (a brother is a male) and relations (both people have the same parents, which means that someone is a brother of one person but not of another). The analytical meaning is defined by the conventional rules of the English kinship system: if A (a male) is brother of B, they share both parents; if A and B only share one parent, A is a half-brother of B. Once the conventional rules are understood, they allow for inferences that generate new knowledge: if one is told that A is brother of B and that B is sister of C, one can infer that A is brother of C. Such inference rests on thinking based on classes as well as on relations defined by the kinship system; if the word "brother" were only connected to specific people (my brother, my friend's brother), it would not be possible to draw inferences.

Research shows that young children find it difficult to use words to refer to classes rather than to specific instances. When children were exposed to artificial concepts in laboratory studies, 3-year-olds tended not to show progress in attaining a class-based use of the words, even though they could identify specific instances; 4- and 5-year-olds showed some progress in using words to refer to a class, if they were supported by adults in making comparisons between classes and in establishing a connection between the word and the criteria used in the classification; 6-year-olds performed above chance level in the assessments (Edwards et al., 2019). Young children have still greater difficulty in learning relational concepts, even if the concepts seem simple, such as "passenger" (Gentner et al., 2011; Hollich et al., 2000; Parish-Morris et al., 2010).

Thus, research results support Vygotsky's and Piaget's theories of how children learn word meanings and contradict the associationist view that word meanings are based on co-occurrences between the sounds of words and referents. However, some psychologists still approach the learning of number words as based on the association between number words and the recognition of numerosities (see Piantadosi et al., 2012, and Ni & Zhu, 2005, for reviews), but such theories are not considered in this chapter.

2 Giving Meaning to Number Words

Like other words in natural language, number words have a referential and an analytical meaning, but with number words there is a procedure for telling which number word to use for a collection of items: counting. Counting in order to attribute the correct number word to a collection depends on the schema of one-to-one correspondence (i.e., each number word must be paired with a single object) and on saying the number words in the right order. Knowing how to count allows children to use number words correctly to indicate how many items are in a group of objects, but it does not guarantee that the children use the words with the same meaning as adults do. For example, many 3- and 4-year-olds who can count to ten and say correctly how many items are in a collection do not give exactly five cookies to a doll when asked by an adult to do so; they just give some cookies without counting. Children who came up with the exact number of objects requested by an adult were classified by Sarnecka and Carey (2008) as "cardinal knowers", but their criterion only indicates that the children used number words in a way that was functionally similar to the adults' use. Later research (Davidson et al., 2012) showed that some children, who were "cardinal knowers" by this criterion, did not make inferences justified by the representational and by the analytical meanings of number words: for example, they did not know which box had more items inside, a box with the label 5 or another box with the label 7. It seems that the children did not treat the numbers as measures of the quantities inside the box.

According to Thompson, "a person constitutes a quantity by conceiving of a quality of an object in such a way that he or she understands the possibility of measuring it." (Thompson, 1993, p. 165). "Measurement, in the broadest sense, is defined as the assignment of numerals to objects or events according to rules" (Stevens, 1945, p. 677). The fact that numerals can be assigned under different rules leads to different types of analytical meaning of numbers.

Helmholtz (1887), one of the founders of the science of psychology, wrote extensively about number and measurement and argued that measurement is the basis for the representational meaning of number. According to him, cardinal number is based on the idea that, as items are counted, they are added to the collection of already counted items; it is this idea of additive relations between counting words (i.e., five is four plus one) that makes counting into a measure of a quantity and allows one to say that there are, for example, five objects in a collection, and that all collections with five objects are equivalent in number.

According to Helmholtz, ordinal numbers are based on the sequence of number words and are rather different from cardinal numbers. In contrast, Piaget (1952a, 1952b) argued that addition is the source for understanding both cardinal and ordinal number. Piaget's writings (Piaget, 1952a, 1952b, 1953) make it clear that his focus was on the action schemas of adding and its inverse, taking away, and not on addition and subtraction as arithmetic operations. When a child counts, for example, four items and realizes that counting another item means adding it to the already-counted items, the child also realizes that addition increases the quantity and, therefore, five is more than four. Piaget's studies focused on the referential meaning of number words, i.e., how number words represent quantities. He investigated what children thought was implied about quantities when the quantities were represented by numbers, and vice-versa: what inferences children made about number when they knew the relations between the quantities. In his well-known conservation studies, the children saw the items in one of two equivalent sets being spread apart or pushed together; some children argued that, after the items in one quantity had been spread apart while those in the other, equivalent quantity remained unchanged, the number of items had not changed, but the quantities were no longer the same (Gréco, 1962); other children argued that quantities only change when something is added or taken away and were confident that both the quantities and the number of items remained the same. Piaget's (1952b) studies, and several subsequent replications (for a list of replications in the Journal for Research in Mathematics Education, see Nunes et al., 2016), support the idea that connecting the action schemas of adding and taking away with the conception of quantities provides the basis for the referential meaning of number.

The action schemas of addition and subtraction also support children's understanding of the analytical meaning of natural number words because, according to the rules of the natural number system, each number is equivalent to the previous number in the counting sequence plus 1. The coordination of the action schema of addition with its inverse, the schema of subtraction, forms the basis for children's thinking about part-whole relations and for the idea that any number is composed of other numbers, a concept known as additive composition. In a more general form, additive composition describes any number as having an infinite number of analytical meanings: 5, for example, means $1 + 1 + 1 + 1 + 1$, $2 + 1 + 1 + 1$, $2 + 2 + 1$, $2 + 3$, $6 - 1$, 7-1-1, 7-2, 8-3 and so on. However, many children may not understand additive composition on their own, simply from applying action schemas to objects without interacting with adults to guide them to think about how number words are related to addition. Research in different countries, such as Brazil, Greece, Hong Kong, Malaysia, India and the UK (see Nunes & Bryant, 2022a) showed that a significant proportion of children have not mastered the concept of additive composition at the age of six. Once children understand the additive composition of number, they can make many inferences: for example, $4 + 1 = 5$ means that 5 is greater 4, an inference about an order relation; it also means that the difference between 5 and 4 is exactly 1. Such inferences elude many "cardinal knowers"; in fact, Sarnecka and Carey (2008) recognized that children need to take further steps beyond counting to learn how natural numbers represent quantities. Later research showed that children's understanding of additive composition and of part-whole relations is a strong longitudinal predictor of their mathematics learning in school (Ching & Nunes, 2017; Nunes et al., 2012).

Summary. When young children count correctly and say how many items are in a collection, they are using number words in a way that is functionally similar to the way adults use them. However, the representational meaning of a number word is not a collection in itself, but rather the children's thinking about quantities and about relations between quantities represented in a counting system. Establishing one-to-one correspondence between number words and items when counting is not sufficient for understanding number: one-to-one correspondence has to be coordinated with thinking about how the action schemas of adding and taking away relate to the meaning of number words. Together, one-to-one correspondence and the action schemas of addition and subtraction create a framework for the children to understand how relations between numbers represent relations between quantities. These developments in thought are coordinated with number words by means of social interactions, during which adults direct the children's attention to relevant relations between the children's own activity and number words.

The same theoretical perspective can be used to think about how students learn the representational meaning of rational numbers. The representational meaning of a fraction, for example a half, is not a referent such as one of the pieces of a chocolate which was cut in two equal parts. The representational meaning of rational numbers is provided by the action schemas of one-to-many correspondence, sharing and partitioning, which support thinking about multiplicative relations between quantities.

Different aims and activities are met by thinking about the relevant relations between quantities. Consider the following questions.

Imagine that in a school there are 32 children in first grade and 8 copies of a book that the teacher wants the children to read.

(1) The first grade teacher wants to know how many copies of the book she needs to buy so that each child can have a copy.
(2) There are no funds in the school for buying lots of copies of the book; the head teacher wants to know how many children would have to share a copy in order to assess the situation and decide how many more books to buy.

There is nothing in the sets of children and books that requires thinking in terms of additive or multiplicative relations. One needs to establish a relation between the number of books and the number of children that is relevant to what one wants to know. In order to answer the first question, the relevant relation between the quantities is additive: the assumption is of one-to-one correspondence between children and books and the difference between 32 and 8 indicates how many books the teacher needs to buy. In order to answer the second question, the relevant relation is multiplicative: the assumption is of one-to-many correspondence and the ratio of books to children, 1 book for 4 children, is what will help the head teacher to assess if there will be too many children sharing one copy of the book for the teaching to work well.

According to Kieren (1976) "rational numbers are 'ratio' numbers … elements of an infinite ordered quotient field" (p. 103: inverted commas in the original). He also maintained that natural number and some aspects of the concept of addition arise out of children's natural activity, and pondered if "in working with rational numbers, children are dealing with mathematical structures which do not have an obvious basis in natural thought" (p. 110).

In this chapter, we suggest that there are action schemas that provide a basis for thinking about ratios; the most important one is the schema of one-to-many correspondence. Past research (e.g., Becker, 1993; Kornilaki & Nunes, 2005; Kouba, 1989; Mamede, 2016; Nunes & Bryant, 1996; Nunes et al., 2008, 2010) has shown that even primary school students can use the one-to-many correspondence action schema to establish multiplicative relations between two quantities. When they do so, they can be taught to use numbers alongside other signs in order to represent two quantities and a ratio between them: for example, 1:4 can be used to represent "1 book for 4 children". Streefland and his colleagues (Middleton & Van den Heuvel-Panhuizen, 1995; Streefland, 1984, 1985) proposed that a useful mathematical representation of the relations between quantities in such situations is the ratio table, in which the measure of each quantity is placed in correspondence with the measure of the other quantity at several points on different sides of a line, yielding a representation of several ordered pairs (see also Brinker, 1998; Dole, 2008).

Teaching of rational numbers often starts with a focus on fractions rather than on ratios and children are asked to share continuous quantities amongst recipients, which requires them to carry out partitioning of wholes. When discrete quantities are used, sharing takes the form of dealing out items. The schema of one-to-many correspondence is the action schema used by children when they share items to recipients: for example, if a child is sharing cookies to recipients, the cookies are placed in correspondence with recipients sequentially until all the cookies have been shared out. When a quantity is continuous, this action schema is still relevant because it is possible to establish correspondences between the quantities without partitioning them: for example, students can place a cake in correspondence with 3 children or in correspondence with 4 children without actually partitioning the cake and can think about which share would be larger.

In the next section, we review research in order to consider how different pairs of teaching situation/action schemas are used as starting points for students to learn about rational numbers. Thompson and Saldanha (2003) argued that understanding the rational numbers system in the way that mathematicians use it is far beyond the grasp of school students, a point that is not disputed in this chapter. We espouse Vygotsky's (1962) view that spontaneous concepts arise from children's activities and are precursors to the scientific concepts, which are elaborated over time and culturally transmitted; students' activities in school can create the opportunity for teachers to help them to transform their spontaneous concepts into concepts that are closer to the way in which mathematicians use rational numbers.

In the remaining sections of this chapter, we analyze situations created by teachers and researchers to teach students about rational numbers as well as the action schemas typically used by students in these situations and how the action schemas become connected to relations between quantities and to numerical signs.

3 Rational Numbers and the Situation/Action-Schema Pair

Teachers create teaching situations using different means: they can ask students to solve problems using objects, to explain their solutions, to draw diagrams, to talk about relations between quantities in a problem, to use mathematical conventions, such as numbers and signs for operations, to represent relations between quantities. The content of the problems used in the teaching situations can also vary, and this variation can prompt students to use different action schemas. Different terminologies have been used to describe this facet of rational numbers. Kieren (1976) described the variations as "interpretations", "situations" and "settings"; Mack (1990) used the expression "real-world situations"; Steffe (2002) used the expression "experiential situations"; Empson et al. (2006) used simply "situations"; the terms "applications", "embodiments", "concepts" and "constructs" have also been used (Behr et al., 1983, 1984; Domoney, 2002; Olive, 2000; Post et al., 1985; Simon, 2017). There is no consensus yet on the terminology nor on a particular classification. The criteria for the classifications and the number of classes described in the literature also differ (see, for example, Behr et al., 1992, 1993; Kieren, 1976, 1988; Mack, 2001; Nunes et al., 2007; Ohlsson, 1988).

Classifications are established by people for specific purposes. In this chapter we present a classification of teaching situations described in the literature in order to examine which action schemas used by primary school students are relevant in each class of situations. For reasons of space, this is a selective review and does not cover all different types of situations. Because the aim is to analyze students' quantitative reasoning, the criteria used in our classification are the type and number of quantities involved in the situations. The labels used for the categories are drawn from previous research. Three classes of teaching situations are distinguished in this chapter.

(1) **Part-whole teaching situations** involve one extensive quantity (often an area or a line segment) divided in equal parts. The numbers are used to represent a part-whole relation: for example, if one chocolate is divided into four equal parts and someone eats three, the portion of the chocolate eaten can be represented by the number 3/4. The action schema most relevant in this situation is partitioning.

(2) **Ratio teaching situations** involve two different extensive quantities and a multiplicative relation between them (Streefland, 1984, 1985). For example, if 3 chocolates are shared fairly among 4 children, the relation between chocolates and children can be represented by the number 3/4: the numerator represents the number of chocolates, the denominator represents the number of children, the slash/represents a division; the number 3/4 represents the amount of a chocolate bar that each child receives. The representation 3:4 can also be used to signify the ratio 3 chocolates to 4 children. The action schema most relevant in this situation is one-to-many correspondence.

(3) **Intensive quantities teaching situations** involve an intensive quantity measured by a ratio between two extensive quantities (Schwartz, 1988; Tolman, 1917); thus, three quantities are involved in this type of situation, whereas in

ratio teaching situations two extensive quantities are involved.[3] Some intensive quantities can be represented either by a ratio or by a fraction. For example, the concentration of a solution of fertilizer can be described as 1 ml of fertilizer to 3 ml of water or as 1/4 of fertilizer and 3/4 water. Other intensive quantities are most often described in ratio language: for example, miles per hour or price per unit. The action schema most relevant in this situation is also one-to-many correspondence. Each of these teaching situations is discussed in the rest of this section.

3.1 *Part-Whole Situations and Relevant Action Schemas*

The widespread use of the part-whole situation in association with the schema of equipartitioning to introduce students to rational numbers makes the pair part-whole/partitioning a good starting point for this analysis. Equipartitioning (henceforth referred to simply as partitioning) is the action schema of cutting a whole into equal parts; in teaching and research situations, the number of parts is often specified by the teacher. The image of a whole divided into equal parts seems to be the most prevalent, and sometimes the only one (Kerslake, 1986; Silver, 1981), that students relate to rational numbers in the form 3/4. However, students often need to use different action schemas in any teaching situation (Vergnaud, 2009); in part-whole situations, partitioning is used in conjunction with counting the parts and adding them to reconstruct the whole.

We summarize in the next sections research findings about the difficulties in carrying out equipartitioning, how the relation between the parts and the whole is established, how teachers and researchers guide students to connect relations between quantities in the situation with signs that represent rational numbers, and logical inferences that can be drawn from equipartitioning that help students to think about rational numbers.

Part-whole Situations and the Schema of Partitioning. Since the pioneering work of Piaget et al. (1960), many researchers have confirmed that partitioning is not easy to carry out. Piaget and colleagues suggested that, when children are asked to cut a

[3] There is not complete agreement regarding the definition of intensive quantities. Some authors suggest that any ratio between two variables constitutes an intensive quantity; for example, in the ratio situations described in the preceding paragraph, one could say that there is an intensive quantity "chocolates per child" or indeed the opposite, "children per chocolate". Others suggest that there must be a quality or an intuition of a quantity that is measureable by the ratio between two extensive quantities for an intensive quantity to be defined. In the latter approach, a ratio is conceived as a mathematical relation between two extensive quantities conceived by someone to measure the intensive quantity, but not as a quantity in itself. When one speaks of the taste of lemonade, for example, taste is a quality that can be conceived as measurable by the relation between amount of water and lemon juice; this relation can be represented by a ratio, but people experience a taste, which is the quality they are trying to measure. We do not presume to solve this issue and refer the interested reader to a collection of papers edited by Steffe et al. (2014b) for further discussion.

whole into a pre-established number of parts, the children need to think about part-whole relations in order to anticipate how many cuts as well as where the partitioning should be done in order to succeed. Piaget and colleagues noted that, when attempting to cut a whole into a specific number of parts, young children sometimes just cut out pieces and stopped when they had reached the required number of parts, without exhausting the whole; sometimes young children confused the number of parts with the number of cuts and, for example, when attempting to cut a rectangle in three equal parts, they made three cuts, ending up with four parts. Piaget and colleagues interpreted this behavior as suggestive of a focus on the number of parts without thinking about part-whole relations.

The level of difficulty of partitioning is influenced by the shape of the whole and the number of parts: rectangles are easier to divide into equal parts because of the symmetry lines that can be used in folding, and it is easier to find a way of folding something in half and then in half again, arriving at 4 parts, than to fold something in three equal parts. Piaget's results have been replicated by researchers in other countries (e.g., Charles & Nason, 2000; Davis & Hunting, 1990; Lima, 1982; Pothier & Sawada, 1983). For example, Pothier and Sawada (1983), working in Canada, asked children to cut cakes into specific number of parts and replicated Piaget et al.'s finding that kindergarten children, aged 4 or 5 years, did not ensure that partitioning resulted in two equal parts even when attempting to cut a rectangle in half. At this age level, the children used the word "half" to mean "cut", without necessarily thinking that the number of parts had to be two. The children used the word "half" "in expressions like *break it [the cake] in half four pieces* and *split it [the cake] in half three pieces*" (Pothier & Sawada, 1983, p. 311, italics in the original). Lima (1982) made similar observations with Brazilian children. When the children were asked what each part would be called, typical answers were "pieces", "a broken cookie" or "I don't know". When asked to cut a circle into three parts, many children used horizontal cuts across a circle, producing unequal parts; this observation was replicated by Charles and Nason (2000) in Australia, who worked with children aged 7 and 8 years.

Pothier and Sawada (1983) further replicated Piaget and colleagues' observation of the use of symmetry lines and successive halving; children in the age range 8–9 years achieved good partitioning of figures in two, four and eight parts, but found it difficult to divide figures in three, five, seven and nine parts. The children themselves remarked that it was easy to divide a figure in an even but not in an odd number of parts (but note that division in six parts is also difficult); they realized that they needed to find a different first move than dividing the whole in two parts, but could not implement such a move. In their study, which included children up to the age of 9 years and 8 months, in one of the tasks the children were asked to divide a whole into nine equal parts; no child succeeded in finding the strategy that Pothier and Sawada (1983) termed *composition*, which would involve trisecting a figure twice in order to generate nine equal parts (see also Maloney & Confrey, 2010, for descriptions of difficulties in partitioning).

We have also observed such difficulties with partitioning amongst English children. Figure 1 displays an English 9-year-old child's attempts to represent 2/6 of a pizza in drawing (Nunes et al., 2006): after two initial attempts based on successive halving of the circle, the child attempted to trisect the two halves, and remained unaware that the resulting figure was divided into seven parts.

In brief, partitioning wholes into equal parts is not an easy task because children need to anticipate how their action of cutting will produce a specific number of same-size parts. The difficulty is not restricted to partitioning areas and is observed also when children are asked to partition lines (see Steffe, 2002, for equipartitioning a stick). Some teaching programs attempt to circumvent the difficulty of partitioning by providing students with already cut materials: for example, same size rectangles of different colors are cut into pieces that can be represented by different fractions (half, quarters, thirds, sixths etc.); students can verify if two fractions are equivalent by superimposing the pieces. Whereas this approach circumvents the difficulty of carrying out the partitioning, it might direct students' attention to visual comparisons as students superimpose the pieces and might not impact students' thinking about part-whole relations.

Quite apart from the difficulty of cutting something into equal parts, tasks that seek to analyze students' understanding of part-whole relations reveal other difficulties in reasoning about part-whole relations. Many researchers have used tasks in which children are invited to compare quantities that result from partitioning the same or same-sized wholes. The wholes are often rectangles that represent chocolate bars or cakes, which are being shared by dolls; the students are asked if the dolls would eat the same amount of chocolate/cake after these had been cut in different ways. Sometimes the researchers cut the rectangles as they pose the questions; sometimes they ask the children to imagine that the chocolates/cakes were cut. These results are summarized here.

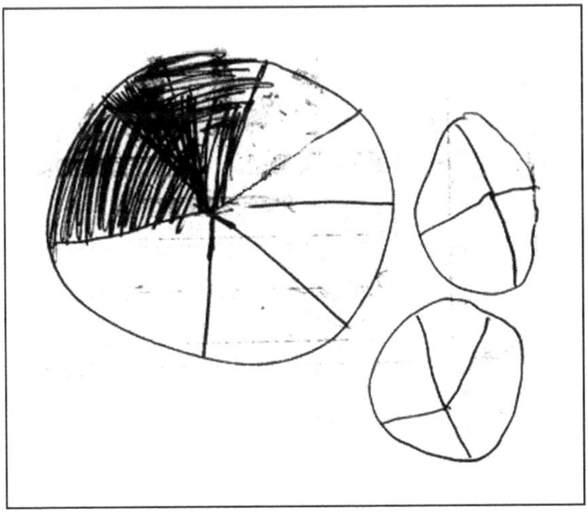

Fig. 1 One 9-year-old's attempts to represent the fraction 2/6 of a pizza

- Many young children (up to about 7 years) did not realize that eating one chocolate bar that was not partitioned and eating two halves of a same-size chocolate bar meant eating equivalent amounts of chocolate (Lima, 1982; Piaget et al., 1960).
- If two dolls ate each one half of same-size chocolate bars and the bars had been cut along different lines of symmetry (e.g., horizontally vs. diagonally), many children thought that the dolls would eat different amounts of chocolate (Kamii & Clark, 1995; Lima, 1982; Piaget et al., 1960).
- If two same-size chocolate bars were cut in different ways in more than two parts, resulting in parts that looked rather different but that would be represented by equivalent fractions, many children believed that the dolls would eat different amounts of chocolate; this conclusion was noted even among children who had been taught about fractions (Kamii & Clark, 1995; Lima, 1982). A similar result was reported when the children were asked to imagine the partitioning without any marks made on the rectangles (Nunes et al., 2006).
- Successive divisions of the same chocolate produce slightly better results than comparing parts of two same-size chocolates (Lima, 1982; Piaget et al., 1960).

Although many students up to the age of nine years struggle to make inferences based on part-whole relations in the context of partitioning, this does not mean that they cannot be supported to reflect about it. Vygotsky's (1962) concept of the zone of proximal development has inspired much research (e.g., Allal & Ducrey, 2000; Brown & Ferrara, 1985; Wood et al., 1976) that shows that, with support, children can do and can learn to do much that they do not accomplish without support. This is the point of teaching.

In concluding this section, we make two final points: the first is about the types of inferences about relations between quantities afforded by part-whole situations and the second about its current use in teaching. Two types of inference about quantities are supported by reasoning about partitioning and part-whole relations: equivalence and order by magnitude. The tasks described in the previous paragraphs exemplify inferences about equivalence. Students can also be invited to think about order relations in part-whole situations: for example, students can be guided to reflect on the idea that the more parts into which they cut a chocolate, the smaller the parts are. This insight is relevant to understanding the inverse relation between the divisor and the quotient. However, one must recognize that this insight is based on the assumption that the whole is constant and that the whole is equal to the sum of the parts. The literature documents that, from the students' perspective, improper fractions violate this additive principle (Hackenberg, 2007; Olive & Steffe, 2002; Tian et al., 2021; Tzur, 1999; Vamvakoussi & Vosniadou, 2004).

Finally, part-whole situations and the schema of partitioning can be used in teaching for different purposes. The previous paragraph considered their use to promote quantitative reasoning and to provide a referential meaning for rational numbers; the focus was on relations between quantities. However, according to Thompson (1990), textbooks and curricula do not promote quantitative reasoning and it would be surprising to find many teachers promoting it in the classroom. In the English National Curriculum (Department for Education, 2013), for example, the aims of teaching about fractions in the first year in school are described as to: "recognise, find and name a half as 1 of 2 equal parts of an object, shape or quantity; recognise, find and name a quarter as 1 of 4 equal parts of an object, shape or quantity" (Department for Education, 2013, p. 8). In the second year, pupils should be taught to: "recognise, find, name and write fractions 1/3, 1/4, 2/4 and 3/4 of a length, shape, set of objects or quantity; write simple fractions, for example 1/2 of 6 = 3 and recognise the equivalence of 2/4 and 1/2" (Department for Education, 2013, p. 13). The assumption underlying these aims seems to be that specific fractions acquire meaning by being connected to a specific referent, independently of their relation to other fractions; this is analogous to the idea that a natural number has meaning on its own, by reference to a numerosity associated with it, and without reference to the number system, a theoretical standpoint that was rejected in earlier sections of this chapter. Alajmi (2012) suggested that this way of teaching does not seem to acknowledge that "it is not possible to learn 'every single fraction'" (p. 251, inverted commas in the original) because the set of fractions is infinite.

Using Rational Numbers to Represent Thinking About Part-whole Situations. Teaching rational number in the context of part-whole situations cannot stop at thinking about relations between quantities after partitioning: students still need to learn to use mathematical signs to talk and think about quantities and relations between them. Two different approaches can be identified in how teaching promotes the connection between the parts and the whole, on the one hand, and numerical representations, on the other hand.

In the most common approach, a fraction is introduced as representing the number of parts taken from a whole (the numerator) and the total number of parts into which the whole was partitioned (the denominator): 1/4, for example, means 1 of 4 parts of a cake divided in 4 parts (see Behr et al., 1983, for a review of teaching approaches). Thus, students are guided to connect a fraction to a referential meaning by means of a double-counting procedure, counting the parts eaten and the total number of parts, a procedure that can be executed as if these were independent natural numbers. After dividing a cake into four parts, the fractions can be added to compose the whole, as if the pieces of the cake were brought close together. It has been suggested that this teaching approach is at the root of students' tendency to interpret fractions as two whole numbers (Kerslake, 1986; Mack, 1990; Pitkethly & Hunting, 1996; Tzur, 1999) and to think of improper fractions as truly inappropriate.

A different approach employed in teaching children how to use fractions to represent quantities and multiplicative relations between them is to guide students to think about a fraction as a unit of measurement by coordinating two different action schemas, partitioning and iteration. Partitioning defines the fraction, usually a unit fraction: once the unit fraction is defined, it is applied repeatedly to objects being measured (Olive & Steffe, 2002; Steffe, 2002). Steffe (2002) argued that, by coordinating the two schemas, partitioning (which he terms *equi-portioning*) and iteration, students build a meaning for fractions that is "at a learning level above the learning level that is made possible by equipartitioning" on its own (p. 204).

In Steffe's (2002) problems, the students were asked to share fairly a candy, represented by a line on a computer screen, into a pre-specified number of parts. In order to do so, they had to estimate the size of each share. The difficulties of estimating the size of the parts in order to divide a whole into a pre-specified number of parts, described in the previous section, were replicated in Steffe's (2002) study, even though in this study the children were able to use resources in the software that facilitated the task. After succeeding with partitioning, the researchers created a unit stick to be used iteratively to measure other sticks and encouraged the students to describe the sticks multiplicatively: a 2-stick was one two times the length of the unit, a 3-stick was three times the unit and so on. However, it is not clear that the students were thinking in these terms: for example, a 9-stick was created by adding one unit to the 8-stick and a 10-stick was made by joining a 3-stick with a 7-stick. It can be argued that the students were thinking in additive terms and continued to think in terms of part-whole relations. Olive and Steffe (2002) reported that, even after the students had themselves produced a written representation of an improper fraction in their teaching study, they still queried its sense: a 4th Grade student produced the fraction 10/7 and said: "I still don't understand how you could do it. *How can a fraction be bigger than itself*? (p. 428, italics in original).

Of the two approaches described in the literature to help students to connect numerical signs for fractions with meanings constructed in part-whole situations, the most common seems to be the counting of parts to define the numerator and the denominator. A review about the teaching of fractions in textbooks used in the United States, Kuwait and Japan (Alajmi, 2012) revealed that the measure approach described by Steffe (2002) did not appear in the textbooks in the US and Kuwait, but it was used in Japan. In contrast with Steffe's (2002) teaching study, in the Japanese textbooks fractions as representations of measures were not introduced in the context of partitioning. One widely used textbook, for example, introduced fractions to 3rd graders after they had been taught about decimals using conventional measures, such as meter and centimeter. In order to introduce fractions, a meter stick without centimeter marks was used to measure a child's arm span, which was a bit longer than one meter; the children in the class were asked to think about how to describe this extra length as part of a meter, and the concept of fraction was introduced by trying to figure out how many times the extra length of the arm would fit onto the meter stick. This question introduced the fractional notation for the unit fraction as involving a multiplicative relation between the segment of the child's arm which extends beyond a meter and the meter stick. Watanabe's (2006) review of Japanese textbooks concurred with this description and further stressed the multiplicative relation between the unit fraction and other fractions: for example, 2 times the unit fraction 1/3 equals 2/3; 3 times the unit fraction equals 3/3 and equivalent to the meter length and 4 times the unit fraction equals 4/3 and describes a length that is more than a meter. Thus, the object being measured (the arm span) and the unit of measure in these examples are not part of a whole; it is unlikely that students using this textbook would see improper fractions as problematic, but one cannot know if this is the case from the review of textbooks.

Simon et al. (2018) provided a similar contrast between Steffe's and the Japanese approach in teaching the connection between the representational meaning of fractions and the numerical fraction signs. They considered the approach suggested by Steffe (2002) a hybrid between part-whole and measurement, which distinguished it from the teaching approach used in Japan and in the Elkonin-Davydov (E-D) curriculum. Simon et al. (2018) noted that in the E-D curriculum students are taught from the outset that all numbers represent measures, so students are taught from the outset a different concept of number than the one taught in current practice in the US. They carried out a case study to investigate what happens if US students are taught about fractions using the measurement approach from the E-D curriculum, even though the students had not been taught about whole numbers in the same way as the E-D curriculum assumes. Their case study of how one student learned about fractions as measurement offers valuable suggestions for further research, but it is not reviewed here for reasons of space.

Summary. This situation/schema analysis illustrates how teachers can use part-whole situations and partitioning to explore the logic of part-whole relations, thereby promoting a form of quantitative reasoning that can give meaning to fractions. Reasoning about part-whole as well as part-part relations can support children's understanding of equivalence and order of magnitude of quantities resulting from partitioning. A weakness of teaching in this context is that part-whole relations are additive and this seems to create an obstacle to understanding improper fractions. Students can be taught to represent their thinking about part-whole relations by using fractions in two ways, but both seemed to interfere with students' understanding of improper fractions.

3.2 Ratio Situations and Relevant Action Schemas

Ratio teaching situations involve two different extensive quantities and a multiplicative relation (a fixed ratio) between them (in Vergnaud's, 1983, terminology: isomorphism of measures). So much has been written about children's understanding of ratio and its connection to rational numbers (e.g., Behr et al., 1992; Brousseau et al., 2004; Hart, 1984, 1988; Inhelder & Piaget, 1958; Kieren, 1992, 1993; Noelting, 1980a, 1980b; Resnick & Singer, 1993) that it is not possible to review this research here. We focus on the work of Streefland (1984), who asserted that ratio is one of the basic concepts in mathematics education because it forms "a basis for fractions, percentages and decimal numbers" (Streefland, 1984, p. 338). His approach provides a clear contrast with teaching children about rational numbers in part-whole situations. In his view, problems that involve fractions represent only one of the "many aspects of the phenomenon of ratio" (p. 339). Streefland suggested that, in sharing problems, the idea of invariance can be intuitively experienced (because the division is expected to be fair and exhaustive) and can be explored by students during teaching, supporting students' reflection about invariance. However, ratio situations are not restricted to sharing problems. For example, Streefland (1984) told students that a man had to take 32 steps to cover the same distance that a giant covered in 8 steps and then asked students how many steps the man would have to take to keep up with a giant when the giant took different numbers of steps. Other examples include research by Inhelder and Piaget (1958) and Singer et al. (1997), who presented students with problems that required establishing a ratio between the length of a fish and the number of food pellets to be fed to the fish a day and by Nunes et al. (2015a, 2015b), who asked students to think about the ratio between the number of scoops of food a hamster eats and the number of days the food lasts.

Ratio situations can be connected to different mathematical representations: for example, 3 chocolates to be shared by 4 children can be represented as 3:4 or 3/4. Students can be guided to think of the numbers in this example as representing the ratio of chocolates to children rather than part-whole relations. Problems involving sharing have been termed "quotient situations" (e.g., Empson et al., 2006; Flores et al., 2006; Mamede et al., 2005; Toluk & Middleton, 2001) because sharing is commonly used to teach students about division, and quotient is the term that refers to the result of a division. Following Streefland (1984), the term "ratio" is used in this chapter because of the wider variation of problems that can be created to teach students to represent the relations between quantities using rational numbers. In the next sections we summarize some of the research that has been carried out about students' use of the schema of one-to-many correspondence in ratio situations; research that invited students to use partitioning is not included in this section, as it was already covered in the previous one.

Ratio Situations and the Schema of One-to-many Correspondence. Streefland (1987, 1991, 1993, 1997) was undoubtedly the most prominent advocate of the use of ratio situations in teaching rational numbers. His work has had considerable international impact on research (see Presmeg & Van den Heuvel-Panhuizen, 2003, for a comment) as indicated by the number of citations of his work and the different countries in which it has been used (e.g., Brinker, 1998, United States; Campos et al., 2013, Brazil; Flores et al., 2006, United States; Mamede et al., 2005, Portugal; Middleton & Van den Heuvel-Panhuizen, 1995, United States; Naik & Subramaniam, 2008, India; Toluk & Middleton, 2001, Turkey), but it does not seem to have had so far a similar impact on teaching.

The aim of teaching students to solve problems in ratio situations is to guide them to realize that they can establish an invariant correspondence between two quantities. The schema of one-to-many correspondence establishes a multiplicative relation between two quantities, but it is a theorem in action (Vergnaud, 2009) in the sense that it only represents a ratio in action, not explicitly by means of words or other external signs. From about age six, many students are able to use the one-to-many correspondence schema to solve multiplicative reasoning problems (Becker, 1993; Kouba, 1989; Nunes et al., 2010; Park & Nunes, 2001; Piaget, 1952b). For example, students were asked how many sandwiches are required in order to fill 8 plates with 3 sandwiches in each plate; the majority of 5- and 6-year-old children provided a correct answer when they had different types of manipulatives to represent the plates and the sandwiches (Nunes et al., 2015a, 2015b). Ellis (2015) investigated the impact on performance in multiplicative reasoning problems of providing students with two types of manipulatives, such as blocks and cut-out shapes, to represent each of the quantities *versus* providing students with only one type of manipulative, such as blocks. She found that the rate of correct responses was higher and the quality of the students' explanations about their answers was significantly better when two types of manipulatives were used than when just one type was used.

Other researchers (Battista & Borrow, 1995; Kaput & West, 1994; Lamon, 1993; Langrall & Swafford, 2000; Steffe, 1992) have used the expressions "composite unit" and "linked composites" to describe students' concepts in the course of learning multiplicative reasoning in teaching experiments. These expressions do not emphasize the action schema of one-to-many correspondence, which students use to solve ratio problems, but they do imply that students are able to find the fixed ratio between two quantities in problem situations without clarifying the process by which the students identify the composite units.

When two quantities are placed in a fixed ratio, two relations between the quantities can be explored: a functional relation (e.g., if for every group of three children, there are two cakes, the number of cakes divided by the number of children is constant) and a scalar relation (if the number of children is doubled, the number of cakes must be doubled for the correspondence to remain the same). Both scalar and functional relations between quantities can be explored when children use the one-to-many correspondence schema and this renders ratio situations quite distinct from part-whole situations.

Streefland's studies on teaching students about ratio (1982, 1984, 1985, 1987, 1991, 1993, 1997) included situations related to fractions as well as situations that are not typically represented using fractions. Many of his studies used sharing as a context because of the intuitive agreement that sharing has to be fair and exhaustive. As Streefland did not provide details of students' answers, we describe some answers observed in our own research in order to explore the role of the action schema of one-to-many correspondence.

Streefland's (1997) teaching sequence in sharing situations aimed to direct students' attention to the relation between the two quantities. This is illustrated in the questions below, which were proposed in the first lesson:

1. Imagine a closed package of cookies and six girls. If the girls receive one cookie each, how many cookies were in the package?
2. If the girls receive a half cookie each, how many cookies were in the package?
3. If some more girls come and now they all share the same package of cookies fairly, will each girl receive more, less or the same amount as before?

These questions were posed to students in a teaching study (Campos et al., 2014) carried out with Brazilian students in 4th and 5th grade, aged 9–10 years. The teacher asked the questions to the whole class; the students first answered individually, then discussed their answers in small groups, and finally presented an agreed solution to the whole class. Figure 2 presents one student's drawings, done during the individual answer phase of the lesson. In the drawing used to answer the first question, the student combines numbers to represent cookies with drawings to represent children, establishing a one-to-one correspondence between cookies and children, which is in line with the question. In the second drawing, the student drew three pairs of half-cookies and established a correspondence between each pair of half-cookies and a child. In the third drawing, the student drew two possible numbers of children, two and three, as recipients of the same package of six cookies; from this comparison, she arrived at her written answer: "if more people come, we have to divide again, there will be fewer cookies for each one". The sequence of questions may have guided the student to use the schema of correspondence and may have helped her to arrive at the correct answers.

Not all the students in the class realized from the start that the relation between the divisor and the quotient is an inverse one but, as this was a teaching experiment, during the group discussions students explained their answers and sometimes reached new insights and changed their minds as they attempted to explain their thinking. For example, one student initially said: "if the girls shared the cookies fairly, they would continue to get the same amount as before", but changed her mind as she explained that "each girl would have to take some of the cookies they had to give to the girls who arrived later".

In another study, Nunes et al. (2006) asked students the same initial questions, which were followed by questions about equivalence. In the equivalence questions, the quantities in the situations differed but the ratios were equivalent. For example, four girls had a party and had one pie to share fairly amongst themselves; eight boys had a party and shared fairly two pies that were identical to the pie that the girls had. Which one of these alternatives is correct: (a) each girl eats more than each boy; (b) each boy eats more than each girl; (c) each girl eats as much as each boy? The percentage of correct answers was 76% for the 9-year-olds and 87% for the 10-year-olds, which is considerably higher than those observed in other studies of equivalence in part-whole situations (e.g., Kamii & Clark, 1995, report 13% correct responses for 8-year-olds and 32% correct responses for 9-year-olds when the questions were presented in part-whole situations). Some of the students' justifications

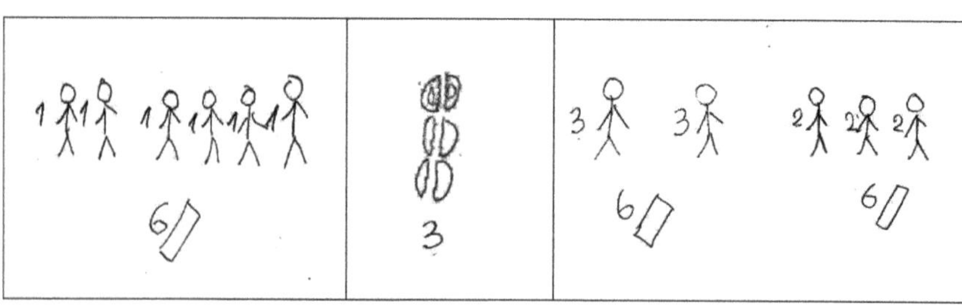

Fig. 2 One student's drawings used to answer Streefland's task: imagine a closed package of cookies and six girls. If the girls receive one cookie each, how many cookies were in the package? If the girls receive a half cookie each, how many cookies were in the package? If some more girls come and now they all share the same package of cookies fairly, will each girl receive more, less or the same amount as before?

for their answers consisted of simply showing that in the boys' party there was also a correspondence of four children to one pie.

We (Campos et al., 2014; Nunes et al., 2006) used two further tasks designed by Streefland (1997) in two teaching experiments to investigate students' use of one-to-many correspondence in reasoning about continuous quantities. In this task, the students were asked to imagine that four children were sharing three chocolates; each child was given a paper with drawings of four children and the chocolates (Fig. 3). The children were asked:

1. Can each child receive a whole chocolate?
2. Can each child receive at least a half chocolate?
3. Show how the chocolates could be shared and write what fraction of a chocolate bar each child would receive.
4. After the students produced drawings of different ways of sharing, the teacher asked: would the children receive the same amount of chocolate if they shared the chocolate bars differently?

Figure 3 shows two ways of sharing the chocolates that typically emerged in all taught groups, which were also described by Streefland (1997). The student's drawing on the left shows the first sharing; on the right, the student drew the children and the chocolates and showed a different way of sharing. The students argued that it does not matter how the chocolates are shared, each girl will eat the same amount. The main argument was summarized by one child: "It is the same number of children and the same number of chocolates, and they share fairly. It does not matter how you cut the chocolates" (Nunes et al., 2006).

The logical implications of one-to-many correspondence were explored further in the subsequent task (adapted from Streefland, 1997), in which the researchers used the language of fractions (i.e., oral language, two sixths, one third; written symbols, 2/6, 1/3) to talk about the problem (the way in which the signs were introduced is described in the section that follows). The sharing task below was presented:

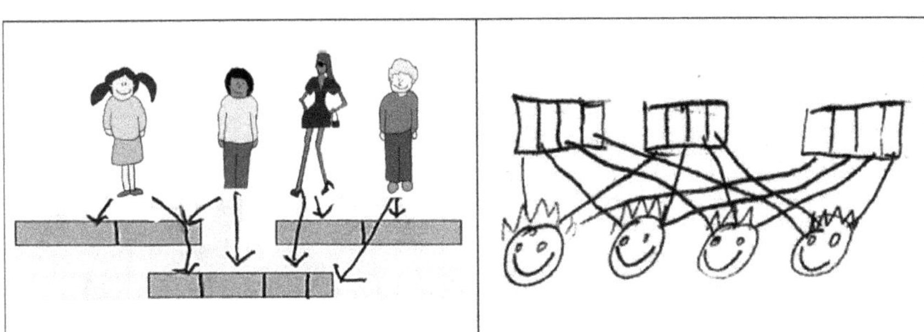

Fig. 3 Different ways of sharing 3 chocolate bars among 4 children. The student's first answer was that "they each get a half and a quarter" and the second answer was that "they each get three quarters"

1. Imagine that six children went to a pizzeria and ordered two pizzas; the waiter brought one pizza at a time so that the pizzas would not get cold; how could they share the pizza?
2. How many sixths would each one receive?
3. Could they share the pizzas differently if the waiter brought both pizzas at the same time?
4. If the children shared the pizzas in thirds, would they eat different amounts of pizza than when they shared in sixths?

This was the first time that the researcher asked questions using fraction language. Most (but not all) students argued that two sixths and one third are equivalent; their justifications continued to rely on the quantities and the relations between them; for example: "it is the same amount of pizzas and the same amount of children"; "they shared it fairly and all the pizzas were gone both times"; "it doesn't matter **how** they shared it" (bold used to represent the child's emphasis).

Most of the students' drawings registered the correspondences between pizzas and children and many showed no concern with equipartitioning (see Fig. 4); when the parts were marked on the pizzas, the marks were used for establishing correspondences. However, a few students attempted to divide circles in 3 or 6 parts before answering and found the task very difficult (see Fig. 1, produced by one of these students); these students often abandoned attempts to draw the parts and turned to using drawings that represented correspondences.

The students' explanations for the equivalence of the amount of pizza that the children would eat when each ate two-sixths *versus* one third of pizza were often not clearly articulated in words but could be understood by taking into account what they were pointing at. Nunes and Bryant (2022b) classified the students' explanations as including four logical arguments:

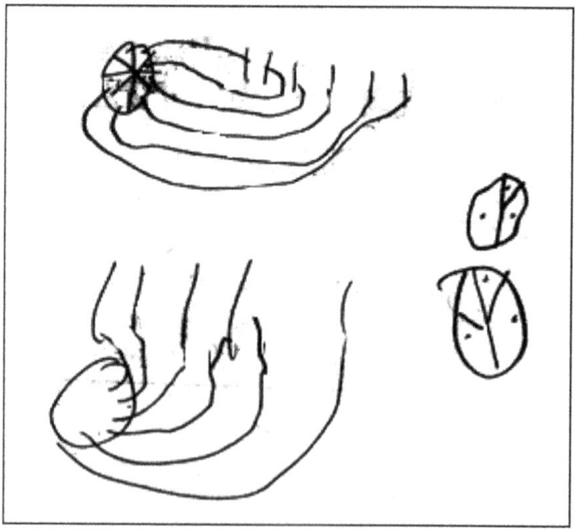

Fig. 4 One student's drawing showing correspondences between pizzas and children; there is no evidence of a focus on equipartitioning

(1) Identity: the quantities did not change, so the relation between the quantities did not change (e.g., "Because it's the same amount of people and the same amount of pizzas, so they get the same amount of pizza. … They're the same amount of people, the same amount of pizzas, and that means the same amount of fractions.")
(2) Inverse relation between the divisor and the quotient: the larger the number of pieces, the smaller the pieces (e.g., "I think, because one third, and one sixth and one sixth, is actually a different way in fractions, and it doubled here [pointing to the pizza cut into sixths] to make it littler, and halving the number [pointing to the pizza cut into thirds] makes it bigger").
(3) Fair and exhaustive division: the division was complete and fair (e.g., "If they have two pizzas, then they could give the first pizza to three girls and then the next one to another three girls. If they all get one piece of that each, and they all get the same amount, they all get the same amount as before. Because the pizzas are all gone.")
(4) Scalar explanation: if the denominator is twice as large, the numerator is also twice as large (e.g., Two sixths is the same as one third "because it's double the one of that [pointing to the denominator] and it's double the one of that [pointing to the numerator].")

The results of this classification do not imply that there are different types of students, because students often combined the different types of justification. The equivalence was not obvious to all the students at first, but the discussions after their individual answers afforded insight into the equivalence of 1/3 and 2/6; most students used the idea of equivalence in subsequent tasks that required comparing other fractions (e.g., 1/3 and 3/9).

Identity arguments have been associated with better performance in different tasks (e.g., conservation of liquids, Elkind, 1967; Arcidiacono & Perret-Clermont, 2009; understanding of the inverse relation between addition and subtraction, Bryant et al., 1999). However, understanding the equivalence of ratios must go beyond problems in which the quantities are the same and have simply been rearranged. Thus, we tested students' use of the schema of one-to-many correspondences in problems in which different children were sharing different pizzas. We used the following task, based on Streefland's (1997) work:

1. Imagine that a whole class of 4th graders, 36 children, went to a pizzeria and ordered pizzas. There wasn't a table where all the children could sit together so they had to sit at separate tables. There were four tables for six people so 24 children sat at these four tables. The 12 children left found a long table and sat together around it. (The children were asked to draw the tables and the children in correspondence for each table).
2. The waiter took two pizzas to each of these five tables and shared them fairly among the children at each table. What fraction of a pizza did the children sitting at the tables for six get? What fraction of a pizza did the 12 children sitting at the long table get? Did all the children in the class get the same amount of pizza?

(The students quickly realized that the sharing was not fair because there were more children at the long table and the same number of pizzas).

3. They need to order more pizzas for the children sitting at the long table. How many more pizzas do they need? What fraction of a pizza will the children at the long table get? Did all the children get the same amount of pizza?

This problem was presented in the last lesson of an 8-lesson teaching study (Nunes & Bryant, 2022b) and by then the students had adopted the use of drawings to represent one-to-many correspondences to support their reasoning. They were able to describe the 2–6 correspondence between pizzas and children sitting at the smaller tables and to realize that they needed two more pizzas for children sitting at the long table, where the pizza to children ratio was 1–6. Because they had thought that 2 pizzas for 6 children and 4 pizzas for 12 children meant that all the children around the different tables would receive the same amount of pizza, when they were asked to compare the fractions 2/6 and 4/12 they concluded that these are equivalent fractions.

Summary. Streefland's pioneering work, although influential in research, starts from a perspective that differs so fundamentally from traditional ideas in curriculum design and in teaching elementary school students about rational numbers that so far its impact on teaching practice has been limited. Research detailing the use of the action schema of one-to-many correspondence in teaching in ratio situations is scarce, and so we drew on our own past research to describe how students use the correspondence schema. Evidence from at least four teaching studies (Campos et al., 2014; Mamede et al., 2005; Nunes & Bryant, 2022b; Streefland, 1997) shows that students are able to connect one-to-many correspondence to ratio situations and that, when they do so, they demonstrate significant insight into the relations between quantities as well as relations between ordinary fractions.

Connecting Ratio Situations to Rational Numbers. When students have materials to represent the two quantities in a ratio situation or when they use drawings that illustrate the correspondences between the two quantities, they typically also speak about the correspondences between the quantities (e.g., 2 pizzas for 6 children). Streefland (1987, 1997) proposed a transitional notation based on drawings before the use of numerical symbols such as 2/6 or 2:6. For example, when the problem was about some children sharing pizzas at a table, the table was represented by a circle and the number of pizzas was written inside the circle; outside the circle, students were told to write the number of children. He did not describe how this transitional notation combining drawings and numbers was subsequently replaced by numerical signs only.

In our studies, we introduced the conventional notation in the context of the task in which four children were sharing three chocolates. English students are able to use the words "half" and "quarter" when talking about fractions. Rather than telling the children how to write fractions, we asked them questions with the aim of extending their knowledge of 1/2 to other fractions. We asked the students: "If you were sharing a chocolate between two people, how much would each one get? Do you know how to write half using numbers?" As there was always at least one student in the group who produced a correct fractional notation, we used the student's answer to connect the notation with the situation represented and to extend it to other fractional numbers. A summary of how the teaching proceeded is presented below.

- The number 1 shows the number of chocolates; the slash shows that this is a division; the number 2 shows how many people are sharing the chocolate. One divided by two means one chocolate divided by two children, and half is what each one gets.
- How do you write a fraction to show that one chocolate is being shared by four children? The students' answers were discussed so that the interpretation of the numerical signs was similar to the explanation we provided previously about half.
- What about one chocolate shared by three children?
- What about a pizza shared by six children?
- What about two pizzas shared by six children?
- What about two pizzas shared by eight children?

After each question, the children wrote their numerical notations, discussed them in their small groups, showed their answers to the rest of the class, and justified their notations. After this teaching episode, the students solved the question about six children sharing two pizzas, presented in the last section of this paper. When the numerical representations for fractions are taught in this way, division is explicitly represented; the numbers do not refer to the double counting of parts nor to the iteration of a part as a measure of the whole, but to the ratio between the quantity being shared and the number of people sharing. Our conjecture is that introducing students to the notation p/q in this way helps them to talk and think about different fractions using multiplicative reasoning based on the one-to-many correspondence action schema. We also hypothesize that ratio situations can provide a basis for students to think about fractions as measures by means of the one-to-many correspondence schema, which could be used to verify the number of times a unit fits onto the measured object. This would not be a hybrid between part-whole reasoning and measurement reasoning because ratio situations start from two quantities in a fixed ratio. Instead of two quantities in a fixed ratio, students could reason about one quantity and a measurement unit.

To our knowledge, only one study (Mamede, 2016) has investigated how children perform in order and equivalence tasks about fractions after having been taught how to use numerical representations in the context of part-whole situations versus being taught in the context of ratio situations. Teaching of fraction notation in ratio situations followed the script described in the previous paragraph; teaching in the part-whole situation connected the fraction to the notion of p parts out of q using counting. Mamede's study included four groups of Portuguese students, three taught about fractions for a total of approximately 70 min and one comparison group who received no teaching. The 1st graders (mean age 6.6 years) had not been taught yet about fractions in school. The group taught in ratio situations made significant gains from pre- to post-test in representing quantities using fractional notation as well as in tasks about order and equivalence of fractions. This group did not show progress in using fractional notation to represent part-whole relations.

Two groups of students were taught to use fractional representation in part-whole situations: one group was taught to think about sharing continuous quantities and the other group about sharing discrete quantities. Both groups showed significant gains in using fraction notation irrespective of whether the quantities were continuous or discrete. However, they made no measurable progress in order and equivalence tasks nor in the use of fraction notation in ratio situations.

The comparison, untaught group made no measurable progress from pre- to post-test in the use of fraction notation nor in tasks about equivalence and order of fractions. It is remarkable that, after relatively little teaching, the students taught in the context of ratio situations made measurable progress in reasoning about ratios and in using numbers to represent ratios. However, no transfer to part-whole problems was observed in this study.

Summary. the situation/schema analysis illustrates how ratio situations can be approached using the schema of one-to-many correspondence, which students employ from about the age 6–7 years when solving multiplicative reasoning problems. Ratio situations can be used to promote the use of language and of numerical notations to represent the ratio between the quantities; students seem to find it relatively easy to use the numerical notations and to compare fractions by thinking about the quantities that they represent. However, research about ratio situations is so far rather limited and much more research is needed to investigate how students' thinking progresses and how it can be expanded to connect to other aspects of rational number.

3.3 *Intensive Quantities and Relevant Action Schemas*

Intensive quantities are inextricably related to multiplicative reasoning and rational numbers. According to Tolman, because intensive quantities "do not have an additive nature, their measurement must be brought about by some device in which the magnitudes to be measured are put into a one-to-one correspondence with a series of quantities having extensive magnitude" (Tolman, 1917, p. 241). Tolman's examples of extensive quantities (magnitudes, in his terminology) include length, volume, and mass, which can be regarded as composed of smaller units of the same kind. These are contrasted with intensive quantities, such as temperature and density, which cannot be regarded in the same way as composed of smaller units. Tolman offered this comparison: if one adds 10 g of platinum to 10 g of platinum, this will give 20 g of platinum, but if one puts together two 10-g pieces of platinum, each with a density of 21 g per cubic centimeter, the larger piece of platinum of 20 g does not have a density of 42 g per cubic centimeter; its density remains at 21 g per cubic centimeter. The two extensive quantities, mass and volume, placed in relation to each other, are used to measure a third quantity, which is the intensive quantity, density.

There is a large amount of research on intensive quantities in mathematics education, which cannot be summarized here. It includes theoretical analyses as well as studies of children's understanding of intensive quantities (e.g., Abrahamson, 2012; Howe et al., 2010a, 2010b; 2011; Johnson, 2014; Kaput, 1985; Nunes & Bryant, 2008; Nunes et al., 2003; Schwartz, 1988, 1996; Simon & Placa, 2012; Singer et al., 1997; Stavy & Tirosh, 1996; Steffe et al., 2014a, 2014b; Thompson, 1994, 2011; Thompson & Saldanha, 2003). Intensive quantities cause great difficulty for students, even when the contexts seem familiar, and the tasks appear very easy. For example, students are familiar with buying popcorn and chocolate in different amounts and paying different prices; however, when they need to consider rate cost, because different amounts of chocolate were bought by the same amount of money, they focus on the amount of money rather than on rate cost (Nunes et al., 2003). For brevity, research on intensive quantities is not reviewed here, but we highlight some of the findings.

- Reasoning about the relations between the two extensive quantities and the intensive quantities is difficult even when no calculation is required to solve a problem: e.g., if two children bought different amounts of popcorn for the same amount of money, students' performance was not at ceiling when they were asked if one purchase was a better value [i.e., rate cost] than the other (Nunes et al., 2003).
- Students find problems about intensive quantities more difficult than problems about extensive quantities, even when the relation between the extensive quantities is inverse (e.g., the more cats one has, the fewer days the same amount of cat food will last: Nunes et al., 2003).
- Explicit teaching about intensive quantities cannot be taken for granted in primary and secondary school curricula in mathematics or in chemistry (Canagaratna, 1992). A review of textbooks in Kuwait, US and Japan (Alajmi, 2012) and a comparison of teachers' practices in teaching fractions in the US and Japan (Moseley et al., 2007) make no reference to intensive quantities nor to per-unit quantities.
- Even when an intensive quantity is particularly relevant in work settings, mathematics teaching does not explicitly focus on intensive quantities and may include the use of cultural tools (such as formulae) to circumvent reasoning about the intensive quantities (Noss et al., 2002).

The Schema of Correspondence and Composite Units in Situations that Involve Intensive Quantities. When a composite unit measures an intensive quantity, two extensive quantities are often mentioned explicitly: for example, speed is measured as kilometers (or miles) per hour; the concentration of a fertilizer can be measured as grams per liter of water. Research has shown that children and adults (Hoyles et al., 2001; Kaput & West, 1994; Noss et al., 2002; Steffe, 1992; Thompson, 1994) often solve problems about intensive quantities by using approaches similar to those described for ratio situations, such as replication of pairs in correspondence and scalar reasoning. Thus, for reasons of space, the details of research on intensive quantities are not reviewed here.

Connecting Intensive Quantities with Rational Numbers. Some intensive quantities, such as the concentration of orange juice or the probability of an event, can be represented in two different rational number formats: ratios or fractions. Some textbooks (see review by Alajmi, 2012) use examples of intensive quantities in chapters about fractions (e.g., a juice made with 1/3 orange concentrate and 2/3 water) but the fractions were not connected to the ratios in these chapters. Although transcoding numerical descriptions of quantities from ratios to fractions (*or* vice versa) might seem obvious and easy, we have found that some primary school teachers were surprised by it; for example, when the fraction of orange concentrate in the juice was 1/3, the teachers' immediate reaction was to transcode this as a ratio of 1 measure of concentrate to 3 of water. Gabriel et al. (2013) acknowledged the potential difficulty of such tasks, but unfortunately did not include relevant items in their measure of rational number knowledge. Duval (1999, 2006) emphasizes that one must not confuse rational number with any of the particular ways in which it is represented, and so such transcoding tasks might offer a good contribution to students' understanding of rational number.

Research about the representation of intensive quantities by using ratios or fractions seems scarce, but the three results below have been replicated at least once.

- When intensive quantities are described as ratios, students can use the schema of one-to-many correspondence in action and are more successful in solving problems than when they do not use this action schema for the comparison (Nunes et al., 2014; Piaget & Inhelder, 1975).
- When students answer questions about intensive quantities, performance is better if the problem is presented in ratio than in fraction language, irrespective of whether the question requires calculation or reasoning without calculation (Desli, 1999; Nunes et al., 2006). Gigerenzer (2002) reports a similar result in the context of understanding probabilities, but he refers to this comparison as language based on frequencies (one in a hundred) or on proportions (0.01).
- When students are taught about intensive quantities using ratio language, they show better post-test performance than when fraction language is used during teaching (Howe et al., 2011; Nunes et al., 2004a, 2004b), possibly because ratio language is more easily connected to the schema of one-to-many correspondences.

Summary. In order to think and talk about intensive quantities using numbers, one must resort to rational numbers, either in ratio or in fraction formats. Explicit teaching about intensive quantities currently does not seem to be included in many mathematics curricula. A search through the English National Curriculum for Mathematics (Department for Education, 2013; updated 2021) in primary and in secondary school did not find the expression "intensive quantity" (or "magnitude"), even though arguments for the inclusion of intensive quantities in the curriculum were made more than 10 years ago on the basis of a project supported by the government of the United Kingdom (Howe et al., 2010a, 2010b). A similar search through the US Common Core State Standards for Mathematics (2022) did not find these expressions either. The concept is certainly not new (Tolman, 1917) and neither is the concern for its absence in the curriculum (Travis, 1937, argued for its inclusion in engineers' education).

4 Concluding Remarks

In the introduction to this chapter, we distinguished between the representational and the analytical meaning of number, and presented a brief summary of theories and research about how the meanings that children give to words in natural language develop. Our aim was to show that children first learn many words when adults direct their attention to referents, but that word meanings in natural language are not just associations between words and specific referents. Vygotsky and Piaget concur in rejecting associationist explanations for word meanings: both argued that the (representational) meanings of words are provided by thought. This notion applies to words whose meanings are related to classes, such as "cat", and to words whose meanings are based on relations, such as "brother of". Psychologists (e.g., Helmholtz and Piaget) have anchored the meaning of whole numbers on two basic schemas of action: one-to-one correspondence (between items in different sets and between items and number words in a counting system) and addition. Neither schema is sufficient for understanding whole numbers: counting supports the understanding of ordinal number and addition the understanding of both ordinal and cardinal number. The combination of these two schemas provides the representational meaning for whole numbers.

Theories about the psychological basis for understanding rational numbers are fragmented so far, but there is a recognition that there is a gap between students' understanding of natural and rational numbers (Van Dooren et al., 2015). Because rational number words are not encountered often in natural language, their meanings have to be taught and learned at school. In this chapter, we analyzed the representational meanings for rational numbers using the unit of analysis situation/action schema. Three classes of situation that can be used in teaching were considered: part-whole, ratio, and intensive quantities. We argue that these classes of situations are distinguishable because they involve different quantities and relations between quantities. We identified a set of schemas that are used to solve problems in these situations. In part-whole situations students are stimulated to carry out partitioning and successive partitioning and to think about part-whole relations when the parts are equal. In situations that involve ratios between two different quantities and intensive quantities students can be invited to use one-to-many correspondences and to think about composite units. Schemas of action contain theorems in action, which can be made explicit in order to promote logical reasoning about situations that involve rational numbers: for example, in part-whole situations, the greater the number of parts, the smaller the parts; in ratio situations, the larger the divisor, the smaller the quotient. However, the schemas that support part-whole logic are additive whereas the schema of one-to-many correspondence used in ratio and in intensive quantities situations represents multiplicative relations in action.

Learning to connect the different situations and schemas to numerical representations of rational number draws on different ways of thinking (see also Moseley, 2005; Post et al., 1993). In part-whole situations, students can learn to think about a part as *p* out of *q* or as a unit iterated *p* times to measure a whole; the relation between the part and the whole is additive and the quantity is extensive. In ratio and in intensive quantities situations, the relations between the quantities are multiplicative. In ratio situations, students can learn to think about one extensive quantity divided by another; this situation can give meaning to the slash [/] used in fractions as "division" and to the colon [:] used in ratios as the "correspondences" between two quantities that form a composite unit. Similar connections can be made in the context of intensive quantities situations, which require thinking about a new concept, the intensive quantity per se, which cannot be measured without recourse to rational numbers.

We conclude by suggesting that reasoning about relations between quantities in all three situations should be the source for the representational meaning of rational numbers. We interpret the research findings as indicating that it is better to start teaching students about rational numbers in ratio situations, which afford reasoning about multiplicative relations by means of the one-to-many correspondence schema. Students successfully use one-to-many correspondences to solve multiplicative reasoning problems from about six or seven years of age. The one-to-many correspondence schema can be used to teach the concept of composite units and to promote scalar reasoning. However, because there is no evidence to show transfer of learning from ratio to part-whole situations, the latter should be included in rational number teaching as well as intensive quantities.

References

Abrahamson, D. (2012). Rethinking intensive quantities via guided mediated abduction. *Journal of the Learning Sciences, 21*(4), 626–649.

Alajmi, A. H. (2012). How do elementary textbooks address fractions? A review of mathematics textbooks in the USA, Japan, and Kuwait. *Educational Studies in Mathematics, 79*(2), 239–261.

Allal, L., & Ducrey, G. P. (2000). Assessment of—or in—The zone of proximal development. *Learning and Instruction, 10*(2), 137–152.

Arcidiacono, F., & Perret-Clermont, A.-N. (2009). Revisiting the Piagetian test of conservation of quantities of liquid: Argumentation within the adult-child interaction. *Cultural-Historical Psychology, 3*, 25–33.

Battista, M. T., & Borrow, C. V. A. (1995). *A proposed constructive itinerary from iterating composite units to ratio and proportion concepts.* In Paper presented at the annual meeting of the north American chapter of the international group for the psychology of mathematics education (PME-NA), Columbus, OH, USA. Last accessed April 2021 from: https://files.eric.ed.gov/fulltext/ED389582.pdf

Becker, J. (1993). Young children's numerical use of number words: Counting in many-to-one situations. *Developmental Psychology, 29*(3), 458–465.

Behr, M., Harel, G., Post, T., & Lesh, R. (1992). Rational number, ratio, proportion. In D. A. Grouws (Ed.), *Handbook of research on mathematics teaching and learning* (pp. 296–333). Macmillan.

Behr, M., Harel, G., Post, T., & Lesh, R. (1993). Rational number: Toward a semantic analysis. In T. Carpenter, E. Fennema, & T. A. Romberg (Eds.), *Rational numbers: An integration of research* (pp. 13–47). Lawrence Erlbaum.

Behr, M. J., Lesh, R., Post, T. R., & Silver, E. A. (1983). Rational-number concepts. In R. Lesh & M. Landau (Eds.), *Aquisition of mathematics concepts and processes* (pp. 91–126). Academic Press Inc.

Behr, M. J., Wachsmuth, I., Post, T. R., & Lesh, R. (1984). Order and equivalence of rational numbers: A clinical teaching experiment. *Journal for Research in Mathematics Education, 15*(5), 323–341.

Brinker, L. (1998). Using recipes and ratio tables to build on students' understanding of fractions. *Teaching Children Mathematics, 5*(4), 218–224.

Brousseau, G., Brousseau, N., & Warfield, V. (2004). Rationals and decimals as required in the school curriculum. Part 1: Rationals as measurement. *Journal of Mathematical Behavior, 23*, 1–20.

Brown, A. L., & Ferrara, R. A. (1985). Diagnosing zones of proximal development. In J. V. Wertsch (Ed.), *Culture, communication and cognition* (pp. 273–305). Cambridge University Press.

Bryant, P., Christie, C., & Rendu, A. (1999). Children's understanding of the relation between addition and subtraction: Inversion, identity and decomposition. *Journal of Experimental Child Psychology, 74*, 194–212.

Campos, T. M. M., Nunes, T., Bryant, P., Silva, A. D. F. G., Canova, R. F., & Cervantes, P. D. B. M. (2014). Uso de situações quociente no ensino de frações. *Jornal Internacional de Estudos em Educação Matemática, 7*(3), 102–128.

Campos, T. M. M., Nunes, T., Costa, N. M. L. D., & Ceragioli, L. (2013). A representação de quantidades menores do que uma unidade. *Acta Scientiae, 14*, 363–373.

Canagaratna, S. G. (1992). Intensive and extensive: Underused concepts. *Journal of Chemical Education, 69*(12), 957.

Charles, K., & Nason, R. (2000). Young children's partitioning strategies. *Educational Studies in Mathematics, 43*, 191–221.

Ching, B.H.-H., & Nunes, T. (2017). The importance of additive reasoning in children's mathematical achievement: A longitudinal study. *Journal of Educational Psychology, 109*(4), 477.

Common Core State Standards for Mathematics. http://www.corestandards.org/wp-content/uploads/Math_Standards1.pdf. Last accessed in 2022.

Davidson, K., Eng, K., & Barner, D. (2012). Does learning to count involve a semantic induction? *Cognition, 123*(1), 162–173.

Davis, G. E., & Hunting, R. P. (1990). Spontaneous partitioning: Preschoolers and discrete items. *Educational Studies in Mathematics, 21*, 367–374.

Department for Education (2013, updated 21 January 2021). *Mathematics programmes of study: key stages 1 and 2. National curriculum in England.* DFE-00180-2013. Retrieved on April 2021 from: https://www.gov.uk/government/publications/national-curriculum-in-england-mathematics-programmes-of-study/national-curriculum-in-england-mathematics-programmes-of-study#key-stage-1---years-1-and-2

Desli, D. (1999). *Children's understanding of intensive quantities* (Ph.D.), University of London, Unpublished Ph.D. Thesis.

Dole, S. (2008). Ratio tables to promote proportional reasoning in the primary classroom. *Australian Primary Mathematics Classroom, 13*(2), 18–22.

Domoney, B. (2002). Student teachers' understanding of rational number: Part-whole and numerical constructs. *Research in Mathematics Education, 4*(1), 53–67.

Duval, R. (1999) Representation, vision and visualization: Cognitive functions in mathematical thinking. Basic issues for learning. In F. Hitt & M. Santos (Eds.), *Proceedings of the annual meeting of the North American chapter of the international group for the psychology of mathematics education* (Vol. 1, pp. 3–26), Cuernavaca, Mexico. https://eric.ed.gov/?id=ED466379

Duval, R. (2006). A cognitive analysis of problems of comprehension in a learning of mathematics. *Educational Studies in Mathematics, 61*(1), 103–131.

Edwards, B. J., Williams, J. J., Gentner, D., & Lombrozo, T. (2019). Explanation recruits comparison in a category-learning task. *Cognition, 185*, 21–38.

Elkind, D. (1967). Piaget's conservation problems. *Child Development*, 15–27.

Ellis, L. (2015). *How does the provision of materials in multiplicative reasoning word problems affect year one children's problem solving? An exploration of the impact of representations on children's ability to reason multiplicatively prior to formal instruction.* Unpublished M.Sc. Dissertation, Department of Education, University of Oxford.

Empson, S. B., Junk, D., Dominguez, H., & Turner, E. (2006). Fractions as the coordination of multiplicatively related quantities: A cross-sectional study of children's thinking. *Educational Studies in Mathematics, 63*, 1–28.

Flores, A., Samson, J., & Yanik, H. B. (2006). Research, reflection, practice: Quotient and measurement interpretations of rational numbers. *Teaching Children Mathematics, 13*(1), 34–39.

Gabriel, F. C., Coché, F., Szucs, D., Carette, V., Rey, B., & Content, A. (2013). A componential view of children's difficulties in learning fractions. *Frontiers in Psychology, 4*, 715.

Gentner, D., Anggoro, F. K., & Klibanoff, R. S. (2011). Structure mapping and relational language support children's learning of relational categories. *Child Development, 82*(4), 1173–1188.

Gigerenzer, G. (2002). *Reckoning with risk. Learning to live with uncertainty.* Penguin Books.

Gréco, P. (1962). Quantité et quotité: nouvelles recherches sur la correspondance terme-a-terme et la conservation des ensembles. In P. Gréco & A. Morf (Eds.), *Structures numeriques elementaires: Etudes d'Epistemologie Genetique* (Vol. 13, pp. 35–52). Presses Universitaires de France.

Guedj, D. (1998). *Numbers. A universal language.* Thame and Hudson.

Hackenberg, A. J. (2007). Units coordination and the construction of improper fractions: A revision of the splitting hypothesis. *The Journal of Mathematical Behavior, 26*(1), 27–47.

Hart, K. (1984). *Ratio: children's strategies and errors. A report of the strategies and errors in secondary mathematics project.* NFER-Nelson.

Hart, K. (1988). Ratio and proportion. In J. Hiebert & M. Behr (Eds.), *Number concepts and operations in middle grades* (pp. 198–219). Erlbaum.

Hollich, G. J., Hirsh-Pasek, K., & Golinkoff, R. M. (2000). *Breaking the language barrier: An emergentist coalition model for the origins of word learning.* University of Chicago Press.

Howe, C., Nunes, T., & Bryant, P. (2010a). Intensive quantities: Why they matter to developmental research. *British Journal of Developmental Psychology, 28*(2), 307–329.

Howe, C., Nunes, T., & Bryant, P. (2011). Rational number and proportional reasoning: Using intensive quantities to promote achievement in mathematics and science. *International Journal of Science and Mathematics Education, 9*, 391–417. https://doi.org/10.1007/s10763-010-9249-9

Howe, C., Nunes, T., Bryant, P., Bell, D., & Desli, D. (2010b). Intensive quantities: Towards their recognition at primary school level. In *BJEP monograph series II, number 7-understanding number development and difficulties* (Vol. 101, pp. 101–118). British Psychological Society.

Hoyles, C., Noss, R., & Pozzi, S. (2001). Proportional reasoning in nursing practice. *Journal for Research in Mathematics Education, 32*(1), 4–27.

Inhelder, B., & Piaget, J. (1958). *The growth of logical thinking from childhood to adolescence.* Basic Books.

Inhelder, B., & Piaget, J. (1964). *The early growth of logic in the child.* Routledge and Kegan Paul.

Johnson, H. L. (2014). Images of intensive and extensive quantity: A framework for reasoning about change in covarying quantities. In L. P. Steffe, K. C. Moore, & L. L. Hatfield (Eds.), *Epistemic algebraic students: Emerging models of students' algebraic knowing, WISDOMe monographs* (Vol. 4, pp. 267–280). College of Education, University of Wyoming.

Kamii, C., & Clark, F. B. (1995). Equivalent fractions: Their difficulty and educational implications. *Journal of Mathematical Behavior, 14*, 365–378.

Kaput, J. (1985). *Multiplicative word problems and intensive quantities: An integrated software response.* Technical Report 85–19. Harvard University, Educational Technology Center. Last retrieved on April 2021 from: https://files.eric.ed.gov/fulltext/ED295787.pdf

Kaput, J., & West, M. M. (1994). Missing-value proportional reasoning problems: Factors affecting informal reasoning patterns. In G. Harel & J. Confrey (Eds.), *The development of multiplicative reasoning in the learning of mathematics* (pp. 237–292). State University of New York Press.
Kerslake, D. (1986). *Fractions: Children's strategies and errors: A report of the strategies and errors in secondary mathematics project*. NFER-Nelson.
Kieren, T. E. (1976). *On the mathematical, cognitive and instructional.* Paper presented at the Number and measurement. Papers from a research workshop. Eric 120 027. Last retrieved on April 08, 2022, from https://citeseerx.ist.psu.edu/viewdoc/download?doi=10.1.1.994.9838&rep=rep1&type=pdf#page=108
Kieren, T. E. (1988). Personal knowledge of rational numbers: Its intuitive and formal development. In J. Hiebert & M. Behr (Eds.), *Number concepts and operations in the middle grades* (pp. 162–180). Erlbaum.
Kieren, T. E. (1992). Rational and fractional numbers as mathematical and personal knowledge: Implications for curriculum and instruction. In G. Lienhardt, R. Putnam, & R. A. Hattrup (Eds.), *Analysis of arithmetic for mathematics teaching* (pp. 323–372). LEA.
Kieren, T. E. (1993). Rational and fractional numbers: From quotient fields to recursive understanding. In T. Carpenter, E. Fennema, & T. A. Romberg (Eds.), *Rational numbers: An integration of research* (pp. 49–84). Lawrence Erlbaum Associates.
Kornilaki, E., & Nunes, T. (2005). Generalising principles in spite of procedural differences: Children's understanding of division. *Cognitive Development, 20*, 388–406.
Kouba, V. (1989). Children's solution strategies for equivalent set multiplication and division word problems. *Journal for Research in Mathematics Education, 20*, 147–158.
Lamon, S. J. (1993). Ratio and proportion: Connecting content and children's thinking. *Journal for Research in Mathematics Education, 24*(1), 41–61.
Langrall, C. W., & Swafford, J. (2000). Three balloons for two dollars: Developing proportional reasoning. *Mathematics Teaching in the Middle School, 6*(4), 254–261.
Lima, M. F. (1982). Iniciação ao conceito de fração e o desenvolvimento de conservação de quantitades. In T. N. Carraher (Ed.), *Aprender Pensando* (pp. 81–127). Editora Vozes.
Mack, N. K. (1990). Learning fractions with understanding: Building on informal knowledge. *Journal for Research in Mathematics Education, 21*(1), 16–32.
Mack, N. K. (2001). Building on informal knowledge through instruction in a complex content domain: Partitioning, units, and understanding multiplication of fractions. *Journal for Research in Mathematics Education, 32*(3), 267–295.
Maloney, A., & Confrey, J. (2010). The construction, refinement, and early validation of the equipartitioning learning trajectory. Retreived on February 11, 2022, from https://repository.isls.org/bitstream/1/2780/1/968-975.pdf
Mamede, E. (2016). *Young children can learn to reason and to name fractions.* In Paper presented at 13th International Congress on Mathematics Education ICME), Hamburg, 24–31 July. Last retrieved on April 2021 from: https://repositorium.sdum.uminho.pt/bitstream/1822/58620/1/Mamede%20CME%202018.pdf
Mamede, E., Nunes, T., & Bryant, P. (2005). *The equivalence and ordering of fractions in part-whole and quotient situations.* In Paper presented at the 29th conference of the international group for the psychology of mathematics education, Melbourne.
Middleton, J. A., & Van den Heuvel-Panhuizen, M. (1995). The ratio table. *Mathematics Teaching in the Middle School, 1*, 282–288.
Moseley, B. (2005). Students' early mathematical representation knowledge: The effects of emphasizing single or multiple perspectives of the rational number domain in problem solving. *Educational Studies in Mathematics, 60*(1), 37–69.
Moseley, B. J., Okamoto, Y., & Ishida, J. (2007). Comparing US and Japanese elementary school teachers' facility for linking rational number representations. *International Journal of Science and Mathematics Education, 5*(1), 165–185. https://doi.org/10.1007/s10763-006-9040-0
Naik, S., & Subramaniam, K. (2008). Integrating the measure and quotient interpretation of fractions. In O. Figueras, J. L. Cortina, S. Alatorre, T. Rojano, & A. Sepúlveda (Eds.), *Proceedings of the*

32nd meeting of the international group of the psychology of mathematics education: PME (Vol. 4, pp. 17–24). CINVESTAV-UMSNH.

Ni, Y., & Zhou, Y.-D. (2005). Teaching and learning fraction and rational numbers: The origins and implications of whole number bias. *Educational Psychologist, 40*(1), 27–52.

Noelting, G. (1980a). The development of proportional reasoning and the ratio concept part I—Differentiation of stages. *Educational Studies in Mathematics, 11*, 217–253.

Noelting, G. (1980b). The development of proportional reasoning and the ratio concept part II—Problem-structure at successive stages: Problem-solving strategies and the mechanism of adaptative restructuring. *Educational Studies in Mathematics, 11*, 331–363.

Noss, R., Hoyles, C., & Pozzi, S. (2002). Abstraction in expertise: A study of nurses' conceptions of concentration. *Journal for Research in Mathematics Education, 33*, 204–229.

Nunes, T., & Bryant, P. (1996). *Children doing mathematics*. Blackwell.

Nunes, T., & Bryant, P. (2008). Rational numbers and intensive quantities: Challenges and insights to pupils' implicit knowledge. *Anales De Psicologia, 24*, 262–270.

Nunes, T., & Bryant, P. (2022a). *Using mathematics to understand the world. How culture promotes children's mathematics*. Taylor & Francis.

Nunes, T., & Bryant, P. (2022b). Using the pair ratio situation/correspondence schema to teach fractions. Manuscript under review.

Nunes, T., Bryant, P., & Hurry, J. (2004a). *The role of awareness in teaching and learning literacy and numeracy in key stage 2*: Report presented to the ESRC-TLRP, Award #L139251015

Nunes, T., Bryant, P., Barros, R., & Sylva, K. (2012). The relative importance of two different mathematical abilities to mathematical achievement. *British Journal of Educational Psychology, 82*(1), 136–156.

Nunes, T., Bryant, P., Burman, D., Bell, D., Evans, D., & Hallett, D. (2008). Deaf children's informal knowledge of multiplicative reasoning. *Journal of Deaf Studies and Deaf Education, 14*, 260–277.

Nunes, T., Bryant, P., Evans, D., & Barros, R. (2015a). Assessing quantitative reasoning in young children. *Mathematics Teaching and Learning, 17*(2–3), 178–196. https://doi.org/10.1080/10986065.2015.1016815

Nunes, T., Bryant, P., Evans, D., & Bell, D. (2010). The scheme of correspondence and its role in children's mathematics. *British Journal of Educational Psychology—Monograph Series II, Number 7—Understanding Number Development and Difficulties*, 83–99.

Nunes, T., Bryant, P., Evans, D., Gottardis, L., & Terlektsi, M.-E. (2014). The cognitive demands of understanding the sample space. *ZDM Mathematics Education, 46*(3), 437–448.

Nunes, T., Bryant, P., Gottardis, L., Terlektsi, M.-E., & Evans, D. (2015b). Can we really teach problem solving in primary school? *Mathematics Teaching, 246*, 44–48.

Nunes, T., Bryant, P., & Hurry, J. (2004b). *The role of awareness in teaching and learning literacy and numeracy in key stage 1*. Report presented to the ESRC-TLRP, Award #L139251015

Nunes, T., Bryant, P., Hurry, J., & Pretzlik, U. (2006). Fractions: Difficult but crucial in mathematics learning. Teaching and Learning: Research Briefing, 13, 1–4. Last accessed on April 2021 from: https://r.search.yahoo.com/_ylt=AwrJO1V..nlgDckAiAt3Bwx.;_ylu=Y29sbwMEcG9zAzEEdnRpZAMEc2VjA3Ny/RV=2/RE=1618635519/RO=10/RU=https%3a%2f%2fhighlandnumeracyblog.files.wordpress.com%2f2015%2f01%2fnunes-et-al-fractions_difficult-but-crucial-in-mathematics-learning.pdf/RK=2/RS=CNlpm0YSskWJniK3L7_nx4BzI5U

Nunes, T., Bryant, P., Pretzlik, U., Bell, D., Evans, D., & Wade, J. (2007). La compréhension des fractions chez les enfants. In M. Merri (Ed.), *Activité Humaine et Conceptualisation* (pp. 255–262). Presses Universitaires du Mirail.

Nunes, T., Desli, D., & Bell, D. (2003). The development of children's understanding of intensive quantities. *International Journal of Educational Research, 39*, 652–675.

Nunes, T., Dorneles, B. V., Lin, P.-J., & Rathgeb-Schnierer, E. (2016). Teaching and learning about whole numbers in primary school. *Teaching and learning about whole numbers in primary school* (pp. 1–50). Springer International Publishing.

Ohlsson, S. (1988). Mathematical meaning and applicational meaning in the semantics of fractions and related concepts. In J. Hiebert & M. Behr (Eds.), *Number concepts and operations in the middle grades* (pp. 53–92). National Council of Mathematics Teachers.
Olive, J. (2000). Children's number sequences: An explanation of Steffe's constructs and an extrapolation to rational numbers of arithmetic. *The Mathematics Educator, 11*(1).
Olive, J., & Steffe, L. P. (2002). The construction of an iterative fractional scheme: The case of Joe. *Journal of Mathematical Behavior, 20*, 413–437.
Parish-Morris, J., Pruden, S. M., Ma, W., Hirsh-Pasek, K., & Golinkoff, R. M. (2010). Relational words. In B. C. Malt & P. Wolff (Eds.), *Words and the mind. How words capture human experience* (pp. 219–244). Oxford University Press.
Park, J., & Nunes, T. (2001). The development of the concept of multiplication. *Cognitive Development, 16*, 1–11.
Piaget, J. (1952a). *The origins of intelligence*. Routledge & Kegan Paul.
Piaget, J. (1952b). *The child's conception of number*. Routledge & Kegan Paul.
Piaget, J. (1953). How children form mathematical concepts. *Scientific American, 189*(5), 74–79.
Piaget, J., & Inhelder, B. (1975). *The origin of the idea of chance in children*. Norton (currently available form Psychology Press)
Piaget, J., Inhelder, B., & Szeminska, A. (1960). *The child's conception of geometry*. Harper & Row.
Piantadosi, S. T., Tenenbaum, J. B., & Goodman, N. D. (2012). Bootstrapping in a language of thought: A formal model of numerical concept learning. *Cognition, 123*(2), 199–217.
Pitkethly, A., & Hunting, R. (1996). A review of recent research in the area of initial fraction concepts. *Educational Studies in Mathematics, 30*(1), 5–38.
Post, T. R., Cramer, K. A., Behr, M., Lesh, R., & Harel, G. (1993). Curriculum implications of research on the learning, teaching, and assessing of rational number concepts. In T. Carpenter P., E. Fennema, & T. A. Romberg (Eds.), *Rational numbers: An integration of research* (pp. 327–362). Lawrence Erlbaum Associates.
Post, T., Wachsmuth, I., Lesh, R., & Behr, M. (1985). Order and equivalence of rational numbers: A cognitive analysis. *Journal for Research in Mathematics Education, 16*, 18–37.
Pothier, Y., & Sawada, D. (1983). Partitioning: The emergence of rational number ideas in young children. *Journal for Research in Mathematics Education, 14*, 307–317.
Presmeg, N., & Van den Heuvel-Panhuizen, M. (2003). Leen Streefland's legacy. *Educational Studies in Mathematics, 54*, 1–4. https://doi.org/10.1007/s11858-020-01138-w
Resnick, L. B., & Singer, J. A. (1993). Protoquantitative origins of ratio reasoning. In T. P. Carpenter, E. Fennema, & T. A. Romberg (Eds.), *Rational numbers: An integration of research* (pp. 107–130). Lawrence Erlbaum Associates Inc.
Sarnecka, B. W., & Carey, S. (2008). How counting represents number: What children must learn and when they learn it. *Cognition, 108*, 662–674.
Schwartz, J. (1988). Intensive quantity and referent transforming arithmetic operations. In J. Hiebert & M. Behr (Eds.), *Number concepts and operations in the middle grades* (pp. 41–52). Erlbaum.
Schwartz, J. L. (1996). Semantic aspects of quantity. *Unpublished manuscript*. Cambridge, MA: MIT and Harvard Graduate School of Education. Last accessed on April 2021 from: http://citeseerx.ist.psu.edu/viewdoc/download?doi=10.1.1.622.7161&rep=rep1&type=pdf
Silver, E. (1981). Young adults' thinking about rational numbers. In T. R. Post & M. Roberts (Eds.), *Proceedings of the Third Annual Meeting of the North American Chapter of the International Group for the Psychology of Mathematics Education* (pp. 149–159). University of Minnesota.
Simon, M. A. (2017). Explicating mathematical concept and mathematical conception as theoretical constructs for mathematics education research. *Educational Studies in Mathematics, 94*(2), 117–137.
Simon, M. A., & Placa, N. (2012). Reasoning about intensive quantities in whole-number multiplication? A possible basis for ratio understanding. *For the Learning of Mathematics, 32*(2), 35–41.

Simon, M. A., Placa, N., Avitzur, A., & Kara, M. (2018). Promoting a concept of fraction-as-measure: A study of the learning through activity research program. *The Journal of Mathematical Behavior, 52*, 122–133.

Singer, J. A., Kohn, A. S., & Resnick, L. B. (1997). Knowing about proportions in different contexts. In T. Nunes & P. Bryant (Eds.), *Learning and teaching mathematics. An international perspective* (pp. 115–132). Psychology Press.

Stavy, R., & Tirosh, D. (1996). Intuitive rules in science and mathematics: The case of 'more of A—more of B.' *International Journal of Science Education, 18*(6), 653–667.

Steffe, L. P. (1992). Schemes of action and operation involving composite units. *Learning and Individual Differences, 4*(3), 259–309.

Steffe, L. P. (2002). A new hypothesis concerning children's fractional knowledge. *Journal of Mathematical Behavior, 102*, 267–307.

Steffe, L. P., Liss II, D. R. L., & Lee, H. Y. (2014a). On the operations that generate intensive quantity. In L. P. Steffe, K. C. Moore, & L. L. Hatfield (Eds.), *Epistemic algebraic students: Emerging models of students' algebraic knowing, WISDOMe monographs* (Vol. 4, pp. 50–79). College of Education, University of Wyoming.

Steffe, L. P., Moore, K. C. & Hatfield, L. L. (2014b, Eds.), *Epistemic algebraic students: Emerging models of students' algebraic knowing, WISDOMe Monographs* (Vol. 4). College of Education, University of Wyoming.

Stevens, S. S. (1946). On the theory of scales of measurement. *Science, 103*, 670–680.

Streefland, L. (1982). Subtracting fractions with different denominators. *Educational Studies in Mathematics, 13*(3), 233–255.

Streefland, L. (1984). Search for the roots of ratio: Some thoughts on the long term learning process (towards...a theory): Part I: Reflections on a teaching experiment. *Educational Studies in Mathematics, 15*, 327–348.

Streefland, L. (1985). Search for the roots of ratio: Some thoughts on the long term learning process (towards...a theory): Part II: The outline of the long term learning process. *Educational Studies in Mathematics, 16*, 75–94.

Streefland, L. (1987). *How to teach fractions so as to be useful.* The State University of Utrecht.

Streefland, L. (1991). Fractions, an integrated perspective. In L. Streefland (Ed.), *Realistic mathematics education in primary school* (pp. 93–118). Freudenthal Institute, Utrecht University.

Streefland, L. (1993). Fractions: a realistic approach. In T. P. Carpenter, E. Fennema, & T. A. Romberg (Eds.), *Rational numbers: An integration of research* (pp. 289–326). Lawrence Erlbaum Associates.

Streefland, L. (1997). Charming fractions or fractions being charmed? In T. Nunes & P. Bryant (Eds.), *Learning and teaching mathematics. An international perspective* (pp. 347–372). Psychology Press.

Thompson, P. (1990). *A theoretical model of quantity-based reasoning in arithmetic and algebra.* Unpublished manuscript based on a paper presented at the Annual Meeting of the American Educational Research Association, San Francisco, CA, 27–31 March 1989

Thompson, P. (1994). The development of the concept of speed and its relationship to concepts of rate. In G. Harel & J. Confrey (Eds.), *The development of multiplicative reasoning in the learning of mathematics* (pp. 181–236). State University of New York Press.

Thompson, P. W. (1993). Quantitative reasoning, complexity, and additive structures. *Educational Studies in Mathematics, 3*, 165–208.

Thompson, P. W. (2011). Quantitative reasoning and mathematical modeling1, 2. In S. A. Chamberlin & L. L. Hatfield (Eds.), *New perspectives and directions for collaborative research in mathematics education* (pp. 33–57). University of Wyoming.

Thompson, P. W., & Saldanha, L. A. (2003). Fractions and multiplicative reasoning. In J. Kilpatrick, W. G. Martin, & D. Schifter (Eds.), *Research companion to the principles and standards for school mathematics* (pp. 95–113). National Council of Teachers of Mathematics.

Tian, J., Bartek, V., Rahman, M. Z., & Gunderson, E. A. (2021). Learning improper fractions with the number line and the area model. *Journal of Cognition and Development, 22*(2), 305–327.

Tolman, R. C. (1917). The measurable quantities of physics. *Physical Review, 9*(3), 237–253.
Toluk, Z., & Middleton, J. A. (2001). *The development of children's understanding of the quotient: A teaching experiment.* Paper presented at the 25th PME CONFERENCE, Utrecht, NL.
Travis, I. (1937). Per-unit quantities. *Transactions of the American Institute of Electrical Engineers, 56*(12), 22–28.
Tzur, R. (1999). An Integrated study of children's construction of improper fractions and the teacher's role in promoting that learning. *Journal for Research in Mathematics Education, 30*, 390–416.
Vamvakoussi, X., & Vosniadou, S. (2004). Understanding the structure of the set of rational numbers: A conceptual change approach. *Learning and Instruction, 14*(5), 453–467.
Van Dooren, W., Lehtinen, E., & Verschaffel, L. (2015). Unraveling the gap between natural and rational numbers. *Learning and Instruction, 37*, 1–4.
Vergnaud, G. (1983). Multiplicative structures. In R. Lesh & M. Landau (Eds.), *Acquisition of mathematics concepts and processes* (pp. 128–175). Academic Press.
Vergnaud, G. (2009). The theory of conceptual fields. *Human Development, 52*, 83–94.
von Helmholtz, H. (1887/translation date 1977). Numbering and measuring from an epistemological viewpoint. In *Epistemological writings* (pp. 72–114). Springer.
Vygotsky, L. (1962). *Thought and language.* M.I.T. Press.
Watanabe, T. (2006). The teaching and learning of fractions: A Japanese perspective. *Teaching Children Mathematics, 12*(7), 368–374.
Wood, D., Bruner, J. S., & Ross, G. (1976). The role of tutoring in problem solving. *Journal of Child Psychology and Psychiatry, 17*(2), 89–100.

Quantitative Reasoning as a Framework to Analyze Mathematics Textbooks

Gülseren Karagöz Akar, Tad Watanabe, and Nurdan Turan

1 Curriculum and Textbooks

The word "curriculum" means different things to different people or even to the same people in different contexts. However, there seems to be a general agreement that it is important to distinguish different meanings of curriculum.

The International Association for the Evaluation of Educational Achievement (IEA) has a model of curriculum. In the IEA model (Valverde et al., 2002), there are three levels of curriculum: *intended curriculum* is a set of expectations and goals for students that are often in the form of curriculum standards or syllabi[1]; *implemented* (or *enacted*) *curriculum* is what actually happens in classroom with respect to those goals and expectations expressed in intended curriculum; and *attained* (or *achieved*) *curriculum* is about goals and expectations actually learned by students.

One important factor in students' opportunities to learn is the textbooks that are used. Textbooks are generally informed by an intended curriculum and play a significant role in how the curriculum is enacted in classrooms. In this sense, textbooks can be considered as a bridge between intended and implemented curricula. In a more refined IEA model, textbooks (and other curriculum resources) are positioned as a potentially implemented curriculum (Valverde et al., 2002). However, with the

[1] Stein et al. (2007) use the phrase "intended curriculum" to describe teachers' plans for instruction, but they recognize that their use of the phrase differs from the IEA model.

G. Karagöz Akar (✉)
Boğaziçi University, İstanbul, Turkey
e-mail: gulseren.akar@boun.edu.tr; gulserenkaragoz@yahoo.com

T. Watanabe
Kennesaw State University, Kennesaw, GA, USA

N. Turan
New York University, New York, NY, USA

G. Karagöz Akar et al. (eds.), *Quantitative Reasoning in Mathematics and Science Education*, Mathematics Education in the Digital Era 21,
https://doi.org/10.1007/978-3-031-14553-7_5

acknowledgement and advances in technology in the digital era, students' need and preference for the use of digital materials is inevitable. On the other hand, "while technology has impacted student behaviours and distinct preferences for learning, one thing that has not changed is the essential need for credible content. Technology is useless without valid content" (Knight & Wang, 2015, p. 4) And, research point that "…textbooks are generally viewed as reliable tools which provide creditable information that supports and enhances students' understanding of critical concepts, and that they present bite-size chunks of information to cement student learning" (a.b.i.d.; p. 1). Kilpatrick et al. (2001) also noted "what is actually taught in classrooms is strongly influenced by the available textbooks" (p. 36). Thus, understanding what is included in textbooks and how specific topics are treated may be informative for examining teaching and learning of those topics in various systems (Fig. 1).

Fan (2013) suggested that research involving textbooks may be placed among three broad areas:

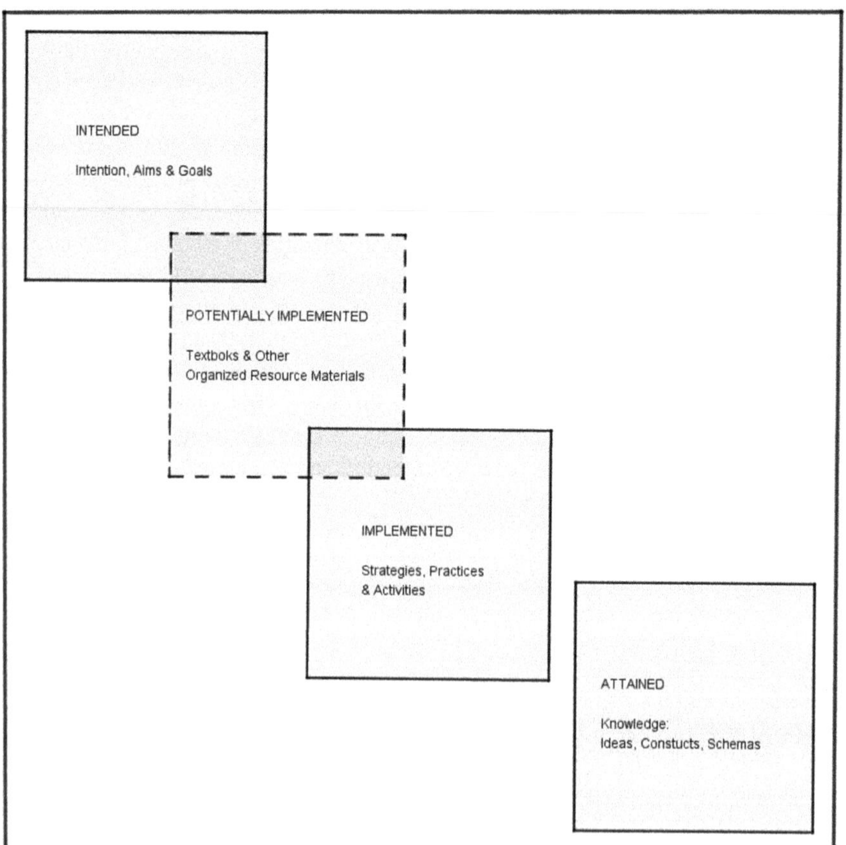

Fig. 1 IEA's the revised tripartite model of curriculum showing mediator role of textbooks (redrawn from Valverde et al., 2002, p. 13)

1. Studies examining the textbooks themselves;
2. Studies examining the factors influencing the textbook development; and
3. Studies exploring how textbooks influence other factors.

Some scholars (e.g., Remillard, 2005) have suggested the need for further studies of how teachers use textbooks and other curriculum resources. Reports like Zembat (2010) might fall into that category. However, at this point, as for Fan (2013), there appear to be more studies that fall into the first two groups.

Content analysis of textbooks as research is a relatively new and still developing field of investigation. Some of the existing research investigated overall structures of textbooks, often focusing on what mathematics is taught at what grade level (e.g., Schmidt et al., 1997). Another group of existing studies examined the treatment of a particular mathematical topic such as fractions (e.g. Son & Senk, 2010; Watanabe et al., 2017) or a mathematical process such as problem solving (e.g., Fan & Zhu, 2007; Li, 2000; Mayer et al., 1995). A few studies examined the historical trends observed in textbooks (e.g., Ross & Pratt-Cotter, 1997; Watanabe, 2014), how teachers evaluate textbooks (Arslan & Özpınar, 2009a) or how the textbooks are aligned with the national mathematics standards (Arslan & Özpınar, 2009b). Charalambous et al. (2010) referred to those textbook studies that examine the structures as horizontal analysis and those that examine the development of a particular idea as vertical analysis. Li et al. (2009) in contrast called the former type "macro analysis" and the latter "micro analysis." Macroanalysis or horizontal analysis involves examining what mathematics is taught at what grade level, whereas microanalysis as a vertical analysis involves examining a particular mathematical topic or process across different grade levels. Although horizontal, or macro, analyses of textbooks give us the general sense of what topics are discussed in what grade level in the textbooks, they do not reveal much about the actual learning opportunities offered by the textbooks. On the other hand, because vertical, or micro, analyses of textbooks focus on a single mathematical topic, they can reveal different approaches taken by different textbooks, possibly from different systems. Thus, some researchers chose to examine textbooks by integrating both the horizontal (or macro) and the vertical (or micro) analysis (e.g., Charalambous et al., 2010; Li et al., 2009).

Often times, international textbook analyses are motivated by the differences in achievements in large scale international comparative studies such as TIMSS (The Trends in International Mathematics and Science Study) and PISA (Programme for International Student Assessment). As a result, textbooks from high achieving systems such as China, Japan, Korea and Singapore are often examined. However, the number of studies examining textbooks internationally is still rather small. Son and Dilleti (2017) reviewed 31 articles that compared textbooks between USA and five high achieving Asian systems published between 1988 and 2015 and found that those studies have examined a variety of aspects of textbooks as well as a wide range of topics. In the 31 articles reviewed, an implication is that no particular topic, perhaps with the exception of fractions, have been studied in depth. Moreover, published studies tended to examine more textbooks for middle and high schools than for early elementary schools. However, none of the studies reviewed by Son and Dilleti

(2017) examined the treatment of whole number multiplication and division. Taking this as a challenge, in this chapter, we examined how the concepts of whole number multiplication and division are developed in the Japanese curricular materials. The reason for choosing the Japanese curriculum will be made clearer in the following section.

Taken together, whole number multiplication and division are arguably the first major challenge young students face in their school mathematics. There is a general agreement among mathematics education researchers that multiplication and division are much more complex than addition and subtraction (e.g., Verschaffel et al., 2007). Studies have shown that students need to extend the meaning of multiplication and division as they progress through their study of whole number multiplication and division and multiplicative conceptual field (Vergnaud, 1983, 1994). Therefore, how international curricular materials, including textbooks, the teachers' guide and the form of curriculum standards or syllabi treat multiplication and division through different grade levels is of interest for this chapter.

In pursuit of this inquiry, in this paper, we offer embedding quantitative reasoning within the blend of both macro and micro analysis. In the following sections, we first explain quantitative reasoning and its constructs by providing a conceptual analysis of multiplication and division. Then, we offer quantitative reasoning as a theoretical model for the examination of curricular materials and provide an analysis of the Japanese curricular materials through the lenses of quantitative reasoning. We start by providing our rationale as to why and how quantitative reasoning might be a possible avenue for examining international curricular materials.

2 Quantitative Reasoning and Whole Number Multiplication and Division

To characterize quantitative reasoning, Thompson (1994) first defined quantity and the quantification process through quantitative and numerical operations. From Thompson's perspective, "quantities are conceptual entities" such that "a person is thinking of a quantity when he or she conceives a quality of an object in such a way that this conception entails the quality's measurability" (p. 184). Emphasizing the relationship between the quantity and the quantification process, Thompson (2011) stated, "Quantification is the process of conceptualizing an object and an attribute of it so that the attribute has a unit and the attribute's measure entails a proportional relationship (linear, bi-linear, or multi-linear) with its unit" (p. 37). For example, think of the problem statement: "how many apples are there if there are 3 groups of 5 apples in each?" One might conceive three quantities involved in this mathematical situation: A total amount of apples, an amount (number) of apples per group, and a number of groups. The *three quantities* involved are either extensive quantities or intensive quantities. Schwartz (1988) stated that all "quantities used in mathematics are derived from the surround by acts either counting or measuring, depending on whether we are quantifying discrete or continuous properties of the surround" (p. 41). He further defined that extensive quantities are directly measurable quantities. Extensive quantities could be discrete or continuous such that once they are combined, the combination is again an extensive quantity (Schwartz, 1988; Thompson, 1990). Intensive quantities, on the contrary, are indirectly measurable quantities derived mostly from two extensive quantities (Schwartz, 1988; Thompson, 1990). So, intensive quantities are a descriptor of a 'quality' of the object, and not of the amount of it. That is, intensive quantities "give no information whatsoever about the number or amount of the relevant extensive quantities…The statement of an intensive quantity is a statement of a relationship between quantities" (Schwartz, 1988, p. 43). So, in our example, *the total amount of apples* or *the number of groups* are extensive quantities. whereas, *the amount of apples per group* is an intensive quantity. All quantities have referents which "…identify the counting unit or unit of measurement to which the number refers" (Simon, 1993, p. 236). So, in our example, "apples", "groups" or "apples per group" are referents.

Moore et al. (2009) stated that, "in order for an individual to comprehend a quantity, the individual must have a mental image of an object and attributes of this object that can be measured" (p. 4) and such image "could be an image interpreted from a problem statement …" (Moore et al., 2009, p. 3). In learner's imagination, one issue is very important: Thompson (1994) pointed out that an individual does not need to know the result of the measurement process to comprehend a quantity. That is, one is not necessarily in need of knowing the numbers associated with the situation. What is important is that the person imagines the quantities involved and how those quantities relate to each other entailing a measurement process. For instance, for the problem situation above, one might have the image that "there is a total amount of apples given that there are so many groups of some apples in each".

When we consider the aforementioned example, the scenario for multiplication in terms of the quantities involved will be:

$$\begin{array}{|l|}\hline \textit{Multiplication}: \\ \textit{number of apples per group} \times \textit{number of groups} = \textit{number of apples (total)} \\ \hline\end{array}$$

Thompson (2011) stated "…quantitative operations are those operations of thought by which one constitutes situations quantitatively" (p. 42) such that they are nonnumerical. In contrast, numerical operations, such as multiplication, are used to evaluate the resultant quantity (Thompson, 1994, 2011). So, in our example, one might consider the quantity, the total amount of apples, as a result of a multiplicative combination (i.e., the quantitative operation) of two quantities such as *some groups* of *some amount in each* (Thompson, 2011).

Similarly, one might consider the problem situation mentioned above reversing the quantification process: Particularly, one might consider the quantity, *so many groups*, as a result of a multiplicative comparison (i.e., the quantitative operation) of two quantities such as *the total amount of apples* and *some amount in each.* In this situation, 'given a total number of apples and some amount in each might result in so many groups'. This scenario is called 'quotitive division'. So, the scenario for quotitive division in terms of the quantities involved will be:

$$\begin{array}{|l|}\hline \textit{Quotitive division:} \\ \textit{number of apples (total)} \div \textit{number of apples per group} = \textit{number of groups} \\ \hline\end{array}$$

By the same token, one might consider the quantity, *some amount in each* as a result of a multiplicative comparison (i.e., the quantitative operation) of two quantities such as *the total amount of apples* and *some groups.* In this second situation 'given a total number of apples and some groups might result in some amount in each', which is called 'partitive division'. So, the scenario for partitive division in terms of the quantities involved will be:

$$\begin{array}{|l|}\hline \textit{Partitive division:} \\ \textit{number of apples(total)} \div \textit{number of groups} = \textit{number of apples per group} \\ \hline\end{array}$$

So, in the division scenarios above, the intensive quantity will be the quotient in the partitive division, whereas it will be the divisor in the quotitive division.

In the aforementioned discussion, although we acknowledged Schwartz's perspective on the notion of quantity especially to refer to the distinctions among extensive and intensive quantities, our analysis of the curricular materials was based on Thompson's (1990, 1994) notion of quantity and quantitative reasoning. In the following section, first we refer to the distinctions between these two perspectives. Then we elaborate on how we consider quantitative reasoning as a framework for examining curricular materials by further examining the different meanings of multiplication and division from the lenses of quantitative reasoning.

3 Quantitative Reasoning: A Theoretical Model for the Examination of Textbooks

Thompson's notion of quantity as a cognitive entity differs from Schwartz's (1988). Particularly, Schwartz (1988) points out that quantities are generated by arithmetic (numerical) operations, which are addition, subtraction, multiplication, and division. Thompson (1994) distinguished numerical operations from quantitative operations. Thompson stated a quantitative operation is "the conception of two quantities being taken to produce a new quantity" (Thompson, 1990, p. 9) such that it "...is nonnumerical; it has to do with the comprehension of a situation" (Thompson, 1994, p. 13). Whereas, numerical operations are used to evaluate the value (numerical result) of a quantity. Thus, from Thompson's (1990, 1994) perspective, quantities are generated from quantitative operations and the numerical value of quantities are determined by the appropriate arithmetic operations. In this paper, our analysis was based on

Thompson's (1990, 1994) notion of quantity and quantitative reasoning for several reasons as detailed below.

First, as Thompson (1990) emphasized, identifying a particular numerical operation with a particular kind of quantity is limited in the following sense: Consider the example, "Jim is 15 cm taller than Sarah. This difference is five times as much as the difference between Abe and Sam's heights. What is the difference between Abe and Sam's heights?" (Thompson, 1990, p. 10). Although "difference" is a quantity which is mostly derived from the quantities compared additively whose value is found by the numerical operation of "subtraction" in a canonical sense, the value of the difference between Abe's and Sam's heights in this problem situation can be found by dividing 15 by 5. Thus, division operation is used to evaluate the relationship between the quantities.

Secondly, Schwartz's (1988) characterization of quantities is based on the ordered pairs of the form: (number, unit). Thompson (1994) argued "To characterize quantities as ordered pairs may be useful formally, but it does not provide insight into what people understand when they reason quantitatively about situations, and it severely confounds notions of number and notions of quantity" (p. 188).

Thirdly, Thompson's point of view takes into consideration the problem situation from an individual's perspective. As we mentioned earlier, textbooks are potentially implemented curricula that are expected to trigger mathematical thinking on the part of learners. In particular, researchers point out that analysis of quantities in problem situations is important. Particularly, "…the central goal is to focus on quantities and how they relate in situations…", and; therefore in any instruction, there is a "… need to open discussions with questions that lead to discussions of *quantities, not numbers*" (Smith III & Thompson, 2007, p. 36). This is significant because learners' imagination of quantities within a mathematical situation can contribute to their ways of thinking about the concepts they are likely to develop (Thompson, 1990, 1994, 2011). Re-examining the problem, "how many apples are there if there are 3 groups of 5 apples in each" in more detail might further elaborate on the discussion.

Learners might think in the following ways: First, they might add five three times. Such understanding is namely the multiplication as repeated addition. Although Fishbein et al. (1985) suggested that the primitive model for multiplication children have is repeated addition, research has shown that this notion of multiplication as repeated addition is a major reason that many students find transitioning from additive to multiplicative reasoning challenging (e.g., Greer, 1992; Thompson & Saldanha, 2003). One way the limitation of the repeated addition model for multiplication becomes obvious is when students must multiply decimal numbers or fractions. In the repeated addition model, at least the number of times the same number is added must be a whole number. Thus, trying to make sense of $3/4 \times 1/3$ or 2.7×3.8 requires something beyond the repeated addition model. Many studies have shown that students often have difficulties in identifying the appropriate calculation when numbers involved in word problems become fractions or decimals as well as developing a common misconception, or overgeneralization, that multiplication always makes bigger, and division always makes smaller (e.g., Fishbein et al., 1985; Graeber & Tirosh, 1990). Similarly, Thompson and Saldanha (2003) argued, "repeated addition is a quantification technique; it is not the thing being quantified" (p. 104).

Thompson and Saldanha (2003) suggested that students need to develop the meaning of multiplication as some number of (or fraction of) some amount. So, they argued that learners need to conceptualize 3 × 5 considering "three fives" since this might trigger on their part the imagination that the product, the result of multiplication, is three times as large as five (or five times as large as three). That is, their focus is taken on to the product:

> We re-emphasize that when a curriculum starts with the idea that "×" means some number of (or fractions of) some amount, it is not starting with the idea that times means to calculate. It is starting with the idea that times means to envision something in a particular way-to think of copies (including parts of copies) of some amount. This is not to suggest that multiplication should not be about calculating. Rather, calculating is just one thing one might do when thinking of a product. (Thompson & Saldanha, 2003, p. 104)

Thus, thinking of 3 × 5 as "three times as large as five" have the potential that the learner envisions the proportionality embedded in multiplication since the learner considers both the product and the factors of the product in relation to it. That is, the learner might think of multiplication multiplicatively such that for instance 3 × 5 can be understood as being in multiple reciprocal relationship to 3 and 5: (3 × 5) is 3 times as large as 5, (3 × 5) is 5 times as large as 3, 3 is 1/5th as large as (3 × 5), and 5 is 1/3rd as large as (3 × 5). In addition, Common Core State Standards in the United States (National Governors Association, 2010) suggests students should understand multiplication as scaling in Grade 5.

Understanding multiplication in this way might be important for also the learning of division multiplicatively since division also need to be understood as relative size comparison (Thompson & Saldanha, 2003). Pointing to partitive and quotitive meanings of division by Simon (1993), Byerley et al. (2012) further argued:

> These two meanings for division do not require multiplicative reasoning. A third model for division, relative size, requires students to reason multiplicatively; the relative size model for division calls upon a comparison between the size of one quantity with respect to another quantity (Thompson & Saldanha, 2003). Division as relative size allows students to be able to reason about non-integer divisors. If division is viewed partitively, it only makes sense to divide a number into *n* equal parts if *n* is an [*sic*] whole number. (p. 359)

We consider such understanding of division as also related with the scalar use of intensive quantities. For instance, for a 13.5 ft length of bookshelf, once considered in yards, it becomes 3.0 ft/yard such that the quantity (length) in feet is understood as relative to the size of one yard. Similarly, the weight of a second grader (52.5 lb, body weight) and an eighth grader whose weight is twice as much could be considered such that 2 represents the measure of the size of the weight of eighth grader relative to the size of the second grader (Shwartz, 1988).

Thompson and Saldanha (2003) provided a detailed description of how someone might conceptualize division as relative size of one quantity with respect to another quantity. For the example we used, someone might reason in the following ways if they understand division multiplicatively: In the partitive division, a total number of apples (3 × 5) and some groups (3 groups) might result in some amount in each (5 apples per group) such that 5 is understood as 1/3rd as large as the total number of apples (15). That is, if we split 15 apples into 3 groups, then each part contains 1/3rd as large as the total number of apples. In the quotitive division, a total number of apples (3 × 5) and some amount in each (3 apples per group) might result in some groups (5 groups) such that 5 is again understood as 1/3rd as large as the total number of apples (15). That is, the number of parts made by putting all apples into three-apple-sized parts is 1/3rd as large as the number of all apples. Thompson and Saldanha (2003) argued that once the learner realizes that in any of these scenarios (i.e., partitive division including sharing; and quotitive division including segmenting (or measuring)), the numerical result is the same, then the learner might make sense of why she uses division for both of them albeit "…at the level of activity they appear to be very different" (p. 107). Such an analysis can also contribute to "an understanding of referential aspects of division" (e.g., apples per group) and "an awareness of and connections between partitive and quotitive division" (Simon, 1993, p. 251) on the part of the learner.

Thus, we argue that quantitative reasoning, albeit a cognitive construct and dependent of the knower, might be useful to examine how curricular materials treat multiplication and division in terms of the possible ways of thinking from the perspective of learners. In any way, curricular materials, especially the textbooks, are mediators of the learning processes educators would like to observe on the part of learners. Particularly, we suspect that Japanese curricular materials might possess the potential to trigger the aforementioned cognitive structures regarding multiplication and division. In fact, Thompson and Carlson (2017) pointed out that "Japanese primary mathematics texts have a clear coherent focus on having students to think about quantities…" (p. 446) and called on inquiring the issue further. In addition, Turan (2021) found that questions given in the Japanese *Mathematics International* textbook series regarding functional relationships "focus on the identification and determination of quantities and relationships between quantities regardless of the grade level." (p. 228). Also, there is an emphasis in the Japanese teacher's guide suggesting teachers to create learning opportunities for students to determine quantities and examine their relationships (Turan, 2021).

Hence, taking such a call as important, in the following section, we provide some possible examples from Japanese Curricular materials, including Tokyo Shosheki's textbooks, to examine the following questions:

> How might the Japanese curricular materials depicted in the concepts of multiplication and division potentially trigger quantitative reasoning? In particular, in what ways the tasks and problem situations might potentially trigger quantitative reasoning in Grades 2 to 4?

4 Analysis of Japanese Curricular Materials

In this section, we first point to some important characteristics of the Japan's Course of Study (COS) and the *Teaching Guide*[2] (Isoda, 2010) in terms of quantities and the operations of multiplication and division. Then, we specifically focus on some examples from the *Mathematics International* Textbook series of Tokyo Shoseki from Grade 2 to Grade 4. Indeed, we purposefully focused on these grade levels because students are expected to understand the meaning of multiplication in the second grade and deepen their understanding of multiplication in the third grade. Additionally, they are expected to understand the meaning of division in the third grade and deepen their understanding of division in the fourth grade. Before we present our findings based on the multiplication and division contents in these grade levels, we explain how quantities and their relationships are introduced in the COS and *Teaching Guide*, which supports our claim that multiplication and division are introduced within quantitative structures.

An overarching objective for all grade levels stated in COS is that "Students will be able to represent and interpret numbers, quantities, and their relationships by using words, numbers, and mathematical expressions as well as diagrams, tables, and graphs" (Takahashi et al., 2008, p. 6). In the *Teaching Guide*, under the content "Quantities and Measurements" it is further stated that "A quantity expresses size of an object" (Isoda, 2010, p. 35). The term "*size*" seems to refer to the *measurable attribute of an object* rather than the actual measure of it. Particularly, length, area, volume, weight, angle, and speed are some examples of quantities that are provided in the *Teaching Guide*. For example, area is defined as "…the size of a surface that extends" (Isoda, 2010, p. 35).

In addition, there is an explicit emphasis on the operations of *counting* and *segmenting* within the quantification process. Particularly, it is stated that the size of some objects can be determined by counting and the size of some objects, such as length of strings or weight of water, should not be expressed by integers since these quantities can be divided infinitely. Also, there is emphasis on the *directly and indirectly measurable quantities with their units*. For instance, in the *Teaching Guide*, it is stated that "a unit of volume, a cubic meter (m^3) is based on a unit of length, meter (m). One cubic meter is the volume of a cube whose edge is 1 m" (Isoda, 2010, p. 36). Similarly, indirectly measurable attributes such as speed and population density are called "…as ratios between two different quantities and are examples of derived units" (Isoda, 2010, p. 36).

Specific to multiplication and division, in the *Teaching Guide*, for the **second grade**, both multiplication in equal group situations and also multiplication as comparison through consideration of relative size are emphasized. It is explicitly stated that "students can understand multiplication as the operation to determine the total number when given the number in a group and the number of groups, or the amount that is so many times as many as the base amount." (Isoda, 2010, p. 31). Also,

[2] *Teaching Guide* is published by the Japanese Ministry of Education, Culture, Sports, Science, and Technology to provide further elaboration of the national course of study.

for the same grade level, it is further stated that students can determine the product by using repeated addition. Together with these, under the content of mathematical relations, in the second grade, "seeing a number as product of other" as well as "How a product changes when the multiplier increases by 1" and in the third grade, "…how a product changes when the multiplier increases or decreases by 1" (Isoda, 2010, p. 44) is emphasized. Interestingly, starting from the second grade, there is also an explicit emphasis on the relative size meaning. For example, it is stated that "To understand the relative size of numbers" means to grasp a numbers' size in terms of units such as tens and hundreds" (Isoda, 2010, p. 62). In particular, it is stated that "students are taught to grasp the entire size of a certain object by using a partial group of that object as a unit and then counting how many units (groups) there are." (Isoda, 2010, p. 62).

Similarly in the *Teaching Guide*, for the **third grade**, it is also stated that "Division includes the case of partitive division and the case of quotitive division. Partitive division finds one part of an equally divided number or quantity. Quotitive division finds how many times one quantity is of another quantity. The methods of calculation for these two are identical and, therefore, are treated as one operation" (Isoda, 2010, p. 31). In addition, pointing to the link between quotitive and partitive division and multiplication in the third grade, it is stated

> Division can be thought of as the inverse of multiplication. Therefore, as it relates to multiplication, it is important to clarify which of the two values is being sought, the one corresponding to the multiplier or the one corresponding to the multiplicand. Partitive division is where □ in 3 × □ = 12 is sought, and quotative division is where □ in □ x 3 = 12 is sought. It is important for students to realize that when we divide in the real world, we can divide things in a partitive way or a quotative way; students should thereby understand that both types of division can be expressed by the same algebraic expression. (Isoda, 2010, p. 83)

Also, it was mentioned that "when comparing quotative division and partitive division, quotative division may be easier to represent with manipulation of concrete materials…" (p. 83).

Furthermore, in the **fourth grade**, the emphasis is put on the meaning of division as relative size comparison. A problem followed by an explanation is provided:

> A ribbon is 96 meters. Another ribbon is 24 meters. How many times as long is the former as the latter?" Here "the base quantity" and "the quantity to be compared" are known and "how many times" is asked. Division is also used in problems such as, "The yellow ribbon is 72 meters long and four times longer than the white ribbon. How many meters long is the white ribbon?" Here the "quantity to be compared" and "how many times" are known and "the unit quantity" is asked. (Isoda, 2010, p. 104)

Similarly, in the fourth grade under the content of decimals, the meaning of multiplication in both equal group situations and the relative size situations is further emphasized. For instance, it is given that "0.1 × 3" could be determined by adding 0.1 three times such as 0.1 + 0.1 + 0.1 through a simple representation of repeated addition. Similarly, "0.1 × 3" could be thought "…as a calculation for finding the quantity corresponding to so many times as much as the base quantity" (Isoda, 2010, p. 107) where the base quantity is thought as 0.1. By the same token, the meaning of division is introduced again as the inverse of multiplication such that it is emphasized that it "can be explained as either finding how many times as much or finding the base quantity" (Isoda, 2010, p. 107).

All these suggest that teachers are directed with a specific and explicit attention to provide their students with opportunities to focus on.

(i) what quantity is, including the nature of different quantities such as length, area, volume, speed etc.
(ii) what operations are involved in the production of quantities such as counting and measuring (segmenting),
(iii) how these quantities relate to each other through different arithmetic operations and
(iv) how students might make sense of multiplication and division through the lenses of reasoning quantitatively on real life situations such as making identical groups as well as comparing the sizes of quantities relative to each other.

In the following paragraphs, using particular examples, we further depict how quantitative reasoning is manifested in the tasks and problem situations in Tokyo Shosheki's *Mathematics International* Textbooks.

5 Whole Number Multiplication in Mathematics International Textbooks

When multiplication is introduced to students for the first time in the second grade, several characteristics of the problem situations seem to be highlighted: Firstly, there is emphasis on the quantities involved in the problem situations. Particularly, focusing on both the numerical measures of the quantities and the referents of the quantities, students are expected to imagine the problem situations quantitatively. Throughout the whole text, extensive and intensive quantities in problem situations are highlighted in different colors to bring to the students' attention explicitly. Secondly, although the main focus is on equal group meaning, times as many/much meaning is also introduced in the second grade. Specifically, while introducing the times as many meaning, although the quantities in problem situations are mostly discrete, both discrete and continuous (e.g., linear) representations are used. In addition, in problem situations, the number of objects in each group is represented and explicitly stated as the *multiplicand*, and the number of groups is represented and explicitly stated as the *multiplier*.[3] Thus, *multiplicand times multiplier* is a shared definition for multiplication both for repeated addition and times as many meanings. In the

[3] The actual Japanese terms used to describe *multiplicand* and *multiplier* are not the formal mathematical terms. Rather, they are more child-friendly terms for which there is no corresponding English words.

1 Investigate how many children are on each ride.

1 How many children are riding **bikes**?

2 How many children are riding in **cars**?

3 How many children are on **tea cups**?

Hiroki: The total number of children in each ride is the same but...

4 Arrange the counters to show how many children are riding on **bikes**

Picture of 6 sets of 2 counters arranged in a line.

On **bike**, there are 2 children on each bicycle and there are 6 bicycles. There are 12 children altogether.

5 Investigate the number of children on the other rides.

☐ children in each airplane ☐ airplanes ☐ children altogether

☐ children in each gondola ☐ gondolas ☐ children altogether

☐ children in each car ☐ cars ☐ children altogether

Fig. 2 Introducing multiplication through a real-life situation (Fujii & Iitaka, 2012, Grade 2, p. B5)

following paragraphs, we specifically exemplify the display of these ideas in Japanese Mathematics International textbooks.[4]

In the second grade, multiplication is introduced by a real life situation in which different numbers of children are riding on different kinds of vehicles and students are asked to reason about the quantities involved (see Fig. 2).

With regards to the picture given in Fujii and Iitaka (2012, Grade 2, pp. B3–B4), several points need to be highlighted. In the situation, not all cases are in an equal number of groups. For example, tea cups have different numbers of children whereas bicycles have equal numbers of children. We consider the picture in Fujii and Iitaka

[4] The publisher, Tokyo Shoseki, does not hold the copyright of *all* drawings and pictures in Japanese Mathematics International textbooks. Therefore, throughout the chapter, we only included the original drawings/pictures for which Tokyo Shoseki had the full copyrights. Otherwise, we either included a brief note in place of the original drawing/picture (e.g., "Picture of 5 sets of 2 counters arranged in a car" as in Fig. 2) or omitted the drawing/picture (e.g., omitted the picture of "whales" as in Fig. 7).

(2012, Grade 2, pp. B3–B4), as important in two ways. First, students' attention is taken to how many children are included in each ride, so that their focus is on quantities depicted in the picture. Second, the idea is to establish the equality of the groups for introducing multiplication. In the following pages of the task, this idea is emphasized when students are expected to put the numbers in the blank boxes and phrase the multiplication sentences, instead of writing expressions and equations, such as "□ children in each airplane, □ airplanes, □ children together" (Fujii & Iitaka, 2012, Grade 2, p. B5). Students are expected to focus not only on the numbers in the situation but also on what the numbers refer to. That is, numbers are given with an emphasis on the referents which enables students to focus on extensive and intensive quantities in the situation. For example, for different rides, quantities are highlighted in different colors (see Fig. 3) and students are expected to think about the number of children on each ride, the number of rides and the total number of children. Although the number of children on each ride can be considered as an intensive quantity, Grade 2 students might conceive the situation as "the children in a ride" as an extensive quantity. Still, it might lay the groundwork for students' ways of reasoning about intensive quantities later on.

Moreover, students are expected to think about how the quantities such as the children in each train, the number of trains and the total number of children are

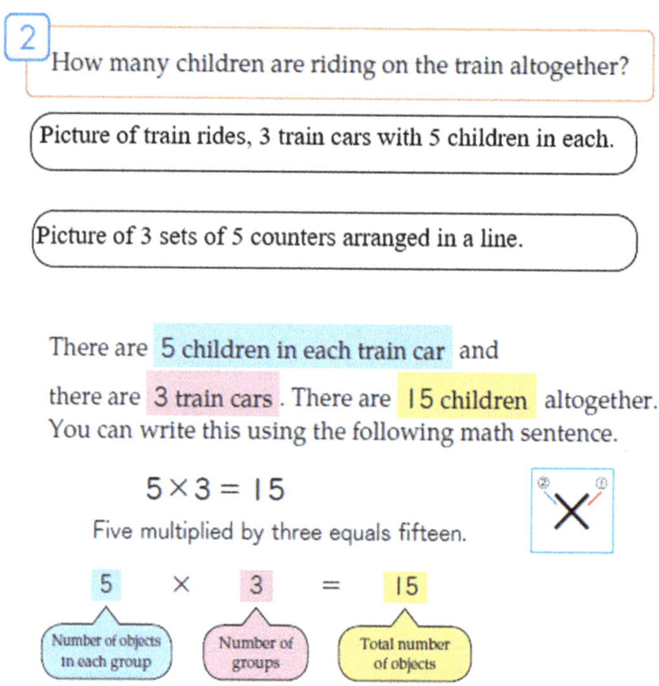

Fig. 3 Color coded representation of multiplication sentences (Fujii & Iitaka, 2012, Grade 2, p. B6)

related simultaneously and focus on both the values of the measures such as 5, 3 and 15 and of the quantities such as children per train, trains and children. Thus, they seem to be triggered to think about quantification processes as well as the numerical results. That is, they are to think of the quantity, the total amount of children, as a result of a multiplicative combination (i.e., the quantitative operation) of two quantities such as *some groups* (the number of trains) of *some amount in each* (the number of children in each train).

In addition, in the second grade, both repeated addition and times as many/much meaning of multiplication is covered prior to students' learning of the multiplication facts (multiplication table) which comes after the lesson given in Fig. 2. Particularly, repeated addition is given as a method of finding the answer for multiplication. Up to this point, students could have found the answers by counting, but the emphasis is understanding situations that can be represented by multiplication expressions (equal groups). As an instance of a situation, a picture of 4 boxes with one including 6 pieces of cake is given and students are asked to find out the total number of cakes, if each box includes an equal number of pieces of cakes. As an answer to this situation, i.e. the result for the multiplication of 6×4, students are given with the calculation of $6 + 6 + 6 + 6$ (Fujii & Iitaka, 2012, Grade 2).

As well as multiplication in equal group situations, in the second grade, students are also introduced to the idea of multiplication as comparison, i.e., times as how many/much. In particular, in a given situation of linear measurement context, students are expected to find "the length of two 3 cm strips of paper put together" (Fujii & Iitaka, 2012, Grade 2, p. B10). The following is explicitly stated as an explanation to the question:

> If a piece of tape is as long as two 3 cm strips of paper put together, we can say the tape is **2 times** as long as the 3 cm tape. You can use the multiplication math sentence 3×2 to find the length that is two times as long as 3 cm. (Fujii & Iitaka, 2012, Grade 2, p. B10)

What is important is that the amount of tape is expressed as "2 **times** as long as the 3 cm tape". That is, the result of the product, i.e. 6, is emphasized relative to the size of one of the quantities, i.e., 3. Also, linear measurement context is used to introduce times as many.

In the following pages of the task, the problems are depicted in school context with the use of discrete quantities, such as, total number of balls on the gym shelves, number of backpacks in the lockers, number of desks in a classroom. That is, as students learn the multiplication facts/table, the book introduces an array representation. Though what is important is that the number of groups represented in rows (e.g., number of shelves) and number of rows are specified to be the multiplier. In particular, students are asked to think 5×4 such that they are invited to think 5, which is the number of balls in each shelf, as a multiplicand and 4, which is the number of shelves, as a multiplier: rather than asking students to think of four groups of five, they are triggered to think five (in each) times four. Also, thinking of "5 balls per shelf" together with "four shelves" might contribute to students' thinking of the quantities (i.e., 5 balls and 4 shelves) simultaneously, which can lay a foundation for rate of change idea later. This is also depicted in the representation which also focuses on the quantities with their referents.

Similarly, in the array diagrams in the following problem situations given in Fig. 4, number of cakes in each group is given as the number of rows and number of groups of cakes is given as the number of columns. In addition, the number of cakes in each group is written as multiplicand and the number of groups of cakes is written as multipliers. Specifically, by taking students' attention even to "what numbers express the quantity in one group?" (Fujii & Iitaka, 2012, Grade 2, p. B21), they are expected to reason on the number sentence [multiplicand × multiplier]. It is even more interesting that in the following pages of the task students are asked to write number sentences such as 4 × 3 and 3 × 4 by thinking about the following: 4 objects in each group and 3 groups of 4 objects; 3 objects in each group and 4 groups of 3 objects. This suggests that there is more emphasis on the distinction between the multiplier and the multiplicand such that students are expected to think about the multiplicand as the number of objects in each group and the multiplier as the number of groups of objects independent from the given problem contexts. That is, *multiplicand* × *multiplier* is given as a structure of (# of objects per group) × (# of groups).

In addition, students are asked to think about how to shift for instance from 3 × 4 to 3 × 5 and from 3 × 5 to 3 × 6 by thinking about the following. If there is one more group, say, increase from 4 to 5, the total number will have to increase by the group size such that one more group of 3 means the total will be 3 more (+ 3). This is important for several reasons. The students' attention is on the change in quantities (i.e., amounts of increase or decrease). That is, students are triggered to think of copies of some amount (i.e., 3) where every time the change in number of groups is one. Specifically, students' attention is on the simultaneous change in the number of groups (from 4 to 5) and the amount of change (the group size). This then might allow students to envision the proportionality embedded in multiplication since they can consider both the product such as 15 and the factors of the product such as 3 and 5 in relation to it. In particular, starting from 3 × 1 and ending at 3 × 5, students might envision that 4 more groups (i.e., 5–1) means 4 times 3 more, so that the total is 5 times as large as 3.

Although these suggest that *multiplicand* × *multiplier* might trigger "times as much" understanding of multiplication, Grade 2 students might not conceive the situation as involving copies of some amount (i.e., the number of cakes in each group) (Thompson & Saldanha, 2003). Still, it might lay the groundwork for students' ways of reasoning about multiplication multiplicatively later on.

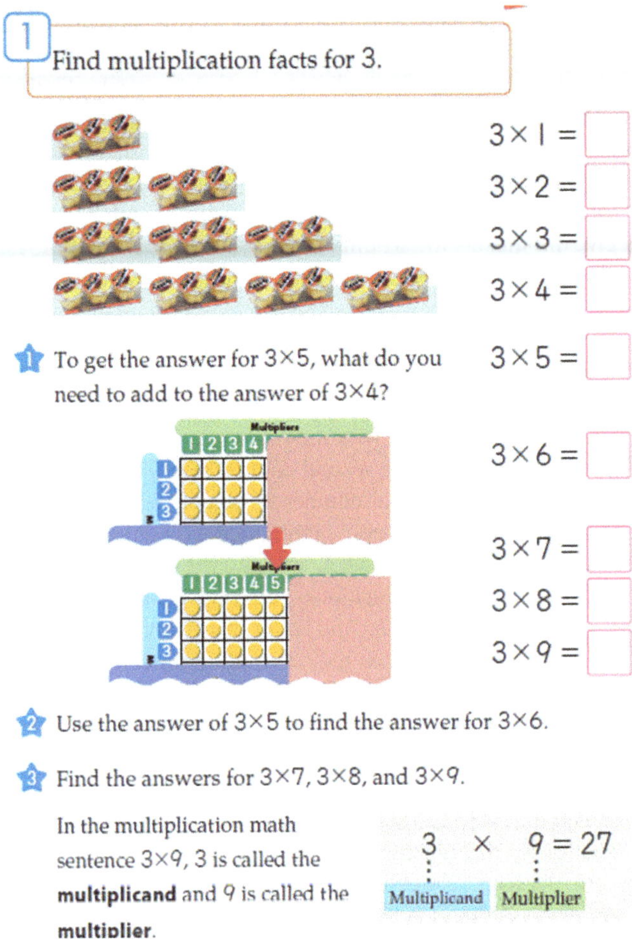

Fig. 4 Multiplicand times multiplier through multiplication facts of 3 (Fujii & Iitaka, 2012, Grade 2, p. B17)

6 Whole Number Division in Mathematics International Textbooks

In Japan, division is introduced the first time in the third grade. Similar to the problem situations introducing the multiplication in the second grade, the third grade problems included both equal and unequal sharing situations. Particularly, as a beginning activity, students are expected to compare two equal and unequal sharing situations to highlight equal sharing as a particular situation. Again, without giving the numerical values, students are asked to think about the quantities and how they are shared in given pictures, such as unequal sharing of noodles and equal sharing of juice.

In the third grade, as an extension of the equal sharing situations, students are first introduced to partitive division, followed by quotitive division. There are several characteristics worth mentioning: First, division situations are explicitly related to multiplication expressions that students established in the second grade. For instance, in the problems, "There are 20 strawberries. If 5 children share the strawberries equally, how many strawberries will each child get?" (Fujii & Iitaka, 2012, Grade 3, p. A27), and "We have 20 stickers. If we give 5 stickers to each person, how many people can we give the stickers to?" (Fujii & Iitaka, 2012, Grade 3, p. A31), students are given the multiplication expression "number of objects for 1 person × number of people = total number of objects" and expected to reason about division through multiplication expressions as illustrated in Table 1. Secondly, the emphasis is on the quantities with their referents. Particularly, the situations are modeled on a table with three columns, where each column explicitly represents the parts of the multiplication expression with a focus on quantities with their referents. The first column refers to "number of objects for 1 person", the second column refers to "number of people" and the third column represents "total number of objects". It is worth mentioning that the result of division is also explicitly highlighted with pointing to the referents (e.g., 4 strawberries, 4 people).

Likewise, in different problem situations, referring back to the multiplication statement as *multiplicand × multiplier*, the result of partitive division is explicitly stated to indicate *multiplicand*; whereas the result of quotitive division is explicitly stated to indicate *multiplier*. Based on all these ideas, finally, the goal for students is to "understand that both types of division can be expressed by the same algebraic expression" (Isoda, 2010, p. 83). Indeed, given the same mathematics sentence, e.g., $6 \div 2 = 3$, students are expected to articulate the two problem situations: One modeling partitive division and the other modeling quotitive division as shown in Fig. 5.

It is worth mentioning that similar to Fig. 5, in most of the pictorial representations in third grade, the idea behind partitive division seems to be multiplication as repeated addition whereas the idea behind quotitive division seems to be repeated subtraction. However, at this grade level, students are also introduced briefly to *times as many/much* meaning within the multiplication and division situations. Especially, times as many/much meaning is modeled in quotitive division situations.

In **the fourth grade**, relative size meaning is emphasized in both partitive and quotitive division situations. Students are first given a partitive division problem in which they are asked to share 80 sheets of colored paper equally among 4 people as given below in Fig. 6. There are several issues to be raised here. First, students are given the division expression "total number of sheets ÷ number of people sharing = number of sheets for each" where the emphasis is on the quantities with their referents. Second, a number line representation is used to represent the quantities in the problem situation, i.e., number of sheets and number of people, albeit it involves discrete quantities. Third, within the linear representation given in Fig. 6, the process of sharing is emphasized to end up with the number of sheets for each person, i.e., the numerical value of the result of the division. A specific attention is given to the relative size idea that when students share 80 (i.e., the total number of sheets) into 4 parts (i.e., among 4 people), each part (i.e., each person) contains 1/4 of the

Table 1 Partitive and quotitive division through multiplication expression

<table>
<tr><th>Partitive division</th><th>Quotitive division</th></tr>
<tr><td>
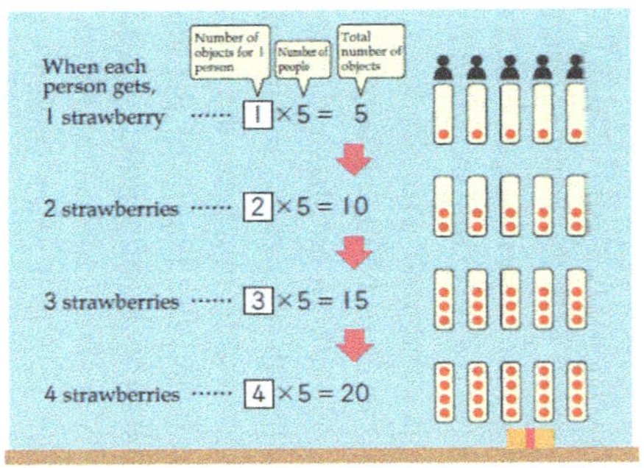

The answer for 20÷5 is the number that goes in the □ of □×5 = 20.
20 ÷ 5 = 4 Answer 4 strawberries

(Fujii & Iitaka, 2012, Grade 3, p. A28)
</td><td>
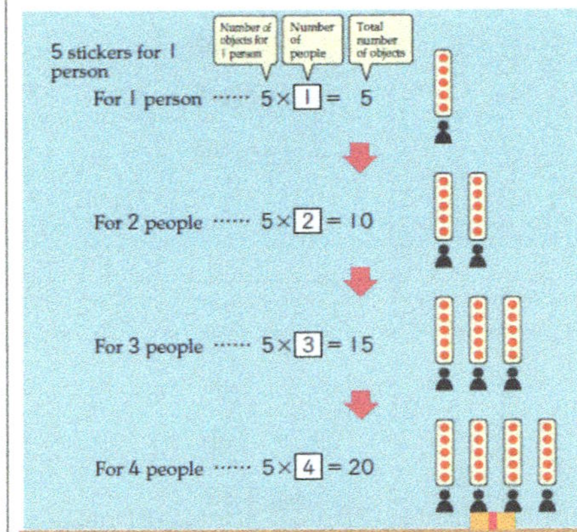

The answer to 20÷5 is the number that goes in the □ of 5×□.
20 ÷ 5 = 4 Answer 4 people

(Fujii & Iitaka, 2012, Grade 3, p. A32)
</td></tr>
</table>

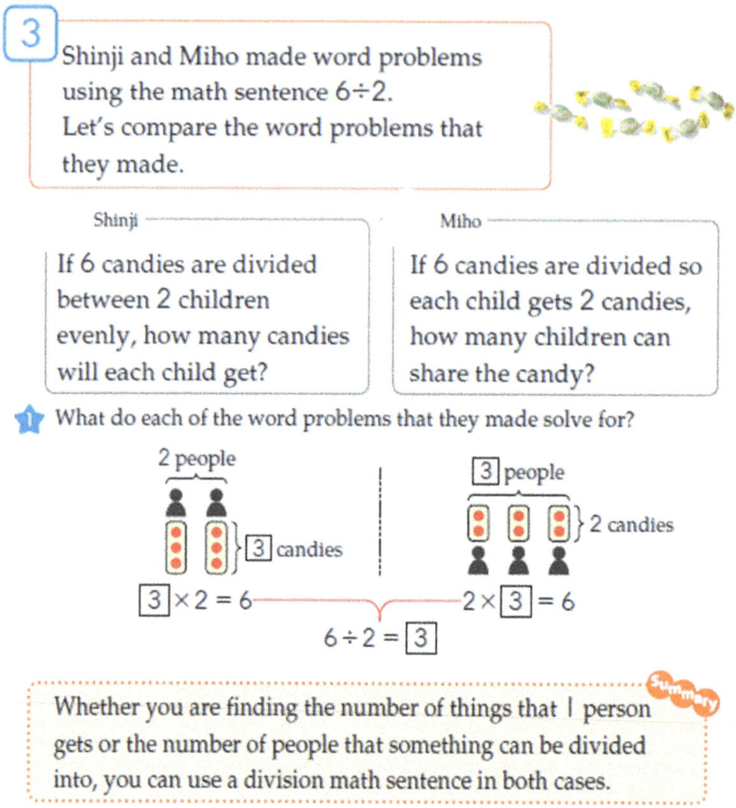

Fig. 5 Modeling partitive and quotitive division for the same mathematical expression (Fujii & Iitaka, 2012, Grade 3, p. A33)

total number of sheets, which is 20. Therefore, even though fraction notation is not explicitly used, the number of sheets for each person is 1/4 as large as the total number of sheets. Indeed, in the rest of the problems used in this grade level, relative size meaning is continually emphasized with the number line representation.

Similarly, there are quotitive division situations given in the problem context of times as many/much. For instance, in the problem shown in Fig. 7, the length of an adult whale, which is 15 m long, is compared with the length of its calf, which is 3 m long. Again, there are several issues to be raised here. First, students are given the mathematics sentence of 15 ÷ 3 with an emphasis on the quantities with their referents. That is, an adult whale's length (in meters) and its calf's length (in meters) are given. Second, a number line is used to represent the quantities in the problem situation, i.e., the length of the calf and the length of the adult whale. Third, the number line in Fig. 7, presents the process of segmenting the length of the adult whale into a number of parts of a given size, which is the length of its calf. A specific attention is given to the relative size idea that when students think of the length of

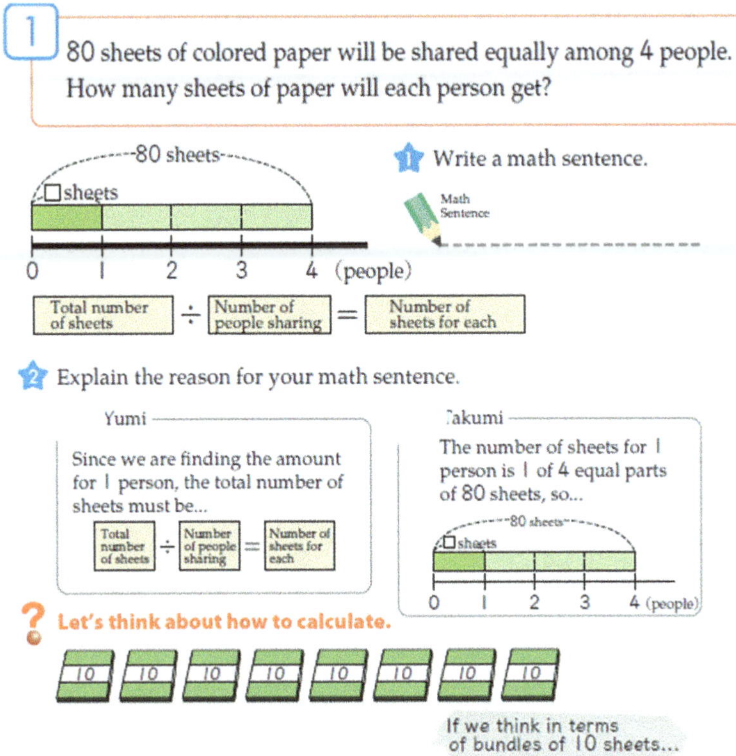

Fig. 6 Emphasis on relatives sizes in a partitive division situation (Fujii & Iitaka, 2012, Grade 4, p. A31)

the calf, 3 m, as a unit of 1, then the length of the adult whale 15 m corresponds to 5 units. Indeed, the line representation seems to explicitly indicate the idea that the length of a calf (i.e., 3 m) is 1/5 times as much as the length of the adult whale (i.e., 15 m). Likewise, the length of the adult whale is 5 times as much as the length of its calf. Moreover, by explicitly stating the meaning of 5 as follows "5 times means that if we consider the 3 m as 1, 15 m corresponds to 5" (Fujii & Iitaka, 2012, Grade 4, p. A44), albeit implicit, it is indicated that 5 is 1/3 of 15 m. That is, the number of parts made by measuring the length of adult whale into three-calf-sized parts is 1/3 as large as the size of the adult whale. All these suggest that students are expected to imagine the problem situation quantitatively in terms of quantities and their relationships. Again, the quantity, *times as many (groups)*, is constructed as a result of a multiplicative comparison (i.e., the quantitative operation) of two quantities such as *the length of the adult whale* and *the length of its calf*. That is, in this situation, 'given the length of an adult whale and the length of its' calf might result in times as many (groups of length of the calf)'. It is also important to state that the textbook also introduces proportionality as an aspect of understanding this situation. Similarly, relative size meaning is continually emphasized with the number

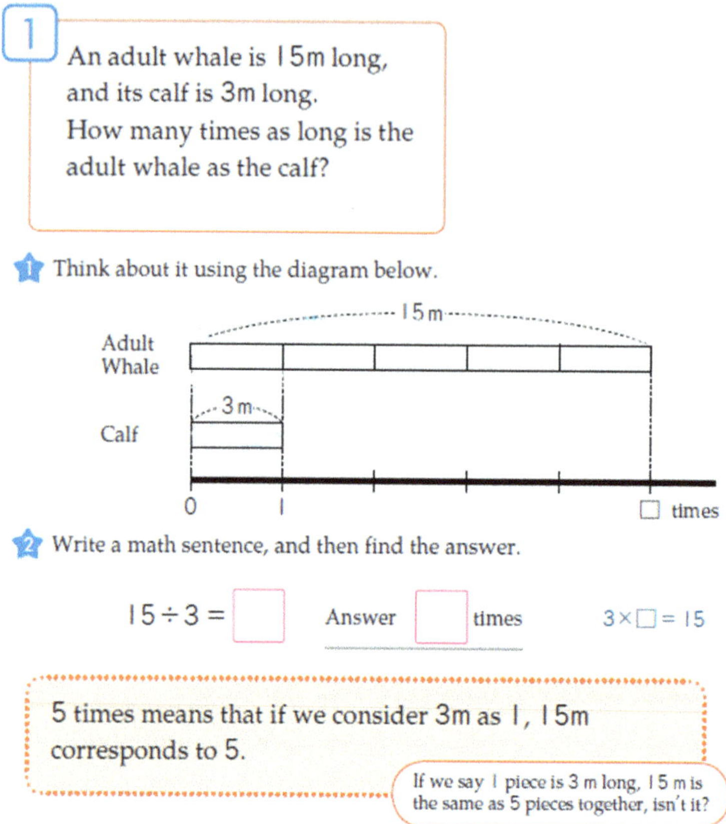

Fig. 7 Emphasis on relatives sizes in a quotitive division situation (Fujii & Iitaka, 2012, Grade 4, p. A44)—Note that there is a picture of a big whale under "15 m" and a picture of a small whale under "3 m" in the original text

line representation in both modeling partitive and quotitive division problems used in the rest of the textbook in this grade level.

7 Discussion

In this chapter, we considered and utilized quantitative reasoning as a framework for textbook analysis. Along with our purpose to determine how curricular materials depicted in the concepts of multiplication and division might potentially trigger quantitative reasoning, we examined Japanese curricular materials including the *Course of Study*, the *Teaching Guide* and the *Mathematics International textbooks* by Tokyo

Shosheki. As the given examples indicated, through all Japanese curriculum materials, quantities with their referents are continually emphasized and used while introducing multiplication and division. Specifically, in the *Teaching Guide* not only the definitions and types of quantities are provided but also the need for learners to think about quantities while learning multiplication and division are highlighted. Indeed, an emphasis on quantities with their referents is important for learners to imagine situations of multiplication and division quantitatively.

Learners are triggered to comprehend the relationships among the quantities through conceptual operations, such as counting and segmenting. In other words, by imagining the result of a relationship between the quantities with a focus on their referents, learners might come to understand the result as a new quantity. This might further allow them to think of multiplication and division not merely as numerical operations but as operations which generate quantities. Specifically, within the second and third grades, multiplication understanding is developed as *multiplicand times multiplier* (Watanabe, 2003) where the multiplicand refers to the number of objects in each group and the multiplier refers to the number of groups. Watanabe noted that teachers' manuals warn teachers not to think of multiplication as repeated addition, rather repeated addition is represented as a way to find the products. So, we argue that the idea of *multiplicand times multiplier* seems to lessen the dependency on repeated addition meaning of multiplication whereas it seems to strengthen times as many/much meaning of multiplication. This is important because researchers emphasized that in order to understand multiplication multiplicatively, students need to develop times as many/much understanding (Schwartz, 1988; Thompson & Saldanha, 2003).

Similarly, division understanding is built upon the multiplication understanding in the third and the fourth grades: both repeated addition and times as much. In particular, the result of partitive division is emphasized to refer to the multiplicand (in multiplication) and the result of quotitive division is emphasized to refer to the multiplier (in multiplication). This idea seems to trigger and strengthen the relative size meaning of division. Byerley et al. (2012) emphasized the importance of learners' thinking of division as relative size. They argued that the two meanings for division, namely partitive and quotitive division, do not require multiplicative reasoning, whereas a relative size meaning which requires learners to compare the size of one quantity with respect to another quantity allow them to reason multiplicatively (Thompson & Saldanha, 2003). Moreover, relative size meaning is depicted through the use of number line (line segment) representation regardless of the nature of the quantities being discrete or continuous. This is also important because although the number line representation is suggested to be used while teaching multiplication, where especially the repeated addition meaning is highlighted (Van de Walle, 2010), the use of number line especially for leveraging the relative size meaning of division on the part of learners seem to be an important characteristic of Japanese Mathematics International textbooks.

All these results together with the explicit attention to both the quantities and their relationships in the *Teaching Guide* suggest that in the Japanese curricular materials quantitative reasoning has a proper place at least for the concepts of multiplication and division. So, we call for further research on examining the depiction of other mathematics concepts through quantitative reasoning in Japanese curricular materials. We also propose that different international curricular materials might be examined with using the lenses of quantitative reasoning. Our call also aligns with the study by Taşova et al. (2018), who provided a framework to analyze the extent and nature of (co)variational and quantitative reasoning in written curriculum, specifically the concept of functions in five U.S. calculus textbooks. Finally, we believe that teachers and teacher educators might use the tasks and problems shared in this chapter while introducing and assessing the different meanings of whole number multiplication and division.

References

Arslan, S., & Özpınar, İ. (2009a). lköğretim 6. sınıf matematik ders kitaplarının öğretrmen görüşleri doğrultusunda değerlendirilmesi. *Dicle Üniversitesi Ziya Gökalp Eğitim Fakültesi Dergisi, 12*, 97–113.

Arslan, S., & Özpınar, İ. (2009b). Yeni ilköğretim 6. sınıf matematik ders kitaplarinin öğretim programına uygunluğunun incelenmesi. *Çukurova Üniversitesi Eğitim Fakültesi Dergisi, 3*(36), 26–38.

Byerley, C., Hatfield, N., & Thompson, P. W. (2012). Calculus students' understandings of division and rate. In S. Brown, S. Larsen, K. Marrongelle, & M. Oehrtman (Eds.), *Proceedings of the 15th Annual Conference on Research in Undergraduate Mathematics Education* (pp. 358–363). SIGMAA/RUME.

Charalambous, C. Y., Delaney, S., Hsu, H. Y., & Mesa, V. (2010). A comparative analysis of the addition and subtraction of fractions in textbooks from three countries. *Mathematical Thinking and Learning, 12*(2), 117–151. https://doi.org/10.1080/10986060903460070

Fan, L. (2013). Textbook research as scientific research: Towards a common ground on issues and methods of research on mathematics textbooks. *ZDM—the International Journal on Mathematics Education, 45*(5), 765–777. https://doi.org/10.1007/s11858-013-0530-6

Fan, L., & Zhu, Y. (2007). Representation of problem-solving procedures: A comparative look at China, Singapore, and US mathematics textbooks. *Educational Studies in Mathematics, 66*(1), 61–75. https://doi.org/10.1007/s10649-006-9069-6

Fishbein, E., Deri, M., Nello, M., & Marino, M. (1985). The role of implicit models in solving problems in multiplication and division. *Journal for Research in Mathematics Education, 16*(1), 3–17. https://doi.org/10.5951/jresematheduc.16.1.0003

Fujii, T., & Iitaka, S. (2012). *Mathematics international* (*Grade 1–Grade 4*). Tokyo Shoseki Co. Ltd.

Graeber, A. O., & Tirosh, D. (1990). Insights fourth and fifth graders bring to multiplication and division with decimals. *Educational Studies in Mathematics, 21*(6), 565–588. https://doi.org/10.1007/BF00315945

Greer, B. (1992). Multiplication and division as models of situations. In D. Grouws (Ed.), *Handbook of research on mathematics teaching and learning* (pp. 276–295). MacMillan.

Isoda, M. (Ed.). (2010). *Elementary school teaching guide for the Japanese course of study: Mathematics (grades 1–6).* http://facultyweb.kennesaw.edu/twatanab/docs/Teaching%20Guide%20Elementary.pdf

Kilpatrick, J., Swafford, J., & Findel, B. (2001). *Adding it up: Helping children learn mathematics.* National Academy Press.

Knight, B. A., & Wang, S. (2015). Teachers' use of textbooks in the digital age. *Cogent Education, 2*(1), 1–11. https://doi.org/10.1080/2331186X.2015.1015812

Li, Y. (2000). A comparison of problems that follow selected content presentations in American and Chinese mathematics textbooks. *Journal for Research in Mathematics Education, 31*(2), 234–241. https://doi.org/10.2307/749754

Li, Y., Chen, X., & An, S. (2009). Conceptualizing and organizing content for teaching and learning in selected Chinese, Japanese and US mathematics textbooks: The case of fraction division. *ZDM—the International Journal on Mathematics Education, 41*(6), 809–826. https://doi.org/10.1007/s11858-009-0177-5

Mayer, R. E., Sims, V., & Tajika, H. (1995). Brief note: A comparison of how textbooks teach mathematical problem solving in Japan and the United States. *American Educational Research Journal, 3*(2), 443–460. https://doi.org/10.3102/00028312032002443

Moore, K. C., Carlson, M. P., & Oehrtman, M. (2009). The role of quantitative reasoning in solving applied precalculus problems. In *Twelfth Annual Special Interest Group of the Mathematical Association of America on Research in Undergraduate Mathematics Education (SIGMAA on RUME) Conference.* North Carolina State University.

National Governors Association Center for Best Practices, Council of Chief State School Officers. (2010). *Common core state standards mathematics.* Author. http://corestandards.org/

Remillard, J. T. (2005). Examining key concepts in research on teachers' use of mathematics curricula. *Review of Educational Research, 75*(2), 211–246. https://doi.org/10.3102/00346543075002211

Ross, S., & Pratt-Cotter, M. (1997). Subtraction in the United States: An historical perspective. *The Mathematics Educator, 8*(1), 4–11.

Schmidt, W. H., McKnight, C. C., & Raizen, S. A. (1997). *A splintered vision: An investigation of U.S. science and mathematics education.* Kluwer Academic Publishers.

Schwartz, J. L. (1988). Intensive quantity and referent transforming arithmetic operations. In J. Hiebert & M. Behr (Eds.), *Number concepts and operations in the middle grades* (pp. 41–52). Lawrence Erlbaum Associates.

Simon, M. A. (1993). Elementary teachers' knowledge of division. *Journal for Research in Mathematics Education, 24*(3), 233–254. https://doi.org/10.5951/jresematheduc.24.3.0233

Smith, J., III., & Thompson, P. W. (2007). Quantitative reasoning and the development of algebraic reasoning. In J. J. Kaput, D. W. Carraher, & M. L. Blanton (Eds.), *Algebra in the early grades* (pp. 95–132). Taylor & Francis Group.

Son, J., & Dilleti, J. (2017). What can we learn from textbook analysis? In J. W. Son, T. Watanabe, & J. J. Lo (Eds.), *What matters? Research trends in International comparative studies in mathematics education* (pp. 3–32). Springer.

Son, J., & Senk, S. (2010). How reform curricula in the USA and Korea present multiplication and division of fractions. *Educational Studies in Mathematics, 74*(2), 117–142. https://doi.org/10.1007/s10649-010-9229-6

Stein, M. K., Remillard, J., & Smith, M. (2007). How curriculum influences student learning. In F. K Lester (Ed.), *Second handbook of research on mathematics teaching and learning* (pp. 319–369). Information Age Publishing.

Takahashi, A., Watanabe, T., & Yoshida, M. (2008). English translation of the Japanese mathematics curricula in the course of study. *Global Education Resources, LLC.*

Tasova, H., Stevens, I. E., & Moore, K. C. (2018). A framework for analyzing written curriculum from a shape-thinking and (co) variational reasoning perspective. In *Proceedings of the Twenty-First Annual Conference on Research in Undergraduate Mathematics Education* (pp. 1527–1533). RUME.

Thompson, P. W. (1990). *A theoretical model of quantity-based reasoning in arithmetic and algebra* (Report No. MDR 87-51381 and MDR 89-50311). Center for Research in Mathematics & Science Education, San Diego State University.

Thompson, P. W. (1994). The development of the concept of speed and its relationship to concepts of rate. In G. Harel & J. Confrey (Eds.), *The development of multiplicative reasoning in the learning of mathematics* (pp. 179–234). SUNY Press. http://pat-thompson.net/PDFversions/1994ConceptSpeedRate.pdf

Thompson, P. W. (2011). Quantitative reasoning and mathematical modeling. In L. L. Hatfield, S. Chamberlain, & S. Belbase (Eds.), *New perspectives and directions for collaborative research in mathematics education*, WISDOMe Monographs (Vol. 1, pp. 33–57). University of Wyoming.

Thompson, P. W., & Carlson, M. P. (2017). Variation, covariation, and functions: Foundational ways of thinking mathematically. In J. Cai (Ed.), *Compendium for research in mathematics education* (pp. 421–456). NCTM. https://www.researchgate.net/profile/Patrick-Thompson-2/publication/302581485_Variation_covariation_and_functions_Foundational_ways_of_thinking_mathematically/links/583c389208ae502a85e3a224/Variation-covariation-and-functions-Foundational-ways-of-thinking-mathematically.pdf

Thompson, P., & Saldanha, L. A. (2003). Fractions and multiplicative reasoning. In J. Kilpatrick, W. G. Martin, & D. Shifter (Eds.), *A research companion to principles and standards for school mathematics* (pp. 95–113). National Council of Teachers of Mathematics.

Turan, N. (2021). *An examination of elementary and lower secondary level Japanese and Turkish mathematics curricula through quantitative and (co)variational reasoning in terms of the treatment of functional relationships*. Master's thesis, Boğaziçi University.

Valverde, G. A., Bianchi, L. J., Wolfe, R. G., Schmidt, W. H., & Houang, R. T. (2002). *According to the book: Using TIMSS to investigate the translation of policy into practice through the world of textbooks*. Kluwer Academic Publishers.

Van de Walle, J. A., Karp, K. S., & Bay-Williams, J. M. (2010). *Elementary and middle school mathematics: Teaching developmentally*. Allyn and Bacon.

Vergnaud, G. (1983). Multiplicative structures. In R. Lesh & M. Landau (Eds.), *Acquisition of mathematics concepts and processes* (pp. 127–174). Academic Press.

Vergnaud, G. (1994). The multiplicative conceptual field: What and why? In G. Harel & J. Confrey (Eds.), *The development of multiplicative reasoning in the learning of mathematics* (pp. 51–39). State University of New York Press.

Verschaffel, L., Greer, G., & De Corte, E. (2007). Whole number concepts and operations. In F. Lester (Ed.), *Second handbook of research on mathematics teaching and learning* (pp. 557–628). Information Age Publishing.

Watanabe, T. (2003). Teaching multiplication: An analysis of elementary school mathematics teachers' manuals from Japan and the United States. *The Elementary School Journal, 104*(2), 111–125. https://doi.org/10.1086/esj.104.2.3202985

Watanabe, T. (2014). Transformation of Japanese elementary mathematics textbooks: 1958–2012. In Y. Li, E. Silver, & S. Li (Eds.), *Transforming mathematics instruction: Multiple approaches and practices* (pp. 199–216). Springer.

Watanabe, T., Lo, J., & Son, J. (2017). Intended treatment of fractions and fraction operations in mathematics curricula from Japan, Korea, and Taiwan. In J. Son, T. Watanabe, & J. Lo (Eds.), *What matters? Research trends in international comparative studies in mathematics education* (pp. 33–61). Springer.

Zembat, I. O. (2010). A micro-curricular analysis of unified mathematics curricula in Turkey. *ZDM—the International Journal on Mathematics Education, 42*(5), 443–455. https://doi.org/10.1007/s11858-010-0236-y

Constructing Covariational Relationships and Distinguishing Nonlinear and Linear Relationships

Teo Paoletti and Madhavi Vishnubhotla

> [T]o ground the development of algebraic thinking on the notion of functions and functional relationships without, in turn, grounding these on understandings of quantities and quantitative reasoning in dynamic situations, is like building a house starting with the second floor. The house will not stand. (Thompson & Thompson, 1995, p. 98)

1 Introduction

Mathematics generally (Crisp et al., 2009; Sass, 2015), and algebra specifically (Loveless, 2013), serve as gatekeepers that have restricted student access to STEM fields. Thus, it is more important than ever that K-12 education supports students in developing foundational knowledge and ways of thinking that support their algebra learning. However, current algebra curricula and teaching often present an abstract, static, symbolic, and largely procedural mathematics (e.g., Hiebert et al., 2005; Litke, 2020). To increase STEM opportunity, pre-algebra and algebra instruction must help students develop ways of thinking that are meaningful, accessible, and applicable broadly across STEM fields. One of these ways of thinking is *covariational reasoning*, the ability to construct and reason about relationships between quantities changing together. Students are not currently being provided sufficient opportunities to reason about covarying quantities (Frank & Thompson, 2021; Smith & Thompson, 2008; Thompson & Harel, 2021). As reflected in the opening quote, the lack of opportunities to reason about covarying quantities may explain much of students'

T. Paoletti (✉)
School of Education, University of Delaware, Willard Hall, Newark, DE 19716, USA
e-mail: teop@udel.edu

M. Vishnubhotla
Department of Mathematics, Virginia Polytechnic Institute and State University, Blacksburg, VA 24061, USA
e-mail: madhavi@vt.edu

G. Karagöz Akar et al. (eds.), *Quantitative Reasoning in Mathematics and Science Education*, Mathematics Education in the Digital Era 21,
https://doi.org/10.1007/978-3-031-14553-7_6

difficulties with algebra; students are not being provided with a foundation on which to build their more formal algebra knowledge.

In this chapter, we propose that middle school students would be better served first having opportunities to reason about dynamically changing quantities to construct basic types of covariational relationships. These covariational relationships can serve as the foundation for students' activity as they begin to use multiple representations (e.g., graphs, tables, equations) to represent such relationships. In the following sections, we first outline our theoretical framework, which details the requisite meanings students need to construct, reason about, and represent covariational relationships between continuously changing quantities. We then outline a task sequence we iteratively designed and tested to support students in constructing, coordinating, and graphically representing covarying quantities. Throughout, we use two students' activity to exemplify the productivity of this task sequence. We conclude by discussing implications for student reasoning and highlighting the potential for such activity to serve as a foundation for students developing meanings for various functional relationships.

2 Theoretical Background

Prior to presenting our task sequence, we describe constructs relevant to how students construct, coordinate, and represent covarying quantities. We then describe how students can leverage their covariational reasoning to characterize basic types of covariational relationships and differentiate between nonlinear and linear relationships. We conclude by characterizing the requisite meanings students must maintain to represent such relationships graphically.

2.1 Foundations of Covariational Reasoning

In this section, we first describe the theoretical framework we use when characterizing students' quantitative and variational reasoning. We then present our coordination of the frameworks from Carlson et al. (2002) and Thompson and Carlson (2017) that we leverage.

2.1.1 Quantitative and Variational Reasoning

Several researchers (see Thompson & Carlson, 2017 for a review) have begun to explore ways in which students' quantitative reasoning (Thompson, 2011) can support their development of productive meanings for various mathematical ideas. Adopting this theory, we contend that quantities are conceptual entities a student constructs to make sense of some phenomenon. A student's quantitative reasoning can involve numerical and non-numerical reasoning (Johnson, 2012; Moore et al., 2019), but the essence of quantitative reasoning is non-numerical (Smith & Thompson, 2008). There are numerous ways students can reason about magnitudes (see Thompson et al., 2014), but we are particularly interested in students conceiving of an increasing or decreasing amount of a measurable attribute of an object or phenomenon. Furthermore, although a student can reason quantitatively about static quantities (e.g., comparing two static lengths to determine the measure of one length in terms of the other), we attend to students' variational reasoning about conceived dynamic quantities.

In this report, we leverage and refine Thompson and Carlson's (2017) variational reasoning framework (Table 1) based on our attempts to analyze student activity using this framework. Specifically, we add smooth variational reasoning, which entails a student reasoning about the variation of a quantity's magnitude or value as changing smoothly across an interval. Such reasoning is not as sophisticated as smooth continuous variation. Smooth continuous variation entails smooth variation with an additional anticipation that any smaller sub-interval would also entail smooth and continuous variation; we only characterize a student as engaging in smooth continuous variation if the student explicitly describes such smaller sub-intervals. In our research, we often observed students engaging in smooth variation without explicitly considering or describing sub-intervals, thereby creating a need for a new level within the framework. We note our characterization of smooth variational reasoning is more sophisticated than gross variation and chunky continuous variation, as the student anticipates the quantity takes on magnitudes or values while changing between intervals of a fixed size (Thompson, personal communication).

We note that while Thompson and Carlson (2017) use the term levels in both their variational and covariational reasoning framework, these levels are not necessarily hierarchical. Students do not need to move sequentially up from the lowest to the highest level. Thompson and Carlson (2017) cautioned researchers not see these levels as

> a learning progression in the sense that one level should be targeted instructionally before the next higher level. As Castillo-Garsow et al. (2013) point out, teachers should emphasize smooth variation in their talk and actions whenever they can. Students will reason at the level they will, and if at some point in time they reason variationally at the highest level, they get all other levels for free. (p. 440)

Consistent with this recommendation, we design tasks we intend to promote smooth variation in students' reasoning.

To exemplify several of the levels in Table 1, and the distinctions we make moving forward, we will use the *Triangle/Rectangle Task*. In the *Triangle/Rectangle Task*, students are presented with a GeoGebra applet which presents an (apparently)

Table 1 An amended version of Thompson and Carlson's (2017) major levels of variational reasoning (p. 440), with changes in bold

Level	Description
Smooth continuous variation	The person thinks of variation of a quantity's or variable's (hereafter, variable's) value as increasing or decreasing (hereafter, changing) by intervals while anticipating that within each interval the variable's value varies smoothly and continuously
Smooth variation (added)	**The person thinks of variation of a variable's value as changing smoothly across an interval without considering sub-intervals within the interval. The person anticipates that the variable changes from *a* to *b* smoothly by taking all values between *a* and *b* without attending to sub-intervals within the interval from *a* to *b***
Chunky continuous variation	The person thinks of variation of a variable's value as changing by intervals of a fixed size...The person imagines, for example, the variable's value varying from 0 to 1, from 1 to 2, from 2 to 3 (and so on), like laying a ruler. Values between 0 and 1, between 1 and 2... and so on, "come along" by virtue of each being part of a chunk...but the person does not envision that the quantity has these values in the same way it has 0, 1, 2, and so on, as values
Gross variation	The person envisions that the value of a variable increases or decreases, but gives little or no thought that it might have values while changing
Discrete variation	The person envisions a variable as taking specific values. The person sees the variable's value changing from a to a_n by taking values a_1, a_2, ..., a_n but does not envision the variable taking any value between a_i and a_{i+1} ($i = 1, 2, \ldots, n$)
No variation	The person envisions a variable as having a fixed value. It could have a different fixed value, but that would be simply to envision another scenario
Variable as symbol	The person understands a variable as being just a symbol that has nothing to do with variation

smoothly[1] growing triangle and rectangle (Fig. 1; https://www.geogebra.org/m/cxeevsyc). The two shapes have equal base lengths (highlighted in pink) defined by the slider value (a), which ranges from 0 to 5. The shorter slider allows students to change the increment with which the a-values change, from apparently smoothly (in increments of 0.01) to larger chunks (in increments of 1.0). We ask students to consider variations in each shape's base length and area as the animations play to support students in conceiving of each shape's base length and area as quantities in the situation.

[1] We acknowledge that, due to the digital nature of the task, all quantities vary according to the discrete parameters set in the applet, which is why we refer to the quantities as varying (apparently) smoothly. Hereafter, we will use smoothly to convey the (apparently) smooth nature of the variations in the applets.

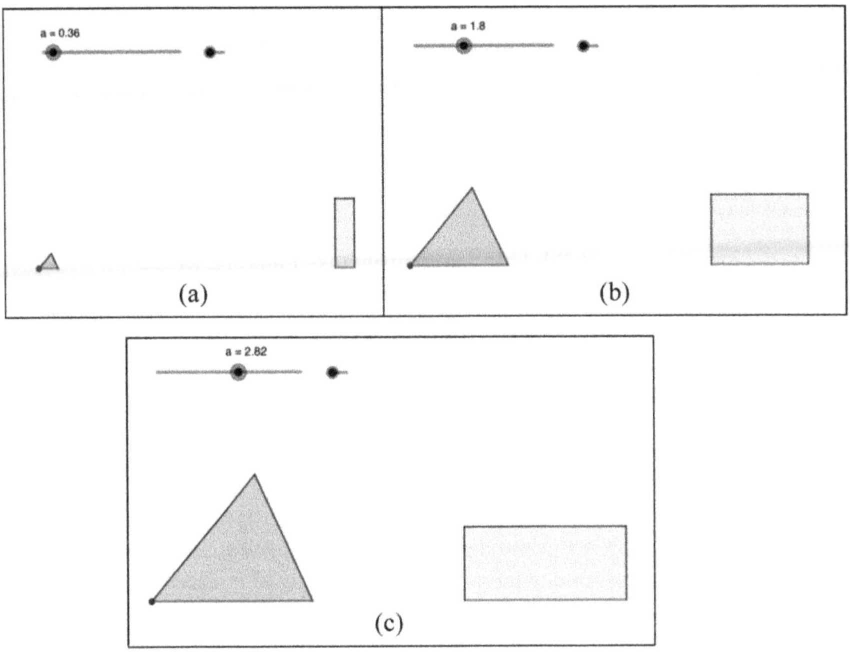

Fig. 1 Screenshots of the triangle/rectangle task

Particular to students' variational reasoning, *gross variational reasoning* is common when students begin to conceive quantities in the situation. For example, a student may initially conceive that each shape has an increasing area as the base length is increasing. If a student conceives area as varying smoothly as the slider increases from 0 to 5 without providing any description of how the area is changing within subintervals, we only classify the student as engaging in *smooth variational reasoning*. Evidence for smooth variational reasoning may entail students using smooth hand motions or active voice to describe how quantities are changing (e.g., motioning to represent an interval of changing *a*-values, describing "the *a*-value starts at 0 and increases until it reaches 5"). We classify a student as engaging in *smooth continuous variational reasoning* if the student anticipates and explicitly describes variations of area within smaller subintervals (e.g., describing what could be happening to area as the base length varies from 1.77 to 1.78).

If a student is constrained to reasoning about incremental changes of a fixed size of the base length (e.g., 0.01, 1.0, or some other value), then we would categorize the student's reasoning as entailing *chunky continuous variation.* It is common for students to exhibit chunky continuous variational reasoning when they describe how the area changes if the short slider is set to integer values (e.g., describing "the area of the triangle jumps by bigger amounts"). Although such reasoning is chunky, chunky reasoning is critical to differentiate between different patterns of growth (Vishnubhotla, 2020). Hence, several tasks are designed to support students in engaging in both smooth and chunky thinking. However, we always start by presenting continuously changing phenomena as we concur with Castillo-Garsow et al. (2013), who contended:

> smooth thinking would entail a capacity to think in chunks (or, at least at its foundation, one chunk). In contrast, in our experience with students, chunky thinking does not seem to entail a capacity to think smoothly, nor does chunky thinking seem to provide a cognitive root for smooth thinking. (p. 36)

With the idea that smooth thinking can entail chunky thinking, we design tasks that allow students to first experience and possibly conceptualize a smoothly changing phenomenon. Only after such opportunities do we introduce features of the applet that allow the quantities to change in chunks.

2.1.2 Covariational Reasoning: Coordinating Frameworks

As our goal is for students to coordinate and represent (at least) *two* varying quantities, we also attend to students' covariational reasoning (Carlson et al., 2002; Thompson & Carlson, 2017). In this section, we first offer an overview of covariational reasoning and multiplicative objects. We then provide the theoretical underpinnings for how we design tasks to support students' covariational reasoning by relating our interpretations of Carlson et al.'s (2002) framework with Thompson and Carlson's (2017) covariational reasoning framework (Table 2).

Researchers (Carlson et al., 2002; Saldanha & Thompson, 1998) have contended covariational reasoning is developmental. Saldanha and Thompson (1998) described that initially a student is likely to coordinate two quantities by thinking, "of one, then the other, then the first, then the second, and so on" (p. 299). Through this process, the student can develop an operative image of covariation that entails a relationship between quantities that results from imaging both quantities being tracked for some duration. Elaborating on their description of covariational reasoning Saldanha and Thompson (1998) stated:

> [Covariational reasoning] entails coupling the two quantities, so that, in one's understanding, a multiplicative object is formed of the two. As a multiplicative object, one tracks either quantity's value with the immediate, explicit, and persistent realization that, at every moment, the other quantity also has a value. (p. 299)

Elaborating on their use of *multiplicative object*, which stems from Piaget's notion of 'and' as a multiplicative operator, Thompson et al. (2017) noted, "A person forms a multiplicative object from two quantities when she mentally unites their attributes to make a new attribute that is, simultaneously, one and the other" (p. 98). Hence, covariational reasoning entails understanding the simultaneity of two quantities' magnitudes or values in relation to each other.

As we are particularly interested in supporting students in conceiving smoothly changing phenomena, we use dynamic applets that could support students in anticipating that quantities covary smoothly, as recommended by others (Castillo-Garsow et al., 2013; Johnson, 2020; Stevens et al., 2017). We agree with Castillo-Garsow

Table 2 An amended version of Thompson and Carlson's (2017) major levels of covariational reasoning (p. 441), with changes in bold

Level	Description
Smooth continuous covariation	Envisioning changes in one quantity's or variable's value as happening simultaneously with changes in another variable's value, and the person envisions both variables varying **via smooth continuous variation**
Smooth covariation (added)	**Envisioning changes in one quantity's or variable's value as happening simultaneously with changes in another variable's value, and the person envisions both variables varying via smooth variation**
Chunky continuous covariation	Envisioning changes in one variable's value as happening simultaneously with changes in another variable's value, with both variables varying with chunky continuous variation
Coordination of values	Coordinating the values of one variable (x) with values of another variable (y) with the anticipation of creating a discrete collection of pairs (x, y)
Gross coordination of values	Forming a gross image of quantities' values varying together. Envisioning a loose link between the overall changes in two quantities' values. The person does not envision that individual values of quantities go together. Instead, the person envisions a loose, non-multiplicative link between the overall changes in two quantities' values
Pre-coordination of values	Envisioning two variables' values varying, but asynchronously—one variable changes, then the second variable changes, then the first, and so on. The person does not anticipate creating pairs of values as multiplicative objects
No coordination	Having no image of variables varying together

et al.'s (2013) assertion that smooth thinking can serve as a cognitive root for chunky thinking, and therefore design tasks to provide students opportunities to first anticipate quantities changing smoothly. Specifically, we design tasks to support students in moving from gross coordination of values to smooth or smooth continuous covariational reasoning, initially bypassing chunky continuous covariational reasoning (Table 2; Thompson & Carlson, 2017). *Gross coordination of values* (Table 2), which Carlson et al. (2002) referred to as coordinating *direction of change*, is common when students are first conceiving a relationship between covarying quantities (e.g., "the triangle's area and base length are both increasing").

After students conceive of the directional changes in two quantities, they can further conceive the relationship via smooth or smooth continuous covariational reasoning. Due to our addition of smooth variational reasoning, we also amend Thompson and Carlson's (2017) covariational framework to differentiate between smooth covariation and smooth continuous covariation (Table 2). The primary distinction we make is to differentiate conceiving each quantity at the smooth variational or smooth continuous variational level. We would characterize a student who consistently attends to simultaneous variations in two quantities across an interval, without describing subintervals, as engaging in smooth covariational reasoning. For example, a student may describe that, until a ball thrown in the air reaches its maximum height, the ball's height increases and its velocity decreases without explicitly describing the ball's height or velocity within any sub-interval. We characterize such reasoning as smooth covariational reasoning unless a student explicitly describes the quantities values within sub-intervals.

After students have had opportunities to engage in smooth (or smooth continuous) covariational reasoning, we provide opportunities for students to engage in *chunky continuous covariational reasoning* (Thompson & Carlson, 2017). Particularly, we are interested in supporting students in reasoning about what Carlson et al. (2002) referred to as *amounts of change* (e.g., the *change* in a triangle's area *increases* as the base length *increases in equal successive amounts*). Although there are ways students may engage in chunky continuous covariation without attending to the amounts of change of one quantity for equal changes in the second quantity, reasoning about amounts of change is a particular form of chunky continuous covariational reasoning (Thompson, personal communication). In this paper, when we refer to a student engaging in chunky continuous covariation, we refer specifically to a student reasoning about amounts of change as described by Carlson et al. (2002). As we describe in the next section, such reasoning can productively interplay with students' smooth reasoning as they conceive of different types of covariational relationships (e.g., Paoletti & Moore, 2017).

2.2 Using Direction and Amounts of Change to Conceive the Basic Types of Covariational Relationships and Distinguish Between Nonlinear and Linear Relationships

Smooth covariational reasoning can support students in conceiving of the directional change of two quantities. A student can conceive that as the first quantity increases or decreases, the second quantity increases, decreases, or remains constant. If both quantities are changing (i.e., the second quantity is not constant), engaging in chunky continuous covariational reasoning is necessary to further characterize a covariational relationship. Specifically, students can examine equal successive changes in the first quantity to explore whether the amounts of change of the second quantity are increasing, decreasing, or remaining constant. For example, in the *Triangle/Rectangle Task*, a student may conceive that, as the base length increases, the triangle's area increases by increasing amounts (as represented by the consecutive trapezoids on the triangle in Fig. 2). For the rectangle, the student may identify that, as the base length increases, the rectangle's area increases by equal amounts (as represented by the five consecutive rectangles in Fig. 2).

Table 3 presents the basic types of covariational relationships that a student can conceive by focusing explicitly on directional and amounts of change. We note that,

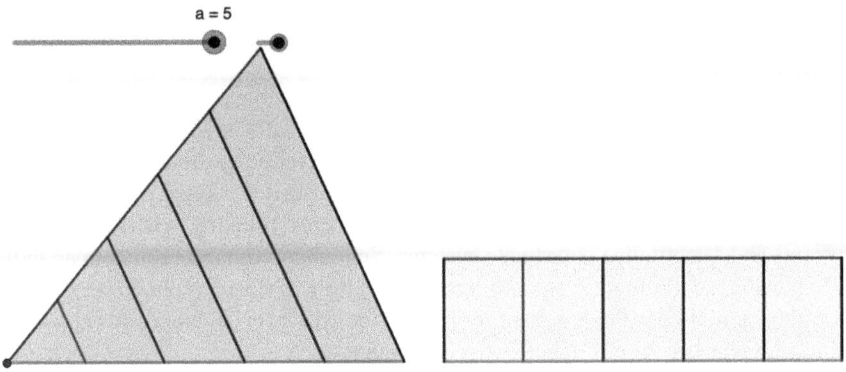

Fig. 2 A screenshot of the triangle/rectangle task showing amounts of change of area for equal integer changes of base length

Table 3 The basic types of covariational relationships

First quantity	Directional change in second quantity	Amounts of change in second quantity
Increasing or decreasing	Constant	N/A
	Increasing	By equal amounts
		By increasing amounts
		By decreasing amounts
	Decreasing	By equal amounts
		By increasing amounts
		By decreasing amounts

due to the prevalence of students' presuming all relationships are linear after learning about linear functions (e.g., De Bock et al., 2007; Esteley et al., 2010), we intentionally provide students repeated opportunities to construct various nonlinear relationships prior to considering a linear relationship. As described in the introduction, we conjecture providing students opportunities to construct these basic types of covariational relationships can serve as the foundation for students' meanings for various nonlinear and linear relationships. Based on prior experiences, and as exemplified in the Faucet Task (Sect. 4.1), we found it important to provide students with repeated opportunities to construct and reason about different directional relationships first. After such experiences, students can experience an intellectual need (Harel, 2008) to further characterize such relationships via the amounts of change of the second quantity with respect to the first.

2.3 Representing Covariational Relationships Graphically

In the above descriptions, we characterize students' reasoning about quantities in situations. To represent a covariational relationship graphically, it is important to attend to students' meanings for the underlying coordinate system. Lee and colleagues (Lee, 2016; Lee et al., 2020; Paoletti et al., 2022) distinguished two types of coordination that result in two uses of coordinate systems in students' thinking: spatial coordinate systems and quantitative coordinate systems. Spatial coordination refers to an individual using a coordinate system to represent a physical space or phenomenon. The resulting spatial coordinate system organizes the space (or an analogous space) in which the phenomenon is conceived (e.g., a map).

Students must construct a quantitative coordinate system to represent two quantities that are not established spatially in a physical space (e.g., temperature, pressure). To construct a quantitative coordinate system, a student must first establish quantitative frames of reference (Joshua et al., 2015) within the situation. They can then disembed the quantities from the situation while maintaining an awareness of the situational quantities (Steffe & Olive, 2010) and project the quantities onto the quantitative coordinate system. Produced graphs in a quantitative coordinate system are not projections of physical phenomena onto the same space containing the original objects or phenomena.

To construct a quantitative coordinate system in the context of area and base length of the triangle in the *Triangle/Rectangle Task* (Fig. 3a–c), a student must first conceive of the triangle's area and base length as quantities. Then, intending to represent the quantities on a coordinate system, the student must consider representing each magnitude (or value) with a corresponding line segment. The student may disembed area and base length from the situation and represent them with a green segment on the vertical axis and a pink segment on the horizontal axis, respectively (e.g., the segments on the axes in Fig. 3a–c). The student can then anticipate that variations in the quantities' magnitudes (or values) correspond to variations in the segments' magnitudes (or values). For example, the student may leverage their situational understanding to argue that if they move the slider to the right, the green segment will go up and the pink segment will go to the right as the area and base length are both increasing. As representing nonlinear quantities using linear segments is non-trivial (Johnson et al., 2017; Paoletti et al., accepted), we provide students repeated opportunities to consider how line segments could be used to represent such quantities, often using tasks similar to the 'Which One' task described by Moore and colleagues (Liang & Moore, 2020; Stevens et al., 2017).

With a quantitative coordinate system in mind, a student can then conceive of a point as a multiplicative object (Lee, 2016; Lee et al., 2020; Thompson, 2011) that simultaneously represents the two segments' magnitudes on the axes. This point reflects the multiplicative object the student conceived of when reasoning covariationally by representing the two quantities' simultaneous values at every moment. For example, a student may argue that moving the slider to the right will result in the

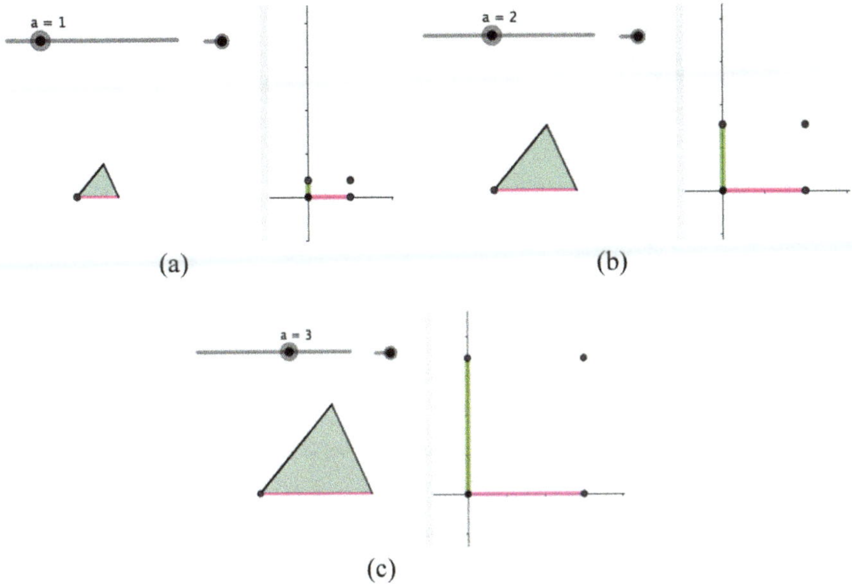

Fig. 3 Screenshots of the triangle/rectangle task applet with a coordinate system shown

point representing (triangle's base length, triangle's area) in Fig. 3 moving diagonally up and to the right because the triangle's area and base length both increase.

2.3.1 Emergent Graphical Shape Thinking

Leveraging the aforementioned descriptions of covariational reasoning, Moore and colleagues have differentiated between students' static and emergent graphical shape thinking (Moore, 2021; Moore & Thompson, 2015). Moore and Thompson (2015) described *emergent graphical shape thinking* (hereafter emergent thinking) as:

> understanding a graph simultaneously as what is made (a trace) and how it is made (covariation)… [E]mergent shape thinking entails assimilating a graph as a trace in progress (or envisioning an already produced graph in terms of replaying its emergence), with the trace being a record of the relationship between covarying quantities. (p. 785)

Students' conceptions of quantities, coordinate systems, and points as multiplicative objects are all critical to their emergent thinking. Prior to conceiving a graph as an emergent trace of covarying quantities, students must construct quantities and consider how such quantities could be represented on a coordinate system.[2] They must then conceive of a point as a multiplicative object simultaneously representing two quantities. With such a conception in mind, a student can conceive of a graph in

[2] Paoletti et al. (2018) characterize ways students could engage in emergent thinking in both spatial and quantitative coordinate systems.

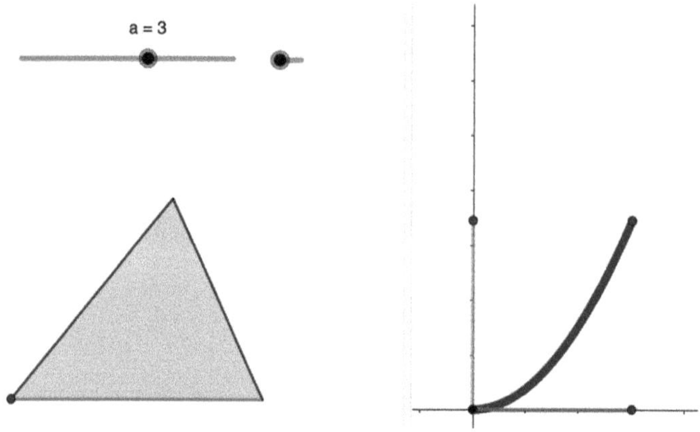

Fig. 4 Trace of a point simultaneously representing the triangle's changing area and side length

terms of an emergent, progressive trace constituted by the point's movement dictated by the covarying quantities' magnitudes represented on the axes.

To support students' emergent thinking, we use GeoGebra's 'trace' feature to trace the point's motion. For example, we have students trace the point in Fig. 3a–c to produce the graph in Fig. 4 representing a record of the relationship between the triangle's base length and area.

In addition to having students observe points producing emergent traces in multiple contexts, we leverage two other techniques to support their emergent thinking. First, as we have contended elsewhere (e.g., Paoletti, 2019; Paoletti & Moore, 2017), students' reasoning about the same final graph as being producible by different traces is a strong indication of a student engaging in emergent thinking; hence, we often provide students with opportunities to engage in such reasoning. For example, in the context of the *Triangle/Rectangle Task*, this opportunity can entail having the animation play in reverse (with a going from 5 to 0).

Second, we often leverage animations, applets, or videos with deliberate pauses. Such pauses provide students opportunities to explicitly attend to the two quantities under consideration (e.g., if the animation pauses and the quantities stop varying, then the point does not move). For example, we use a video of the growing triangle that deliberately pauses for several seconds at two a-values to provide students with opportunities to consider how such pauses impact their graphs. Such opportunities help address a common difficulty students experience with graphs, namely, reasoning univariationally about one quantity with respect to time (e.g., Carlson et al., 2003; Leinhardt et al., 1990; Paoletti, 2015). Students using such reasoning might expect that, if the animation pauses, the graph should contain a straight horizontal line to represent the pause.

2.3.2 Differentiating Between Nonlinear and Linear Relationships Graphically

Particular to differentiating between nonlinear and linear relationships, once a student has conceived of each type of relationship situationally (as described in Sect. 2.2), they can consider how such changes will constrain the movement of the segments and the point in the coordinate system. For example, in the *Triangle/Rectangle Task*, a student may conceive that the increasing changes in the triangle's area will correspond to increasing jumps of the segment representing area along the vertical axis (shown in Fig. 5a). These changes will therefore create points with increasing vertical changes for equal horizontal changes, represented by the large green dots in the coordinate system in Fig. 5a. The student may then leverage their smooth covariational reasoning to anticipate the smooth nature of the increasing quantities to draw a smooth (concave up) curve representing the relationship. The student can engage in similar reasoning for the growing rectangle where the vertical changes are equal, thereby creating a straight graph (Fig. 5b). In both cases, the student understands the shape of the graph is dictated by the relationship between the covarying quantities, which is indicative of emergent thinking.

3 Methods, Participants, and Analysis

The results reported here are a part of a larger design-based research study (Cobb et al., 2003) involving six small group teaching experiments (Steffe & Thompson, 2000). The goal of the study was to examine ways to support middle school students developing various mathematical ideas via their variational and covariational reasoning. We iteratively designed, tested, and redesigned tasks and a task

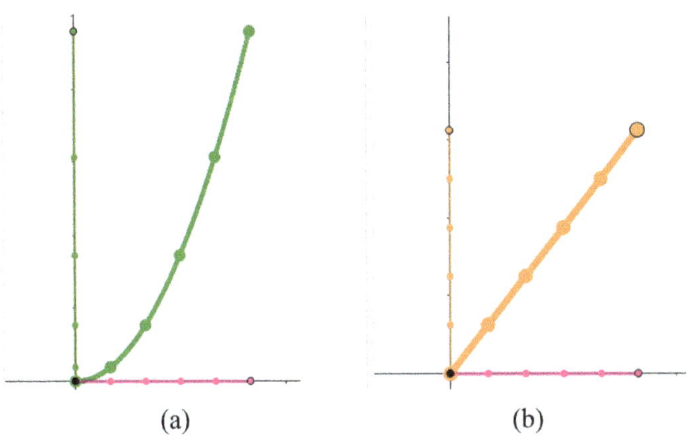

Fig. 5 Graphs representing nonlinear and linear relationships for the *Triangle/Rectangle Task*

sequence that was productive for this goal. In this report, we describe the final task sequence we designed and use student data to exemplify ways students engage with the task sequence. We briefly describe the subjects and data analysis below.

3.1 Subjects and Setting

We conducted the teaching experiments in a school in the Northeastern U.S. that hosts a diverse student population with over 75% students of color and over 67% students who qualify for free or reduced-price lunch. We chose to work with middle school students because they had not taken or completed Algebra I. We asked teachers to recommend students who would be willing to participate and could articulate their thinking.

In this report, we focus on two male students, Vicente (Hispanic) and Lajos (Asian) (pseudonyms), as they engaged with the task sequence. The students participated in 10 teaching episodes that mostly occurred one week apart (February through May), though, due to scheduling constraints (e.g., spring break), some sessions occurred two weeks apart. Each session lasted approximately 40 min. We only report on their activity in the first 8 sessions, as it is critical to their differentiating between linear and nonlinear relationships. Table 4 provides an overview of the 8 sessions including the time span and the task the students were engaged in during each session.

3.2 Data Analysis

We employed on-going and retrospective analyses to characterize models of each student's reasoning. During each phase of analysis, we conducted conceptual analysis—"building models of what students actually know at some specific time and what they comprehend in specific situations" (Thompson, 2008, p. 105). To accomplish this, we analyzed the records using open (generative) and axial (convergent) approaches (Strauss & Corbin, 1998). Specifically, we watched all videos and identified instances that provided insights into each student's reasoning about, coordination of, or representations of varying and covarying quantities. We developed tentative models of each student's mathematics with special attention to the students' covariational reasoning and emergent thinking, including the possibility that they were distinguishing between the different types of relationships (e.g., linear and nonlinear). To test these models, we returned to the previously identified instances, searching for supporting or contradicting instances. When evidence contradicted our models, we revised the models based on interpretation of latter instances. This iterative process resulted in viable models of each student's mathematics. We reiterate that our goal in this paper is not to present complete models of the students' mathematics but to use evidence from these models to exemplify the efficacy of the task sequence.

Table 4 A summary of the teaching experiment sessions

Session	Month	Task	Reasoning supported
1	February	Faucet task	Construct numerous directional relationships
2	March	Faucet task	Construct a quantitative coordinate system and graph directional relationships
3	March	Faucet task	Reason emergently to represent directional relationships
4	March	Growing triangle task	Construct an increasing by increasing amounts relationship
5	March	Pausing triangle task Shrinking triangle task	Explicitly reason emergently Construct a decreasing by decreasing amounts relationship
6	April	Growing trapezoid task	Construct an increasing by decreasing amounts relationship
7	April	Triangle/rectangle task	Construct an increasing by constant amounts relationship Differentiate between nonlinear and linear relationships
8	April	Triangle/rectangle task	Graph nonlinear and linear relationships

4 Building to Nonlinear and Linear Growth: A Task Sequence with Student Work

In this section, we describe most of the final task sequence that supported students reasoning covariationally to construct and graphically represent many of the basic types of covariational relationships (Table 3) and differentiate between nonlinear and linear relationships. For each task, we first describe the task and how it relates to our goals for student learning. We then present results highlighting how Vicente and Lajos engaged in the task and connect their activity back to our goals and theoretical framework. Notably, all tasks were situated in an experientially real situation (Gravemeijer & Doorman, 1999) that entailed quantities we intended to be conceived as varying smoothly (i.e., entail smooth variation). As we are interested in middle school students' initial variational and covariational reasoning, we did not design tasks with eliciting smooth continuous variational reasoning in mind.

4.1 The Faucet Task: Gross Covariational Reasoning and Emergent Thinking

We designed the *Faucet Task* (https://ggbm.at/rdxkrwek; see Fig. 6 for screenshots of initial applet) for two primary purposes. First, the *Faucet Task* provides students repeated opportunities to engage in smooth covariational reasoning in which one quantity (temperature) increased, decreased, or remained constant as the other quantity (amount of water) increased or decreased, reflecting the directional relationships in Table 3. Second, we designed this task to have students consider how to represent two changing quantities as an emergent trace in a quantitative coordinate system.

4.1.1 Students' Quantification and Directional Covariation in the *Faucet Task*

To support students' quantification, at the outset of the *Faucet Task*, we present students with a GeoGebra applet intending to represent a faucet with water coming out (Fig. 6). Students can use red and blue sliders to smoothly turn the hot and cold knobs on and off. As they do so, the rectangle below the faucet smoothly increases or decreases in width to represent the changing amount of water leaving the faucet. Further, the color of the water changes to represent the water's changing temperature (darker red for hotter, darker blue for colder). Initially, our goal is to provide students with the opportunity to construct quantities within the situation.

After Vicente and Lajos explored the applet, the teacher-researcher (TR) asked them what quantities they can measure, with the intention of discussing water temperature and amount of water leaving the faucet (e.g., flow rate, water pressure). Vicente quickly identified "temperature or direction of the knobs" as quantities we could consider, with Lajos adding "the degree of the [knob]". Shortly thereafter, Vicente described the "speed" of the water as another quantity, which we referred to as the "amount of water" or "volume" throughout the rest of the task. Both students had constructed amount of water and water temperature as quantities in the situation.

Fig. 6 Several screenshots for Scenario A of the *Faucet Task* with the cold-water knob being turned all the way on

After students describe water temperature and amount of water, the TR further describes the faucet system in relation to an "engineering problem." The applet reflects that, situationally, if only cold water is turned on then the temperature of the water leaving the faucet is the constant temperature of groundwater. Similarly, if only hot water is turned on, then the temperature of the water is the constant temperature determined by the hot water heater's settings. This conversation includes describing why water feels as if it is warming up when the hot knob is first turned due to the loss of heat of stagnant water in the hot water pipe. By describing the situation, we intend to provide opportunities for students to conceive a situation in which one quantity (temperature) remains constant while the second quantity (amount of water) varies.

After this conversation, we begin to pose questions intended to support students' covariational reasoning. For each prompt, both the hot and cold knobs start halfway on. We ask students to predict what will happen to water temperature and amount of water leaving the faucet for the following prompts, with the directional relationship between (amount of water, temperature) noted in brackets:

(A) they turn the cold knob all the way on [increasing, decreasing] (i.e., Fig. 6),
(B) they turn the cold knob off [decreasing, increasing],
(C) they turn the hot knob all the way on [increasing, increasing], and
(D) they turn the hot knob off [decreasing, decreasing].

Additionally, we ask students to explore how the same two quantities will vary when:

(E) the hot knob stays off and the cold knob is turned on and/or off [increasing/decreasing, constant] and
(F) the cold knob stays off and the hot knob is turned on and/or off [increasing/decreasing, constant].

Our goal is to support students in engaging in directional covariational reasoning with either quantity increasing, decreasing, or remaining constant reflecting each directional relationship in Table 3. Additionally, we intend to foreshadow for students the notion that graphs can be producible in different directions (i.e., support their emergent thinking).

Vicente and Lajos had little difficulty describing how each quantity changes as the knobs are turned. For instance, addressing Prompt B, Vicente quickly described, "I think that it'll only be hot water [left running]. So temperature will increase, but volume will decrease because it's less water." Addressing Prompt D, he described, "it'll be more cold, like the temperature will go down. And I think less water will be pouring out of the faucet." Further, when asked to address Prompts E and F, Vicente identified in each case that temperature would remain the same while the amount of water varied. For instance, addressing Prompt F, Vicente described, "Temperature is going to stay the same, and less water will be coming out as you turn the knob."

There are several notable features from Vicente's activity. First, based on the active nature of his utterances describing changing quantities (e.g., "temperature will increase, but volume will decrease," "less water will be coming out as you turn the knob"), we infer Vicente engaged in (at least) smooth variational reasoning as he

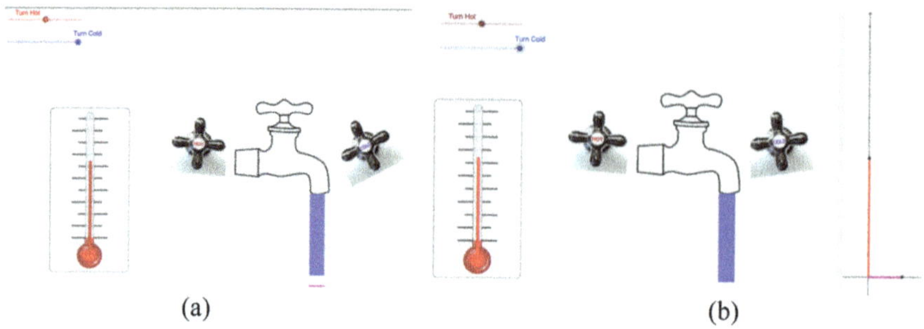

Fig. 7 **a** The Faucet Task applet with an additional thermometer and pink segment below the water and **b** the next applet showing red and pink segments on the vertical and horizontal axis

developed smooth images of change, including varying temperature, amount of water, and the turning of the knobs. We note that since Vicente never explicitly referred to sub-intervals of either quantity, we do not classify his reasoning as smooth continuous variational reasoning; student reasoning compatible with Vicente's motivated our modification of Thompson and Carlson's (2017) variational reasoning framework.

Second, Vicente's quantitative understanding of the situation supported him in reaching numerous (accurate) conclusions regarding the directional relationships between temperature and amount of water. This included situations in which one quantity changed as the other quantity remained constant. Third, Vicente consistently described how both the amount of water and temperature changed as a knob was turned without ever referring to sub-intervals, which is indicative of his engaging in (at least) smooth *co*variational reasoning.

4.1.2 Students Developing Graphing Meanings Via the *Faucet Task*

Once students have described each relationship covariationally, we ask a series of prompts designed to support students in constructing a quantitative coordinate system. First, we present a revised applet that includes a thermometer to gauge the water temperature and a horizontal pink line segment below the rectangle corresponding to the rectangle's width to represent the amount of water leaving the faucet (Fig. 7a).[3] We present these segments for two reasons. First, the red segment provides students a way to describe temperature changing without referring to the color of the water. Second, by using vertical and horizontal linear segments to represent the quantities' magnitudes, we intend to foreshadow a similar representation in the next applet. In that applet, we present students what we intend to be a quantitative coordinate system, with temperature represented by a red segment on the vertical axis and amount of water represented by a pink segment on the horizontal axis (i.e., Fig. 7b).

[3] Depending on time constraints, we sometimes have the thermometer and pink segment as two different tasks and sometimes present them simultaneously, as we did in this case.

Lajos and Vicente interpreted the segment lengths as representing variations in the disembedded quantities. Describing a situation in which the hot knob is turned on, Lajos described, "the temperature will increase [*motioning his finger in an upward direction*], and the pink segment [*putting two fingers together then moving them apart horizontally*] will get wider." Lajos characterized each segments' variation based on his conception of the quantities in the situation. Also, like Vicente, Lajos's words (e.g., "temperature will increase") and actions (e.g., smooth motions with his fingers) are indicative of at least smooth variational reasoning.

After students describe what each segment represents situationally, we move to the next applet showing a red and a pink segment on the vertical and horizontal axis, respectively (Fig. 7b). Hoping to support students in conceiving of a quantitative coordinate system, the TR asks them to describe what will happen to each segment for Prompts A–F described above.

After observing the applet, Lajos and Vicente described how the segments vary based on their understanding of how the quantities change situationally. For example, when tasked with describing how the segments vary for Prompt D, the following conversation ensued:

Lajos The temperature will decrease and [*pause*] the water will decrease.
TR So the amount of water will decrease, and you said, why will the temperature decrease?
Lajos Because since the cold is still on, and temp. The hot water will, ah, you're turning it off, and the cold is still on so it will decrease.

[*The TR asked Lajos to describe what that means for the segments.*]

Lajos Down.
TR Yeah, this one [*pointing to the red segment on the vertical axis*] will definitely move down, but when you say this one [*pointing to the pink segment on the horizontal axis*] will move down, what does that mean?
Lajos Like smaller.
TR Smaller, so it'll go down?
Lajos Like to the left.

We infer Lajos disembodied amount of water and water temperature from the situation as he interpreted the varying segment on each axis, which is critical to his constructing a quantitative coordinate system.

After describing how each segment varies for Prompts A–F, we intend to support students in constructing a point as a multiplicative object simultaneously representing water temperature and amount of water. Hence, we present another applet that now includes a point in the coordinate system with a horizontal magnitude corresponding to the endpoint of the pink segment and a vertical magnitude corresponding to the endpoint of the red segment (Fig. 8a). We first have students turn the knobs and observe the movement of the point.

Constructing a point as a multiplicative object is non-trivial. For example, after Vicente and Lajos turned each knob and observed the point, the following conversation ensued:

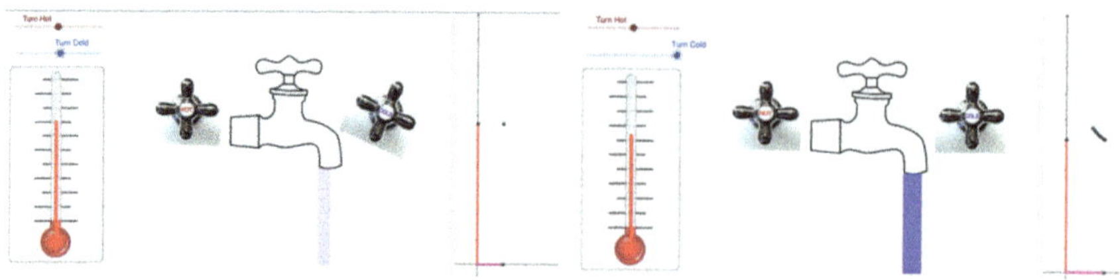

(a) (b)

Fig. 8 **a** The applet with the point shown and **b** the applet with one emergent trace resulting from turning the cold on

TR So, Vicente, what do you think you've got about [the point]?
Vic. I think as, as the water gets warmer [*turning hot knob on*], [the point] moves farther away from [the vertical axis]. It's still like in the same spot but like it goes farther away.
TR Ohhh, why do you think it might be moving to the right?
Vic. Maybe, because of this [*motioning the cursor over the pink segment on the horizontal axis*].

Whereas previously, Vicente always attended to variations in both quantities, when initially making sense of the point's motion, he only attended to the horizontal motion dictated by the pink segment. We often observe such reasoning when students are first considering how to represent two quantities on a quantitative coordinate system.

Immediately after the above interaction, Vicente again attended to only one quantity as he described the point moving left when the amount of water decreased. The TR attempted to draw his attention to the vertical motion of the point by providing a prompt analogous to Prompt C (hot on):

TR There is some cold water right now. Does the temperature go up or down as I turn hot all the way on?
Vic. It's going to go up.
TR It's going to go up. So this red segment is going to go up. So, what do you think is going to happen to that point in terms of moving up, down?
Lajos Go like [*Vicente interjects, Lajos continues*] away from the [*motioning away from the horizontal axis*].
TR It's going to go away because there's more water but will it go away like to the right and down or to the right and up?
Lajos & Vic. [*simultaneously*] To the right and up [*each moving his finger in the air to the right then up*].

After conceiving of the point's movement as being constrained by the varying magnitudes on each axis (i.e., as a multiplicative object), each student described the point's motion in the quantitative coordinate system so that the point represented the covarying magnitudes in this and other cases.

Once students have conceived of the point as a multiplicative object, we support them in engaging in emergent shape thinking. To do this, we use the 'Trace' feature of GeoGebra to trace the point as the students again address (at least a subset of) Prompts A–F (Fig. 8b shows the resulting trace for Prompt A). Our goal is to support students in imagining the graph as being produced by the trace of the point as it moves based on the two quantities. Further, the variety of prompts ensures students have opportunities to engage with graphs tracing in multiple directions.

After having numerous opportunities to observe how graphs are created via the changing quantities in the situation, we present students with several completed graphs (e.g., Fig. 9) and ask them to predict how the knobs began and what action occurred to produce the graph. Our goals are to explore how students interpret a graph representing covarying quantities and to examine if students consider reasoning emergently to describe different scenarios that create the same final graph. If the students do not consider more than one scenario, the TR can raise a second scenario as a hypothetical classmate's solution and ask students to comment on this solution.

Addressing the first of these tasks (Fig. 9a), Vicente accurately argued, "I think the hot water is going to be turned on… because it looks like the temperature is going up… there's more water coming out, it's going to the right." After this, the TR posited that a pair of their classmates had argued the graph was made from turning hot water off and asked Vicente and Lajos if these students could have been correct. Responding to this, and indicative of reasoning about a single graph being traceable in multiple directions, Vicente immediately responded:

> [M]aybe backwards. Maybe they could be thinking about it in reverse because, so you turn hot off right? [*TR agrees*] So that means if you turn it off there will be less water, so you go to the left [*motioning from the top right point leftwards*], and the temperature is going to go down [*tracing along the curve from the top right point down to the left*].... [T]hey're imagining it backwards.

Hence, Vicente, and later Lajos, reasoned about covarying quantities to describe two possible emergent traces producing the same final graph.

Additionally, each student correctly described numerous ways a straight horizontal line could be produced from the situation, with Lajos arguing that "everything

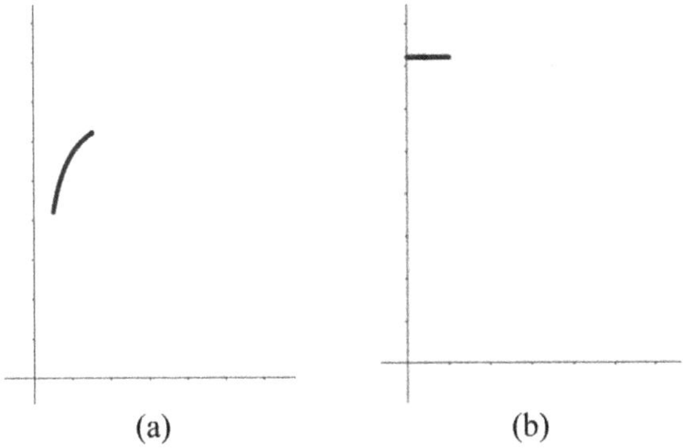

Fig. 9 Two examples of graph interpretation tasks

turned off then turn hot all the way on," would produce the graph in Fig. 9b. Hence, the students repeatedly engaged in smooth covariational reasoning in which temperature increased, decreased, or remained constant as the amount of water increased or decreased, reflecting the directional relationships in Table 3.

4.2 The Growing Triangle Task

After the *Faucet Task*, we have students address the *Growing Triangle Task*. This task provides students additional opportunities to reason emergently about a smoothly changing phenomenon in a quantitative coordinate system. Additionally, we designed the task to extend their covariational reasoning by supporting them in reasoning about amounts of change to construct and accurately represent such a relationship. Specifically, we intended to support students in conceiving that the triangle's area grows by increasing amounts for equal changes in its base length; this relationship is the first type of nonlinear covariational relationship we have students construct.

To support students in imagining and anticipating smooth variation, we first have them interact with a dynamic GeoGebra applet (https://www.geogebra.org/m/yu25d2my) showing a smoothly growing scalene triangle (Fig. 10a). We ask them what quantities they could measure in this situation. To support them in attending to and coordinating area and base length (i.e., to reason covariationally), we highlight the triangle's base length in pink and area in green. After describing the directional change of area and side length, we specifically ask students to identify if, for equal changes in the base length, the area increases by (a) more, (b) less, or (c) the same amount. As described in Sect. 2.1, we included a second smaller slider which allows students to increase the increment by which the pink length increases (e.g., to integer chunks versus smoothly). We have the 'trace' option available so that students can visually identify the increasing amounts of change of area in the applet (i.e., the increasing size of the consecutive trapezoids shown in Fig. 10b). This feature of the task supports students in conceiving of increasing changes in area.

Consistent with constructing quantities in the situation, when asked what quantities they could measure, Vicente quickly described, "all sides are increasing," and, as the base length increases, "the area gets bigger." Later in the session, with the pink length increasing in one unit chunks, Vicente described, "[the area jump] starts with small and… keeps getting bigger and bigger." After conceiving of the area increasing by more, Vicente and Lajos approximated values for each of the amount of change amounts to numerically represent the increasing changes in the triangle's area, as shown in Fig. 10c (i.e., 1, 3, 5, 7, and 9 represent the amounts of change). Hence, while initially leveraging smooth variation (e.g., "area gets bigger"), the students began to leverage chunky variational reasoning as they conceived of and numerically represented the amounts of change of area increasing by more.

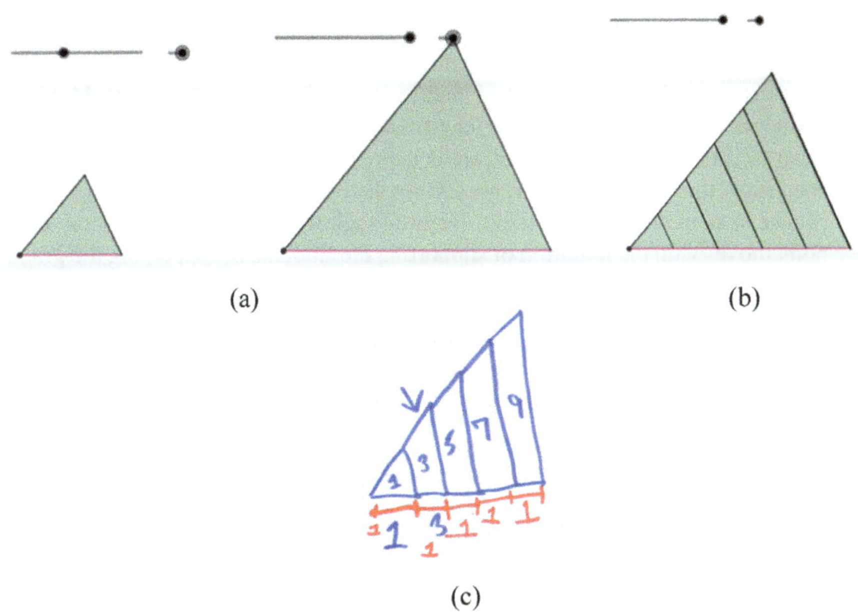

Fig. 10 **a** Two screenshots of the *Growing Triangle Task*, **b** the triangle shown in the applet with chunky changes, and **c** the students' work approximating area values

4.2.1 Graphing the Relationship in the *Growing Triangle Task*

Once students describe the amounts of change in area as increasing, we present them with what we intend to be a quantitative coordinate system with the side length represented by a pink line segment on the horizontal axis. We ask them to describe how the increasing change in the triangle's area will correspond to the movement of a segment representing area on the vertical axis. Our intent is to offer students repeated opportunities to reason about covariational relationships and consider how to represent a quantity's magnitude (or value) with a corresponding line segment.

Indicative of disembedding area from the situation and representing it with a segment's magnitude on the vertical axis, each student described that the increasing changes in the triangle's area will correspond to increasing jumps of the segment on the vertical axis. For instance, referring to the approximated area values, Lajos motioned his finger by three units in an upward direction along the vertical axis from a point representing the area when the side length was one unit. Lajos described that "the area is four…the area would go to nine [*motioning his pointing finger by five units in an upward direction along the vertical axis from the point placed by the TR at (0, 4)*]." While Lajos described how the segment representing area increased by more, Vicente simultaneously motioned his finger by one unit to the right on the horizontal axis, indicative of reasoning about the horizontal segment varying by equal amounts. We infer Lajos engaged in numeric chunky continuous covariational reasoning to describe how the segments varied, and hence conceived of a quantitative coordinate system. Further, the students extended this activity by creating points in the coordinate system that simultaneously represented side length and area.

The next several prompts are designed to support students' emergent thinking. After students represent specific base length and area magnitudes via points as multiplicative objects, we change the small slider to present the triangle growing smoothly. We then show the dynamic point representing the two quantities' magnitudes in the coordinate system and use the 'trace' feature to allow the students to observe how the point moves with the intention of supporting the students in conceiving the graph as being produced by an emergent trace.

When asked to explain why the graph contains more than the five chunky points shown in Fig. 11a, Vicente claimed "it's [the point] just not skipping, it'll go like this [*motioning his pointer finger in a curve that passes through the five points as though sketching a smooth concave up curve*] and Lajos added, "it's [the point is] tracing, tracing slowly up [*motioning his fingers in the air as though sketching a smooth concave up curve*]." After this, each student sketched a smooth curve (Fig. 11b, c) joining the five points on a given handout. Further, Vicente claimed, "[the area] is not going to be just here and here [*pointing to consecutive points shown on the graph*]" and explained that "like [the area] could be at 50, but at some point it has to be smaller than that like 49, 48." Realizing that Vicente spontaneously began to describe smaller sub-intervals of the changing area, the TR asked him if the area's value must ever be 48.5. Possibly indicative of Vicente engaging in smooth continuous variational reasoning, he quickly agreed that the area must take on such a value. However, the TR did not provide additional follow-up questions to allow us to claim definitively if Vicente was engaging in smooth continuous variational reasoning.

4.2.2 The *Pausing* and *Shrinking Triangle Tasks*: Examining Students' Emergent Reasoning

To investigate the extent to which the students are attending to the two intended quantities and to examine their potential emergent thinking, we include two follow-up tasks. In the *Pausing Triangle Task*, we present a video showing the same smoothly growing triangle. However, twice in this video, the triangle's growth pauses. In the *Shrinking Triangle Task*, we present a video showing the same triangle, but with its side length and area decreasing from their maximum values until they are both zero. In each case, we intend to support the students in attending to the two changing quantities in the situation and considering how the new features of the situation (pauses, going in reverse) do or do not influence either their graph or the trace producing their graph. An added affordance of the *Shrinking Triangle Task* is that students have the opportunity to describe a decreasing by decreasing amounts relationship; this relationship is the second type of nonlinear covariational relationship we have students construct.

Addressing these tasks, each student exhibited emergent thinking as they related their original graph to these new situations. For the *Pausing Triangle Task*, Lajos argued this new situation would have a different trace but produce the same graph. He explained, "it [the moving point] would stop for a few seconds here [*marking a*

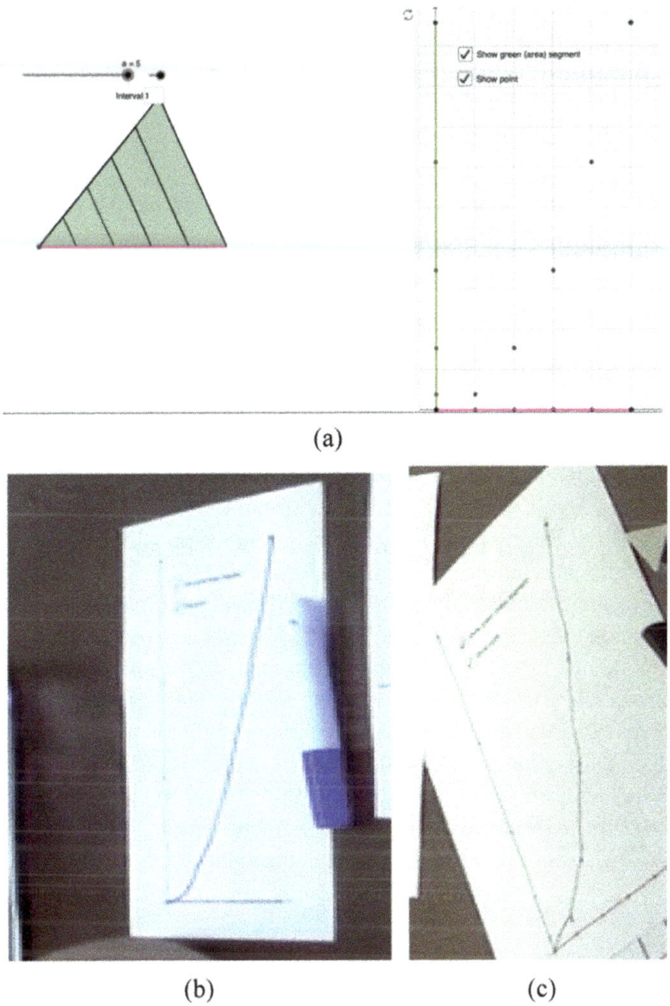

Fig. 11 **a** One screenshot from the *Growing Triangle* applet that includes a graphical representation and points produced by the 5 equal changes of side length, **b**, **c** Lajos's and Vicente's graphs, respectively

point on the curve in Fig. 11b] and then keeps going [*tracing the pen on the curve*] then stops [*stops pen along the curve*] and then keeps going [*moves the pen along the curve*]." Addressing the *Shrinking Triangle Task*, Lajos re-traced the original graph from the top right to the bottom left while claiming, "[the point] would start right up there and it would go reverse and go back down." Likewise, Vicente claimed that the graph "would go backwards." We infer each student was reasoning emergently as he created and interpreted graphs representing the triangle's varying base length and area. Further, when addressing the *Shrinking Triangle Task*, the students explicitly described that the triangle's area decreased by less for equal decreases in the base length. Hence, the students constructed a second type of nonlinear covariational relationship.

4.3 The Growing Trapezoid Task: An Increasing by Less Relationship

After engaging in the prior tasks, we hope students will begin to spontaneously examine the direction and amounts of change of a relationship and will leverage emergent thinking when prompted to graphically represent a relationship. We use the *Growing Trapezoid Task* (https://ggbm.at/jbk6kw8f) to examine this possibility. In this task, we present students with a smoothly growing figure that starts as a line, becomes a trapezoid, and increases until it results in a triangle (Fig. 12a). Although the resultant triangle is the same triangle as in the *Growing Triangle Task*, in this situation, area increases by less for equal changes in the pink length, which is a third type of covariational relationship in the task sequence.

Relevant to the students' quantitative and covariational reasoning, each student quickly responded that the "area gets bigger." Vicente described the area increases by "smaller amounts," and Lajos elaborated the consecutive amounts of change in the area "are smaller." Particular to their graphing activity, the students plotted points on the vertical axis in a way that was indicative of leveraging a quantitative coordinate system. Specifically, with prompting from the TR, Vicente and Lajos worked together to connect the increasing amounts of change represented on the vertical axis in their graph in the *Growing Triangle Task* to corresponding decreasing amounts of change in this task (e.g., approximating the size of the final jump for their graph in the

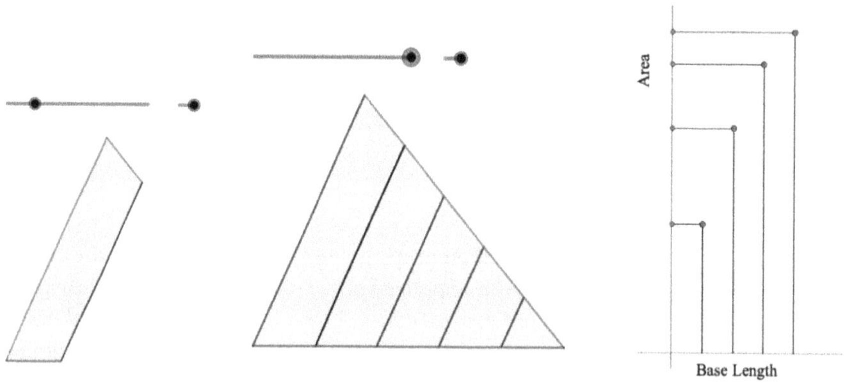

(a) (b)

Fig. 12 **a** The *Growing Trapezoid Task* and **b** a recreation of Vincent's work showing area increasing by smaller amounts and points produced as multiplicative objects

Growing Triangle Task and using this magnitude for the first jump in this task). Leveraging this reasoning, Vicente plotted several points on the vertical axis that he conceived "jump by less." Using these points, Vincente plotted points in the coordinate system representing simultaneously the growing trapezoid's base length and area (Fig. 12b), which we infer represented the trapezoid's area increasing by decreasing amounts for equal changes in base length.

4.4 The Triangle/Rectangle Task

In addition to providing another opportunity to reason emergently, the *Triangle/Rectangle Task* provides students an occasion to construct a linear covariational relationship, a fourth type of covariational relationship in the sequence. Further, the task provides opportunities to compare linear and nonlinear growth and consider how each type of relationship can be represented via graphs as emergent traces. In the *Triangle/Rectangle Task*, we present students with a smoothly growing rectangle next to the original triangle from the *Growing Triangle Task*; both figures have equal pink base lengths. We ask students to describe how each area is changing and to graphically represent the relationships between area and side length for each growing shape.

Once Lajos and Vicente described that each area is increasing, the TR began to pose questions to investigate their covariational reasoning. For instance, after Vicente and Lajos described the area of the rectangle as increasing, the following conversation ensued:

TR How is the area of the rectangle increasing?

Vic. I think for the rectangle, I think that it's increasing by, keeps increasing by the same amount.

TR Increasing by the same amount?

Vic. Yeah, 'cause it keeps adding that one block [*pointing to the smallest amount of change rectangle*] over and over again [*motions hand over successive amounts of change rectangles, shown in* Fig. 2].

We infer Vicente (and later Lajos) engaged in chunky continuous covariational reasoning to describe the area of the rectangle increasing by equal amounts for equal changes in base length.

To investigate the students' emergent shape thinking, we then asked the students to graph the relationship between area and side length for each growing shape on a handout with the area and side length represented on the vertical and horizontal axes, respectively. Watching the applet with side length increasing in increments of 1, Lajos leveraged chunky continuous variational reasoning as he used his fingers to indicate the segment representing the rectangle's area would jump by equal amounts along the vertical axis. Justifying these equal changes along the vertical axis, he described, "Because all of the [smaller] rectangles (shown in Fig. 2) are equal sized, so it [the increase in area] has to be the same amount." After this, Vicente also engaged in chunky continuous covariational reasoning as he plotted points that represented the area changes described by Lajos vertically and corresponding equal side length changes horizontally (Fig. 13a). As Vicente plotted points, Lajos described that the points correspond to "both of them" referring to the rectangle's base length and area. After Vicente plotted the last point, the TR changed the smaller slider to change the applet from playing in chunks to smoothly and asked "and what if I have it playing, sort of, smoothly?" Immediately, and indicative of engaging in smooth covariational reasoning and emergent shape thinking, Lajos picked up the marker, said "it would be like this," and sketched a straight line through the points Vicente plotted (Fig. 13b).

Shortly after this, the TR prompted the students to graphically represent the triangle's area and side length on the same coordinate system. The students recalled their work in the prior sessions to sketch a smooth concave up curve to represent this relationship (Fig. 13c). Hence, the students were able to leverage a combination of their chunky continuous and smooth covariational reasoning to conceive of both nonlinear and linear relationships. Further, they graphically represented each relationship via an emergent trace on a quantitative coordinate system.

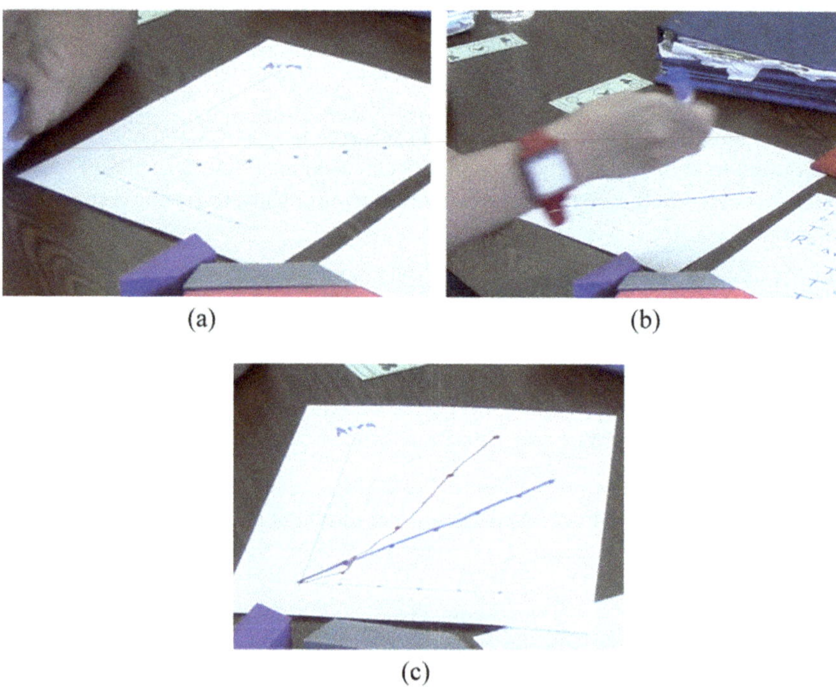

Fig. 13 **a** Vicente's plotted points, **b** Lajos finishing sketching a straight line, and **c** the pair's graph with the curve representing the triangle's base length and area

5 Discussion

We first discuss contributions this chapter provides to the literature on students' covariational reasoning. We then relate our task design to the theoretical framework and provide implications for developing other mathematical ideas. We conclude with areas for future research.

5.1 Middle School Students' Covariational Reasoning

In this chapter, we explored the possibility of middle school students constructing and reasoning about basic types of covariational relationships (Table 3), which supports them in differentiating between nonlinear and linear relationships. Through the task sequence, students had repeated opportunities to construct numerous directional relationships. Such activity was foundational for the students' later activity as they characterized covariational relationships with differing amounts of change; Table 5 presents all of the directional and amounts of change relationships Lajos and Vicente constructed. Further, we described how such covariational reasoning supported students' emergent reasoning as they accurately constructed and interpreted graphs tracing in numerous directions with varying concavities.

Table 5 The directional and amounts of change covariational relationships Vicente and Lajos constructed with the first quantity in the *Faucet Task* being amount of water

First quantity	Directional change in second quantity (task where student experienced)	Amounts of change in second quantity (task where student experienced)
Increasing	Constant (*Faucet Task*, Scenarios E and/or F)	–
	Increasing, (*Faucet Task*, Scenario C; *Growing Triangle Task*; *Growing Trapezoid Task*)	Increasing by the same amount (rectangle's area in *Triangle/Rectangle* task)
		Increasing by more (triangle's area in *Growing Triangle* task)
		Increasing by less (trapezoid's area in *Growing Trapezoid* task)
	Decreasing (*Faucet Task*, Scenario A)	–
Decreasing	Constant (*Faucet Task*, Scenarios E and/or F)	–
	Increasing (*Faucet Task*, Scenario B)	–
	Decreasing (*Faucet Task*, Scenario D; *Shrinking Triangle Task*)	Decreasing by less (triangle's shrinking area in *Shrinking Triangle* task)

Throughout our presentation of the tasks, we explicitly connected our task design to our theoretical framing. Compatible with the Learning Through Activity framework described by Simon and colleagues (e.g., Simon et al., 2018), our goal was to design a task sequence that supported students in gradually developing ways of thinking that would eventually lead to their constructing sophisticated mathematical understandings. We intend for such descriptions to serve as a resource for other researchers' and teachers' efforts at adapting these tasks or designing new tasks that could afford similar shifts in students' reasoning.

In addition to providing empirical examples of middle school students constructing numerous covariational relationships, we extend Thompson and Carlson's (2017) variational and covariational framework to include smooth variational and smooth covariational reasoning. We provide empirical examples of middle school students first engaging in gross and smooth covariational reasoning prior to engaging in chunky continuous covariational reasoning. Consistent with the conjecture of Castillo-Garsow et al. (2013), smooth reasoning seemed to entail a capacity to engage in chunky reasoning, with the latter reasoning supporting students in further characterizing their conceived relationships. These forms of reasoning interplayed productively with the students' meanings for quantitative coordinate systems and points as multiplicative objects as students constructed and interpreted graphs as emergent traces, "with the trace being a record of the relationship between covarying quantities" (Moore & Thompson, 2015, p. 785).

5.2 *Task Design in Relation to Our Theoretical Framework*

We highlight that each part of the *Faucet Task* involved (almost all) of Prompts (A)–(F). We conjecture these repeated opportunities were critical for the students' developing graphing meanings as they considered how to represent a relationship via a point as a multiplicative object constrained by the motions of segments on axes. Further, their directional covariational reasoning in the *Faucet Task* laid the foundation for their later activity discerning the amounts of change of one quantity with respect to a second quantity in the tasks that followed. In these latter tasks, the students leveraged a combination of chunky continuous and smooth covariational reasoning to construct and graphically represent numerous nonlinear and linear relationships.

We contend the ability for students to change the intervals by which an applet's parameter varied from smooth to chunky was critical. By first engaging with smoothly changing phenomena, Vicente and Lajos developed smooth images of change. However, and as we contended elsewhere (Paoletti & Moore, 2017), smooth thinking alone is not sufficient for discerning the amounts of change of one quantity with respect to another. Hence, changing the parameter via the slider also supported students in developing chunky images of the situation, which was critical to them constructing the different covariational relationships in Table 5.

Relatedly, across all students we interviewed in the larger design experiment, using smoothly changing phenomena supported students in developing smooth images of change. In contrast, we conjecture tasks which present a table of values, regardless of the teacher's or researcher's intention, are more likely to elicit (at best) chunky covariational reasoning from students. We contend that creating mental imagery of smoothly changing phenomena from a table of values is possible, but non-trivial; providing students with dynamic representations of smoothly changing phenomena is invaluable to their development of smooth variational and covariational reasoning (Castillo-Garsow et al., 2013; Johnson, 2020; Stevens et al., 2017).

5.3 Implications for Developing Other Mathematical Ideas

An immediate consequence of constructing various nonlinear and linear relationships is that students can experience an intellectual need (Harel, 2008) to further differentiate between types of covariational relationship (or function) classes that exhibit similar change patterns. For example, both quadratic and exponential relationships can exhibit growth such that the second quantity increases by an increasing amount for equal changes of the first quantity. As described by Vishnubhotla (2020), once students identify such a similarity, they may further explore numeric relationships to identify patterns. Hence, once students have repeated experiences constructing and representing covarying quantities, other representations, such as tables of values, can be useful as they further differentiate between various forms of change (beyond linear versus nonlinear).

To exemplify this, we turn to Vicente and Lajos's activity described in Sect. 4.2. Specifically, after identifying numeric values for specific amounts of change (+ 1, + 3, + 5, etc.) the pair identified that these amounts of changes were changing by a constant amount. As our goal in this study did not entail students developing meanings for quadratic relationships, we did not design tasks or prompts to explore this reasoning further. However, such activity supported this pair, and other students (Mohamed et al., 2020), in identifying the defining characteristic of quadratic growth (Ellis, 2011; Lobato et al., 2012). Hence, the presented task sequence has the potential to lay a foundation for students developing meanings for specific nonlinear relationships.

5.4 Concluding Remarks and Areas for Future Research

We contend that providing middle school students opportunities to reason about dynamically changing quantities to construct basic types of covariational relationships can serve as a foundation for their developing meanings for various functional relationships (Thompson & Thompson, 1995). There is a further need to develop or adapt tasks that extend middle school students' covariational and emergent thinking to support them in developing meanings for other relationships as described by other researchers, including quadratic relationships (Ellis, 2011; Lobato et al., 2012), exponential relationships (Confrey & Smith, 1994, 1995; Ellis et al., 2015; Thompson, 2008), and possibly even trigonometric relationships (Moore, 2014).

In addition to designing or adapting tasks to foster students' thinking, there is also a need to investigate ways to scale a task sequence like this one to be effective in larger settings (e.g., whole class). Based on a pilot whole class teaching experiment with 6th-grade students, we conjecture there is a need to provide students with sufficient opportunities to reflect on their activity for them to develop stable meanings that entail covariational reasoning and emergent thinking. Such reflective activities can further support students in developing stable meanings for graphs, relationships, and various relationship classes. Hence, we call for further research on how to make task sequences like the one presented in this chapter both productive in whole class settings and usable by middle school teachers everywhere. Such research has the potential to impact the teaching and learning of middle school mathematics across the world, as such an approach can support all students in developing foundational knowledge and ways of thinking that are critical to their algebra learning.

Acknowledgements This material is based upon work supported by the Spencer Foundation under Grant No. 201900012. We thank Allison Gantt and Julien Corven for their feedback.

References

Carlson, M. P., Jacobs, S., Coe, E., Larsen, S., & Hsu, E. (2002). Applying covariational reasoning while modeling dynamic events: A framework and a study. *Journal for Research in Mathematics Education, 33*(5), 352–378.

Carlson, M. P., Larsen, S., & Lesh, R. (2003). Modeling dynamic events: A study in applying covariational reasoning among high performing university students. In R. Lesh & H. Doerr (Eds.), *Beyond constructivism in mathematics teaching and learning: A models & modeling perspective* (pp. 465–478). Lawrence Erlbaum.

Castillo-Garsow, C., Johnson, H. L., & Moore, K. C. (2013). Chunky and smooth images of change. *For the Learning of Mathematics, 33*(3), 31–37.

Cobb, P., Confrey, J., diSessa, A. A., Lehrer, R., & Schauble, L. (2003). Design experiments in educational research. *Educational Researcher, 32*(1), 9–13.

Confrey, J., & Smith, E. (1994). Exponential functions, rates of change, and the multiplicative unit. *Educational Studies in Mathematics, 26*, 135–164.

Confrey, J., & Smith, E. (1995). Splitting, covariation, and their role in the development of exponential functions. *Journal for Research in Mathematics Education, 26*(1), 66–86.

Crisp, G., Nora, A., & Taggart, A. (2009). Student characteristics, pre-college, college, and environmental factors as predictors of majoring in and earning a STEM degree: An analysis of students attending a Hispanic serving institution. *American Education Research Journal, 46*(4), 924–942.

De Bock, D., Van Dooren, W., Janssens, D., & Verschaffel, L. (2007). *The illusion of linearity: From analysis to improvement* (Vol. 41). Springer Science & Business Media.

Ellis, A. B. (2011). Middle school algebra from a functional perspective: A conceptual analysis of quadratic functions. In L. R. Wiest & T. Lamberg (Eds.), *Proceedings of the 33rd Annual Meeting of the North American Chapter of the International Group for the Psychology of Mathematics Education* (pp. 79–87). University of Nevada Reno.

Ellis, A. B., Özgür, Z., Kulow, T., Williams, C. C., & Amidon, J. (2015). Quantifying exponential growth: Three conceptual shifts in coordinating multiplicative and additive growth. *The Journal of Mathematical Behavior, 39*, 135–155.

Esteley, C. B., Villarreal, M. E., & Alagia, H. R. (2010). The overgeneralization of linear models among university students' mathematical productions: A long-term study. *Mathematical Thinking and Learning, 12*(1), 86–108.

Frank, K., & Thompson, P. W. (2021). School students' preparation for calculus in the United States. *ZDM—Mathematics Education, 53*(3), 549–562. https://doi.org/10.1007/s11858-021-01231-8

Gravemeijer, K., & Doorman, M. (1999). Context problems in realistic mathematics education: A calculus course as an example. *Educational Studies in Mathematics, 39*(1–2), 111–129.

Harel, G. (2008). DNR perspective on mathematics curriculum and instruction, part I: Focus on proving. *ZDM Mathematics Education, 40*, 487–500.

Hiebert, J., Stigler, J. W., Jacobs, J. K., Givvin, K. B., Garnier, H., Smith, M., Hollingsworth, H., Manaster, A., Wearne, D., & Gallimore, R. (2005). Mathematics teaching in the United States today (and tomorrow): Results from the TIMSS 1999 video study. *Educational Evaluation and Policy Analysis, 27*(2), 111–132.

Johnson, H. L. (2012). Reasoning about variation in the intensity of change in covarying quantities involved in rate of change. *The Journal of Mathematical Behavior, 31*(3), 313–330.

Johnson, H. L. (2020). Task design for graphs: Rethink multiple representations with variation theory. *Mathematical Thinking and Learning*, 1–8. https://doi.org/10.1080/10986065.2020.1824056

Johnson, H. L., McClintock, E., & Hornbein, P. (2017). Ferris wheels and filling bottles: A case of a student's transfer of covariational reasoning across tasks with different backgrounds and features. *ZDM, 49*(5).

Joshua, S., Musgrave, S., Hatfield, N., & Thompson, P. W. (2015). Conceptualizing and reasoning with frames of reference. In T. Fukawa-Connelly, N. E. Infante, K. Keene, & M. Zandieh (Eds.), *Proceedings of the 18th Meeting of the MAA Special Interest Group on Research in Undergraduate Mathematics Education* (pp. 31–44). RUME.

Lee, H. (2016). Just go straight: Reasoning within spatial frames of reference. In M. B. Wood, E. E. Turner, M. Civil, & J. A. Eli (Eds.), *Proceedings of the 38th Annual Meeting of the North American Chapter of the International Group for the Psychology of Mathematics Education* (pp. 278–281). The University of Arizona.

Lee, H. Y., Hardison, H. H., & Paoletti, T. (2020). Foregrounding the background: Two uses of coordinate systems. *For the Learning of Mathematics, 40*(1), 32–37.

Leinhardt, G., Zaslavsky, O., & Stein, M. K. (1990). Functions, graphs, and graphing: Tasks, learning, and teaching. *Review of Educational Research, 60*(1), 1–64.

Liang, B., & Moore, K. C. (2020). Figurative and operative partitioning activity: Students' meanings for amounts of change in covarying quantities. *Mathematical Thinking and Learning.*

Litke, E. (2020). The nature and quality of algebra instruction: Using a content-focused observation tool as a lens for understanding and improving instructional practice. *Cognition and Instruction, 38*(1), 57–86.

Lobato, J., Hohensee, C., Rhodehamel, B., & Diamond, J. (2012). Using student reasoning to inform the development of conceptual learning goals: The case of quadratic functions. *Mathematical Thinking and Learning, 14*(2), 85–119.

Loveless, T. (2013). *The algebra imperative: Assessing algebra in a national and international context.* Brookings. Retrieved November 19, 2021 from https://www.brookings.edu/wp-content/uploads/2016/06/Kern-Algebra-paper-8-30_v14.pdf

Mohamed, M. M., Paoletti, T., Vishnubhotla, M., Greenstein, S., & Lim, S. S. (2020). Supporting students' meanings for quadratics: Integrating RME, quantitative reasoning and designing for abstraction. In *Proceedings of the Annual Meeting of the Psychology of Mathematics Education - North America* (pp. 193–201).

Moore, K. C. (2014). Quantitative reasoning and the sine function: The case of Zac. *Journal for Research in Mathematics Education, 45*(1), 102–138.

Moore, K. C. (2021). Graphical shape thinking and transfer. In C. Hohensee & J. Lobato (Eds.), *Transfer of learning: Progressive perspectives for mathematics education and related fields* (pp. 145–171). Springer.

Moore, K. C., Stevens, I. E., Paoletti, T., Hobson, N. L. F., & Liang, B. (2019). Pre-service teachers' figurative and operative graphing actions. *The Journal of Mathematical Behavior, 56*.

Moore, K. C., & Thompson, P. W. (2015). Shape thinking and students' graphing activity. In T. Fukawa-Connelly, N. Infante, K. Keene, & M. Zandieh (Eds.), *Proceedings of the Eighteenth Annual Conference on Research in Undergraduate Mathematics Education* (pp. 782–789). West Virginia University.

Paoletti, T. (2015). Students' reasoning when constructing quantitatively rich situations. In T. Fukawa-Connolly, N. E. Infante, K. Keene, & M. Zandieh (Eds.), *Proceedings of the Eighteenth Annual Conference on Research in Undergraduate Mathematics Education* (pp. 845–852). West Virginia University.

Paoletti, T. (2019). Support students' understanding graphs as emergent traces: The faucet task. In M. Graven, H. Venkat, A. Essien, & P. Vale (Eds.), *Proceedings of the 43rd Conference of the International Group for the Psychology of Mathematics Education* (Vol. 3, pp. 185–192). PME.

Paoletti, T., Lee, H. Y., & Hardison, H. (2018). Static and emergent thinking in spatial and quantitative coordinate systems. In T. E. Hodges, G. J. Roy, & A. M. Tyminski (Eds.), *Proceedings of the 40th Annual Meeting of the North American Chapter of the International Group for the Psychology of Mathematics Education* (pp. 1315–1322). University of South Carolina & Clemson University.

Paoletti, T., & Moore, K. C. (2017). The parametric nature of two students' covariational reasoning. *The Journal of Mathematical Behavior, 48*, 137–151. https://doi.org/10.1016/j.jmathb.2017.08.003

Saldanha, L. A., & Thompson, P. W. (1998). Re-thinking co-variation from a quantitative perspective: Simultaneous continuous variation. In S. B. Berenson & W. N. Coulombe (Eds.), *Proceedings of the Annual Meeting of the Psychology of Mathematics Education—North America* (Vol. 1, pp. 298–304). North Carolina State University.

Sass, T. R. (2015). *Understanding the STEM pipeline*. Georgia State University.

Simon, M. A., Kara, M., Placa, N., & Avitzur, A. (2018). Towards an integrated theory of mathematics conceptual learning and instructional design: The learning through activity theoretical framework. *The Journal of Mathematical Behavior, 52*, 95–112.

Smith, J. P., & Thompson, P. W. (2008). Quantitative reasoning and the development of algebraic reasoning. In J. J. Kaput, D. W. Carraher, & M. L. Blanton (Eds.), *Algebra in the early grades* (pp. 95–132). Lawrence Erlbaum Associates.

Steffe, L. P., & Olive, J. (2010). *Children's fractional knowledge*. Springer.

Steffe, L. P., & Thompson, P. W. (2000). Teaching experiment methodology: Underlying principles and essential elements. In R. A. Lesh & A. E. Kelly (Eds.), *Handbook of research design in mathematics and science education* (pp. 267–307). Erlbaum.

Stevens, I. E., Paoletti, T., Moore, K. C., Liang, B., & Hardison, H. (2017). Principles for designing tasks that promote covariational reasoning. In A. Weinberg, C. Rasmussen, J. Rabin, M. Wawro, & S. Brown (Eds.), *Proceedings of the 20th Annual Conference on Research in Undergraduate Mathematics Education* (pp. 928–936).

Strauss, A. L., & Corbin, J. M. (1998). *Basics of qualitative research: Techniques and procedures for developing grounded theory* (2nd ed.). Sage.

Thompson, A. G., & Thompson, P. W. (1995). A cognitive perspective on the mathematical preparation of teachers: The case of algebra. In C. B. Lacampagne, W. Blair, & J. Kaput (Eds.), *The algebra initiative colloquium* (pp. 95–116). U. S. Government Printing Office.

Thompson, P. W. (2008). Conceptual analysis of mathematical ideas: Some spadework at the foundations of mathematics education. In O. Figueras, J. L. Cortina, S. Alatorre, T. Rojano, & A. Sèpulveda (Eds.), *Proceedings of the Annual Meeting of the International Group for the Psychology of Mathematics Education* (Vol. 1, pp. 31–49).

Thompson, P. W. (2011). Quantitative reasoning and mathematical modeling. In S. A. Chamberlain & L. L. Hatfield (Eds.), *New Perspectives and Directions for Collaborative Research in Mathematics Education: Papers from a Planning Conference for WISDOM^e* (Vol. 1, pp. 33–56). University of Wyoming College of Education.

Thompson, P. W., & Carlson, M. P. (2017). Variation, covariation and functions: Foundational ways of mathematical thinking. In J. Cai (Ed.), *Compendium for research in mathematics education* (pp. 421–456). National Council of Teachers of Mathematics.

Thompson, P. W., Carlson, M. P., Byerley, C., & Hatfield, N. (2014). Schemes for thinking with magnitudes: A hypothesis about foundational reasoning abilities in algebra. In L. P. Steffe, K. C. Moore, L. L. Hatfield, & S. Belbase (Eds.), *Epistemic algebraic students: Emerging models of students' algebraic knowing* (pp. 1–24). University of Wyoming.

Thompson, P. W., & Harel, G. (2021). Ideas foundational to calculus learning and their links to students' difficulties. *ZDM Mathematics Education, 53*(3), 507–519. https://doi.org/10.1007/s11858-021-01270-1

Thompson, P. W., Hatfield, N. J., Yoon, H., Joshua, S., & Byereley, C. (2017). Covariational reasoning among U.S. and South Korean secondary mathematics teachers. *The Journal of Mathematical Behavior, 48*, 95–111. https://doi.org/10.1016/j.jmathb.2017.08.001

Vishnubhotla, M. (2020). *The role of covariational reasoning in pre-service teachers' meanings for quadratic and exponential relationships.* (Unpublished doctoral dissertation). Montclair State University.

A Conceptual Analysis of Early Function Through Quantitative Reasoning

Amy Ellis, Zekiye Özgür, and Muhammed Fatih Doğan

1 Introduction: Functions as Rates of Change

The concept of function is a unifying principle in mathematics instruction (Krüger, 2019; Panaoura et al., 2017; Steele et al., 2013), and recommendations to support students' algebraic reasoning advocate the introduction of function ideas in the middle grades (e.g., Australian Curriculum, Assessment and Reporting Authority [ACARA], 2020; National Governor's Association Center for Best Practices, 2010; Turkish Ministry of National Education [MEB], 2018; U.K. Department for Education, 2009). A strong understanding of function is critical not only for success in algebra, but also for students who intend to study geometry, statistics, calculus, and other content courses essential in many STEM fields (Oehrtman et al., 2008). Despite the instructional emphasis on function throughout secondary school, there is copious evidence of students' challenges in developing a meaningful understanding of function. Students exit secondary school viewing functions in terms of symbolic manipulations, relying on memorized rules and procedures (Carlson & Oehrtman, 2005; Oehrtman et al., 2008; Stephens et al., 2017). Many think a relation is only a function if it is definable by a single algebraic formula (Carlson & Oehrtman, 2005; Clement, 2001; Vinner, 1983), and students can struggle to distinguish variables from parameters (Best & Bikner-Ahsbahs, 2017). Relations that are not continuous and do not have a one-to-one correspondence are often not recognized as functions (Clement, 2001; Leinhardt et al., 1990; Vinner & Dreyfus, 1989), and students

A. Ellis (✉)
The University of Georgia, Athens, GA, USA
e-mail: amyellis@uga.edu

Z. Özgür
Dokuz Eylül University, İzmir, Turkey

M. F. Doğan
Adiyaman University, Adiyaman, Turkey

G. Karagöz Akar et al. (eds.), *Quantitative Reasoning in Mathematics and Science Education*, Mathematics Education in the Digital Era 21,
https://doi.org/10.1007/978-3-031-14553-7_7

often believe that a function's graph should be continuous, symmetric, or constantly increasing or decreasing (Jones, 2006; Vinner, 1983). Students also struggle to find outputs and inputs (McGowen et al., 2000), to generate appropriate definitions of functions (Panaoura et al., 2017), and to meaningfully translate across representations (McGowen et al., 2000).

Many of the above documented challenges are an understandable consequence of instruction that treats function as a static relation, predicated on a set theoretic image, with an emphasis on procedural manipulation and algebraic representations. Set theoretic notions of function are certainly important, particularly for higher mathematics such as analysis, graph theory, and other topics, but we argue this should not be the foundation for introducing function concepts in the middle grades. As Yoon and Thompson (2020) discussed, instruction on function early in the twentieth century emphasized variability and change (Krüger, 2019), and it was only in recent decades that a static treatment of function as object became prominent. Now, in many secondary schools across the globe, function is commonly introduced as a correspondence relation between sets (McCulloch et al., 2020), although some newer curricula have also begun to address rate-of-change approaches. For instance, this can be seen in Turkey (Akkoc & Tall, 2005; Yılmaz et al., 2020), in Israel and the U.K. (Ayalon et al., 2017), and in the U.S. (Thompson & Carlson, 2017).

Using the correspondence definition when introducing function can encourage students to rely on representational cues to determine whether a given relation is a function. Oehrtman et al. (2008) pointed out that this treatment removes the intellectual need to develop functional relationships: "To use the modern definition of function in an introduction to the function concept is to present students with a solution to problems of which they cannot conceive" (p. 4). Consequently, even when students can produce a formal definition of function, they often still hold a restricted concept image (Clements, 2001; Sajka, 2003; Vinner, 1983; Vinner & Dreyfus, 1989). Ayalon et al. (2017) argued that instead we should promote a dual view of function that represents not only the correspondence perspective, but also an emphasis on function as a representation of covariation.

Following this recommendation, we propose that leveraging situations involving covarying quantities that students can observe, manipulate, and meaningfully investigate can foster their abilities to reason flexibly about function ideas as representations of dynamically changing events (Carlson et al., 2002). By quantities, we mean schemes composed of a person's conception of an object, a quality of the object, an appropriate unit or dimension, and a process for assigning a numerical value to the quality (Thompson, 1994). By covariation, we refer to Saldanha and Thompson's (1998) characterization. They extended Confrey and Smith's (1994) notion of covariation—as the coordination of sequences—to consider the possible imagistic foundations for someone's ability to imagine covariation. From this perspective, covariation entails a person visualizing a sustained image of two quantities' values simultaneously. This requires coupling the two quantities to form what is called a multiplicative object, a conceptual object that is a uniting of two or more quantities' magnitudes or values simultaneously (Saldanha & Thompson, 1998). This then enables one to attend to either quantity's changing value with the understanding that at every instance, the other quantity also has a corresponding value (Thompson et al., 2017).

Madison et al. (2015) characterized quantitative and covariational reasoning as "cross-cutting reasoning abilities that enable students to determine, describe, and represent patterns of change" (p. 56). As Kaput (1994) pointed out, function has its deepest conceptual roots in growth and joint variation. Reasoning about variation is crucial for the development of the function concept itself (Thompson & Carlson, 2017). Further, understanding function from a covariation perspective is key to mastering critical concepts in calculus and beyond, including limits, continuity, and average and instantaneous rates of change (Carlson & Oehrtman, 2005; Carlson et al., 2003; Oehrtman et al., 2008; Rasmussen, 2000; Zandieh, 2000). There is also a growing emphasis on a covariation approach to function in some curricula and instructional treatments (Cooney et al., 2010). For instance, one German curriculum refers to functional dependencies connected to the exploration of variation behavior (Krüger, 2019), and there are Israeli curricula that emphasize both covariation and correspondence approaches (Ayalon et al., 2017).

Consequently, covariation is emerging as a meaningful foundation on which to ground functional reasoning. Leinhardt et al. (1990) argued that attending to values changing together is natural for students, which is supported by Confrey and Smith's (1995) findings that students typically first analyze functional situations from a coordinated change perspective. Similarly, Blanton and Kaput (2011) noted that "even as early as kindergarten, children can think about how quantities co-vary and, as early as first grade, can describe how quantities correspond" (p. 14). Greeno (1988, as cited in Leinhardt et al., 1990) reported similar findings for middle-school students. Learning to track two sources of information simultaneously can be useful for constructing ratios and rates (Ellis et al., 2020), for seeing graphs as records of constant and changing rate relationships (Saldanha & Thompson, 1998), and for reasoning flexibly about dynamically changing events (Castillo-Garsow et al., 2013). Emphasizing covarying quantities in rate situations can also support students' understanding of rate as a relationship, rather than as the outcome of a calculation (Herbert & Pierce, 2012).

Following these approaches, we offer a model for the introduction of function that relies on an investigation of covarying quantities. This model is situated at the middle-school level because we are concerned with the introduction of formal function instruction, rather than subsequent investigation with students who have already constructed functional relationships. Using linear and quadratic functions as an example, we describe a sequence of conceptual activities students can undergo to construct initial models of linear and quadratic growth.

2 The Case of Linear and Quadratic Growth: Conceptual Analysis

When considering functions from a covariation perspective, what makes a function linear or quadratic? Linear functions have traditionally been defined by having a graph that is a straight line, by having the form $f(x) = \mathrm{a}x + \mathrm{b}$, or by being a polynomial function of degree zero or one (e.g., McGraw-Hill Education, 2012). Similarly, quadratic functions have also been defined by having parabolic graphs, by having the form $f(x) = \mathrm{a}x^2 + \mathrm{b}x + \mathrm{c}$, or by being a second-degree polynomial function (e.g., McDougal Littel, 2008). These characterizations define the phenomena of functional growth in terms of a function's graph or its algebraic representation, but that presupposes an extant understanding of those representations and their meanings. When initially introducing function families, we advocate for a rate-of-change approach. Specifically, a relation between two co-varying quantities is linear when one quantity changes at a constant rate compared to the other quantity's change, and a relation is quadratic when one quantity changes at a constantly-changing rate compared to the other quantity's change.

Below we introduce five conceptual acts students can undergo in order to construct linear and quadratic growth: (a) identify the attribute to be measured; (b) identify the quantities that affect the attribute; (c) imagine gross coordination; (d) imagine coordination of values; and (e) quantify the covariation. In order to illustrate these conceptual acts, we provide a specific example drawn from the context of growing area. In this context, students explore growing rectangles and triangles that extend in length from left to right (Fig. 1). Students work with a dynamic software program, such as Geogebra or Geometer's Sketchpad, to drag the right side of the figure to increase its length, and they can also adjust each figure's height. The relationships under examination are the comparison of each figure's area to the length swept. These images can be adjusted to either show or hide the length and height measures according to the goals of the particular task.

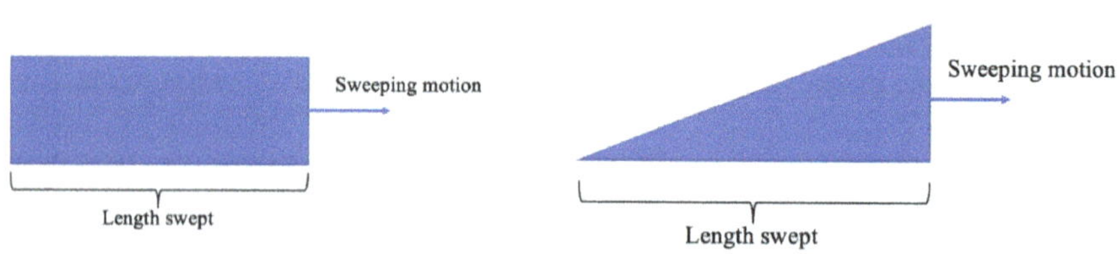

Fig. 1 A sweeping rectangle and a sweeping triangle

2.1 Identify the Attribute to Be Measured

The first conceptual act is that of *identifying the attributes to be measured*. In the linear case, the attributes are the amount of length swept and the amount of area produced as one drags the rectangle to sweep out length, which depends on the height of the rectangle. Students can sweep out different shapes, such rectangles with shorter and taller heights, to determine that the taller the height, the more area will be produced for a given amount of length swept. For the quadratic case, the attributes are also the amount of length swept compared to the amount of area produced, which depends on the height of the triangle. At this stage, determining which attributes affect the area should be explicitly identified, but do not yet need to be quantified.

2.2 Identify the Quantities Affecting the Relevant Attributes

The second conceptual act is that of *identifying the quantities that affect the relevant attributes*. In the linear case, students must quantify the amount of length swept and the rectangle's height in order to produce an amount of area. One way to encourage this can be to ask students to come up with their own measures, such as 6 cm, 2 m, or 1 in. for the rectangle's height (within our data, the students and the teacher-researchers often shifted between units from task to task). Then, they can determine area amounts for various length amounts that they choose, for instance for 1 in., 2 in., 5 in., 12 in., or 24 in. Through this process, students may notice that the area increases by a constant amount for every additional inch of length swept. It may be useful at this stage to encourage students to explore with height values other than 1 in. (or 1 unit of any kind), because relying on a height of 1 may lead to the incorrect assumption that the amount of area swept is always equal to the amount of length swept.

In the quadratic case, students would need to quantify both the height and the length, because both measures change as the triangle grows. For this reason, it may be effective to first provide students with a given triangle that has a 1:1 ratio of height to length. Students would need to be able to determine the area of a triangle, and a triangle with a 1:1 ratio would allow them to generalize that the triangle's area would be one half the height times the length for any given length. In cases in which the ratio of the triangle's height to length is not 1:1, students will benefit from being able to conceive the height as a ratio of the length, which we discuss in more detail below. At this stage, some students may notice a squaring relationship between the triangle's length and area, but others may not. Students might also need support with appropriately determining the triangle's area, and not, for instance, defaulting to determining the area using the formula for finding the area of a rectangle.

2.3 *Imagine Gross Coordination and Coordination of Values*

As students quantify and determine ways to calculate area, the idea of rate may begin to emerge, even if implicitly. Especially when determining area values for length values that increase by 1 unit, which is a common approach, students may use language such as "for every" or "every time". This leads us to the third and fourth conceptual acts, that of imagining *gross coordination* and *coordination of values*. Thompson and Carlson (2017) identified six major levels of covariational reasoning, and as students construct covarying relationships between quantities for the growing figures, we have found that they may begin to do this either through gross coordination of values or coordination of values (Ellis et al., 2020). According to Thompson and Carlson (2017), gross coordination of values entails the formation of a gross image of quantities' values varying together, such as, "the area increases as the length increases." This act may be more prevalent in the quadratic case than in the linear case, due to the complexity of quantifying the relationship between area and length for the triangle. Students might notice in both cases that the area grows larger as the length grows longer, or even, in the quadratic case, that the amount that the area increases is growing larger as the length grows longer.

Coordination of values entails coordinating the values of one variable, in this case length, with values of another variable, in this case, area, with the anticipation of creating a discrete collection of pairs. Thompson and Carlson (2017) indicated that at this level, no thought is given to intermediate values. Thus, in the linear context, students could indicate that the area would grow by the same amount, such as 1 m^2 or 3 m^2 "every time", but the notion of "every time" would be a discrete marker, such as flipping over a card, with each new flip or increase entailing an additional amount of area without any sense of change within the amount of length pulled. Graphs that students produce comparing area and length might be sets of points, without those points connected to form a line. In the quadratic context, coordination of values will be more challenging. One might see students first engaging in a nascent form of coordination of values by quantifying increments of length before fully quantifying corresponding increments of area. For instance, students might recognize that the area grows more for each inch of length swept than it grew the previous time, identifying an implicit idea of a changing rate of change without yet quantifying how the rate of change is changing. Students might also calculate multiple length-area pairs, but those pairs will remain a discrete set without attention to how area and length grows from one pair to the next.

2.4 *Quantify Covariation*

The fifth conceptual act is that of *quantifying the covariation*, which has several constituent parts and differs depending on whether students are constructing linear or quadratic growth. We discuss each in turn.

2.4.1 Quantify Covariation for Linear Growth

The first part of quantifying covariation for linearity is *creating a ratio*, by which we mean either a multiplicative comparison of two quantities, or a joining of two quantities in a composed unit (Lobato & Ellis, 2010). A multiplicative comparison includes a consideration of how many times as large as one thing is than another, for instance, determining for a rectangle with a height of 3 ft and a length of 5 ft that the area, 15 ft^2, is three times as large as the length. A composed unit, in contrast, is the joining of two quantities to create a new unit, which students may then iterate or partition to determine other equivalent ratios (Lamon, 1994). A student could, for example, create a composed unit of an area-length pair, 3 ft^2:1 ft, and then iterate that unit to form equivalent ratios such as 6 ft^2:2 ft, 9 ft^2:3 ft, 18 ft^2:6 ft, and so forth. Forming a composed unit is a more rudimentary act than the development of a multiplicative comparison, but it is often a foundational act that can be leveraged to build up more robust ratio understandings. As Lobato and Ellis (2010) noted, forming a ratio as a measure of a real-world attribute, such as amount of area swept, requires isolating that attribute from other attributes (such as the amount of length swept and the rectangle's height) and understanding the effect of changing each quantity on the attribute of interest.

The second part of quantifying covariation is *developing equivalent ratios*. One way to support the development of equivalent ratios is to have students identify as many area:length pairs as they can for a given rectangle. As we noted above, this can initially occur through the iteration and partitioning of a composed unit. For the example rectangle with a height of 3 ft, for instance, students could be encouraged to create very large pairs (such as 636 ft^2:212 ft or 3,000 ft^2:1000 ft) and very small pairs (such as 1 ft^2:1/2 ft, 0.3 ft^2:0.1 ft, and 0.03 ft^2:0.01 ft). This activity can encourage multiplicative comparisons as well, by multiplying each part of the composed unit to create new ratios, and by attending to the fact that the area is always three times as large as the length for each equivalent ratio. Students can also be encouraged to develop a more general strategy to determine the area for any unspecified length swept, x, which requires the recognition that the corresponding area will be $3x$.

The third part of quantifying covariation is *creating a rate*. By rate, we mean a reflectively abstracted constant ratio, as described by Thompson and Thompson (1992). If a ratio is a multiplicative comparison of two quantities, it requires viewing two such quantities as changing together, and treating the collection of equal ratios they generate as a single quantity of its own. It symbolizes the ratio structure as a whole while giving prominence to the constancy of the result of the multiplicative comparison. For instance, in order to understand the ratio 3 ft^2:1 ft as a rate, one would need to have an image of change such that it represents an equivalence class of ratios, with the unit ratio simply being a convenient measure of expressing the growth in area for a standard unit of length. A student with this conception would be able to imagine any length, including an infinitesimal length, as sweeping out an area amount that was three times as large.

The final conceptual act for constructing linearity is that of *representing the constant rate of change in a general form*, whether that be graphical, algebraic, or another form. As we mentioned above, when reasoning with coordination of values, students' graphs may be collections of discrete points. As students create ratios and ultimately rates, their graphs may also evolve to be smooth lines. That said, given differences in instruction, many students may have already learned to represent any linear relationship as a line, regardless of their own stage of ratio reasoning. A line graph is therefore not a guarantee that students have constructed a rate, but one can ask students to think about what each point on a line represents, as well as what happens in between any two given points, in order to encourage the development of equivalent ratios and rates. Students should also be able to ultimately express area:length relationships for unspecified length amounts by writing algebraic expressions such as A = 3L.

2.4.2 Quantify Covariation for Quadratic Growth

In order to quantify the covariation in the quadratic context, students must first quantify growth across increments. Initially, this may happen through *coordination of values,* by creating a collection of area:length pairs and then by comparing increases across same-size increments. For instance, consider a growing triangle with a height to length ratio of 1:1. Students could create multiple length:area pairs, such as (1, ½), (2, 2), (3, 4.5), (4, 8), and (5, 12.5). By attending to the growth in area for each additional 1-in. length increase, one can notice that the area grows by a constantly-changing amount: First by 1½ in.^{2}, then by 2½ in.^{2}, then by 3½ in.^{2}, then by 4½ in.^{2}. Quantifying the growth across increments can enable students to identify that the area is changing by a constantly changing amount, growing by an additional 1 in.^{2} in area compared to the prior increment's growth. Articulating that the amounts of area increase change at a constant rate is key for understanding quadratic growth.

It may be natural for students to pick a standard unit of increase for the length, such as 1 in., and then not attend to the fact that the corresponding increases in area are specific to that unit of increase. This is particularly easy to do when the unit of increase in length is 1. As a way to combat this, one can encourage students to engage in repeated reasoning (Harel, 2007) by creating other collections of length:area pairs, but for different uniform increments of length, such as 2 in., half an inch, or 0.1 in.. This can also encourage the next conceptual act, which is to *generalize across increment sizes.* By generalizing across increment sizes, one recognizes that the ratio of change is constantly changing, regardless of the increment. What that value is depends on the increment size itself, but the generalizable feature is that the ratio always changes at a constantly-changing rate. It can be challenging for students to develop this generalization. In order to do so, it can be helpful to try to imagine changes between successive values of accumulated area and accumulated length, through drawing pictures and creating graphs. This can support chunky-continuous reasoning (Thompson & Carlson, 2017), in which one envisions simultaneous changes between two quantities by imagining each quantity's value changing by intervals of a fixed size. Ultimately, it will be useful to also support students to reason about very small length increments, to encourage reasoning with smooth continuous covariation (Thompson & Carlson, 2017).

A closely related conceptual act is to *create a constantly-changing rate of change*, in which students must understand that the rate of the rate of change of area, when averaged across some increment, will remain invariant across all increments. That is, students can articulate that the rate of change of area is changing by a constantly-changing rate, regardless of increment size, and that this characterizes the growth in area. Again, reasoning with very small increments could support the development of this rate of change. Further, the rate of change can be understood as dependent on the slope of the triangle's hypotenuse, or the ratio of the triangle's height to length. For instance, for the 1:1 triangle, the rate of increase of the area is changing by 1 in^2 for a 1-unit increase in length; the triangle's slope is a convenient way to express a unit ratio.

The final conceptual act is *representing the constantly-changing rate of change in a general form*. To do so algebraically, one can again rely on the slope of the triangle's hypotenuse in developing a general expression. For instance, for a 1:1 triangle, one can ask students to determine the triangle's area for any given length swept, x. First, one can conceive of the height in relation to the length, in this case $1x$. Then the area is determined in relation to the length: A = ½ base * height. In this case the base and height are both x, so $A = \frac{1}{2}x^2$. In general, for any triangle with a height to length ratio of a:b, $A = \frac{a}{2b}x^2$.

3 Data Examples: Students' Reasoning with Linear and Quadratic Growth

We conducted two teaching experiments, each with a pair of students, in which they explored the growing figures context. The first pair of students, Olivia and Wesley, were both eighth-grade students (age 13) who had not yet taken an algebra course. The second pair of students, Homer and Barney, were siblings. Homer was a rising 9th-grade student (age 14), and Barney was a rising 7th-grade student (age 12). Homer had taken a beginning algebra course, but Barney had not. Both pairs of students engaged with both the linear and quadratic tasks. Wesley and Olivia met 10 times for approximately 60 min each time, and Homer and Barney met 5 times, with each session ranging between 60 and 90 min. We will draw on excerpts from both teaching experiments to illustrate students' conceptual activities for constructing linear and quadratic growth.

3.1 Identifying the Attribute and the Quantities

The initial introduction to the growing rectangle tasks was within the context of a paint roller rolling out paint (Fig. 2). After drawing painted regions, we asked the students to consider the attributes that would affect the area of the painted rectangle. In order to determine the amount of area produced for both the rectangle and the triangle, the students pointed to the amount of length swept, and the height of each figure. For instance, Wesley explained that the amount of area produced depended on "How big it is" (pointing to the height of the rectangle), and Olivia added, "The distance", meaning the total length swept. When determining how the quantities length and height affected the area, we asked the students to assign their own values to the relevant quantities. Wesley said, "I decided I would think the height would be 1 m. So, for every 1 m it's pulled, it [the area] gets 1 by 1 m."

In contrast, Homer decided to make the height of his rectangle 3 m. He explained, "This (pointing to the amount of area swept out with a length of 2 m) is two times as far as this (pointing to the amount of area swept out with a length of 1 m). Three times one is three, three times two is six. Six is twice as much as three." Barney responded, "It's going up the same", which we infer he meant that the rectangle is adding the same amount of area for every additional meter swept. Homer agreed, stating, "If it's 3 m far, three times three, it's going up at the same rate." By this agreement, we infer that Homer, like Barney, was thinking about "the same" from a calculational perspective. He noticed that the rectangle first had 3 m^2 (for the first meter of length), then a total of 6 m^2 (for the second meter of length), and then a total of 9 m^2 (for the third meter of length). He had already emphasized multiplication by

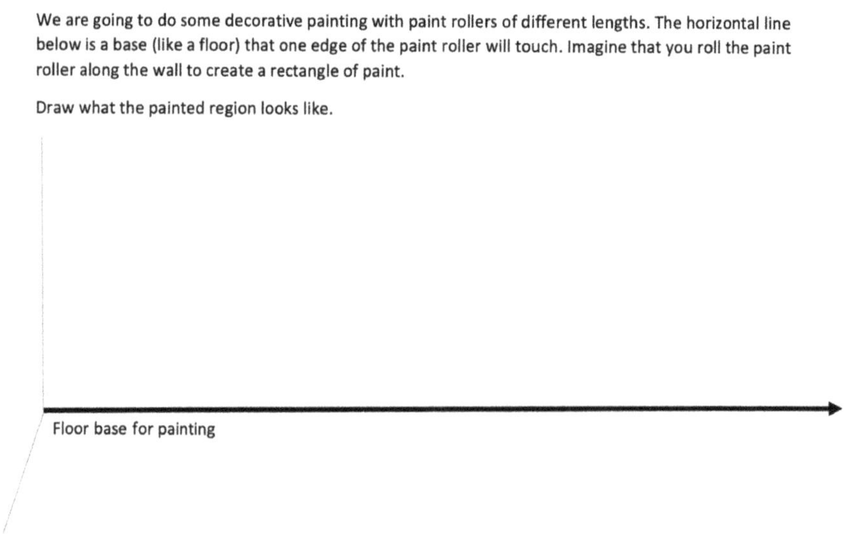

Fig. 2 Drawing the painted region

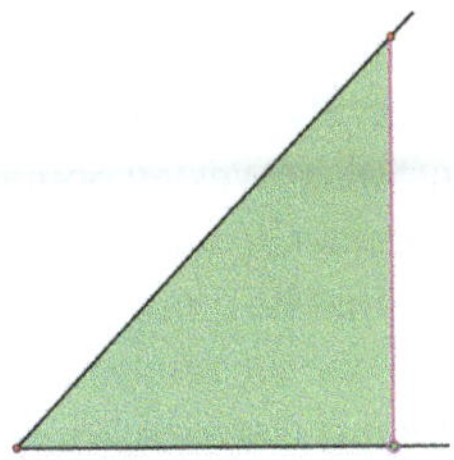

Fig. 3 Screen capture of a movie depicting the 1:1 growing triangle

three ("three times one is three, three times two is six"), and appeared to understand that the rate of increase in area was three for every meter of length pulled.

In the quadratic case, the students also began by exploring a triangle with unspecified dimensions. This triangle looked like it had a 1:1 ratio of height to length (Fig. 3), and Wesley therefore assigned those values without specifying units. He said that he could determine the area for many different lengths by calculating "the height times the length divided by two". In fact, all four students determined ways to calculate the area of the right triangle for lengths and heights they assigned, relying on the triangle area formula, area = ½ base * height. Homer also noticed a squaring relationship between the triangle's length and area, by sketching a square with twice the area of the triangle: "It keeps squaring itself. This (the area of the square he sketched) is one. And then (referring to a triangle with a length and height of (2) it's going to be four total, and then (referring to a triangle with a length and height of (3) it's going to be nine total." Barney replied, "So the square keeps squaring, but the triangle is half the square, so it's (the length) squared divided by two."

3.2 *Imagining Gross Coordination and Coordination of Values*

We saw more evidence of gross coordination in the quadratic context than in the linear context, which is likely due to the fact that quantifying change is more challenging in the quadratic context. As an example of gross coordination, after watching the movie of the growing triangle (Fig. 3), we asked the students to describe how the triangle was growing. Olivia said, "The longer it gets, the more area there is (than there was before)." Wesley explained, "The area grows more as you pull it (pointing to the length) out." Similarly, Barney described the growing area by stating "The amount the triangle is growing is also growing", by which we infer that he meant that the amount of area added on for each amount of length was growing larger as the triangle grew. At this stage, gross coordination is an appropriate way to begin to make sense of the triangle's growth, and it also raises the natural question of how to actually quantify the nature of that growth.

In the linear context, the students were able to reason by coordination of values early on. In fact, notice that in the above conversations, the idea of rate began to emerge, even if implicitly. For instance, Wesley pointed out that the rectangle gained a square meter in area *for every* meter it was swept. When exploring the triangle, Homer identified a series of areas that he compared multiplicatively, by saying "six is twice as much as three", each one in relation to the length increasing by one unit at a time. Here it appeared that the students were coordinating the increases in area with length increases, but there was no evidence that they were attending to intermediate values between each additional meter in length. We also saw evidence that the students' reasoning was discrete in their early graphs. For instance, after having drawn and considered a number of different rectangles, we provided the following graphing task to the students: "Draw a graph that shows how much paint has been painted so far, as the paint roller glides from left to right along the length of the wall." The students created graphs based on individual points (Fig. 4a), and only after discussion did they decide to connect the points into a line. Olivia was the one exception: She created a smooth graph (Fig. 4b), and explained, "I knew it would line up (pointing to the tally marks she placed on each axis) because for every length that you've pulled, it should be the same amount of area."

The students also demonstrated a nascent form of coordination of values in the quadratic context. After watching a movie of a growing triangle (Fig. 3) and making some rough sketches comparing the area to the length, we asked the students, "Make a graph of it on graph paper. Do it accurately, as accurately as you can." All four students plotted points to construct a graph, and then connected the points afterwards; Wesley's is shown in Fig. 5. In explaining why his graph looked like it curved up, Wesley explained, "Every inch it goes (referring to the length), it, it goes, it covers more area for that inch, so it keeps getting steeper and steeper." (Although Wesley had previously referred to meters, here he switched his unit to inches; the students often shifted between units in this manner.) Here he quantified the increments of length, but the corresponding amounts of area were not yet quantified beyond an

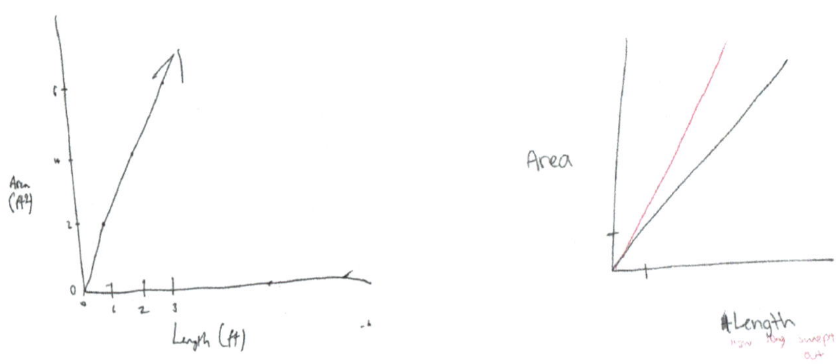

Fig. 4 Homer's **a** and Olivia's **b** first graphs comparing the area and length of growing rectangles

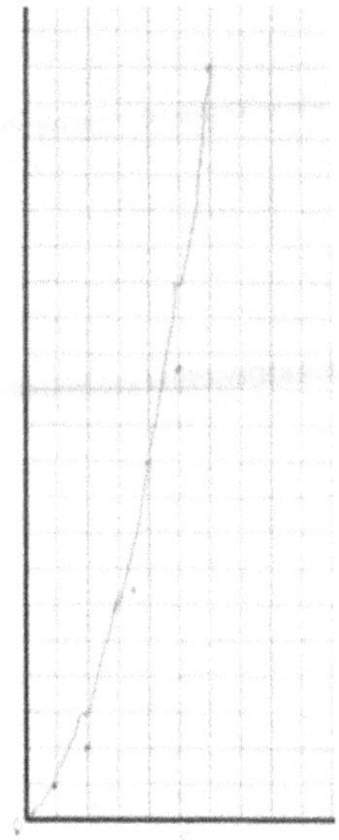

Fig. 5 Wesley's graph comparing area and length of the 1:1 growing triangle

idea that each amount of area (from one point to the next) was greater than the previous amount. We saw more explicit quantification and comparing in the next conceptual act, quantifying covariation.

3.3 *Quantifying Covariation*

3.3.1 Quantifying Covariation for Linear Growth

Recall that the first part of quantifying linear covariation is the creation of a ratio. It may appear that students have developed a ratio if they can articulate a unit ratio, which all four students were able to do early in the teaching experiments. For instance, the teacher-researcher gave Homer and Barney a task to graph the area and length of a rectangle with a height of two ft: "I'm going to give you one that's…2 ft tall, and it's sweeping out a path like this (draws a growing rectangle similar to that shown in Fig. 1)…graph the area per length again for a 2 ft tall paint roller." Both students graphed a linear function (see Fig. 4a for Homer's graph), and they explicitly referenced both area and length for the rectangle:

Barney: Two square ft per ft.
Homer: Two square ft of area per-
Barney: Per length.
Homer: Per ft of length traveled.

Both Barney and Homer articulated a composed unit (Lamon, 1994), i.e., a joining of the two quantities, area and length, to create a 2 ft^2:1 ft unit. Further, when the teacher-researcher asked the students what would happen to the rectangle's area when another foot was added on to any arbitrary length, Barney responded "You're going to add two more", by which we infer he meant 2 ft^2 of area. This suggests that Barney understood that the additional area would always be 2 ft^2, regardless of how much length had already been swept.

However, from this exchange alone, it is not clear whether the students had constructed an iterable composed unit, or could make a multiplicative comparison. One way to encourage iteration and multiplicative comparisons is to introduce rectangles of different heights. For instance, we provided the students with the following task asking them to draw a painted region with two different heights (Fig. 6):

Barney decided that his paint roller would roll out a rectangle with a height of 2 m for 2 s, and then sweep out at a height of 4 m for the next two seconds. He also decided to create another area-length graph for this scenario. Homer, in contrast, chose an initial height of 1 m, and then doubled the height to 2 m. The students explained their thinking:

Barney So the first time, it would go up to 4 m^2 at 2 ft (in length). Then it would go up two (more ft in length), it would add, it would go up to 12 m^2 (total for the entire figure). I need a bigger graph.

Homer Okay, so what I did at the beginning, I said okay. It rolls out at a meter per second for speed, the thing is 1 m tall to start with. So, if it rolls 2 m in 2 s, times 1 m, after that the area would be, after 2 s, it would have rolled 2 m, which would put it at 2 total m^2.

TR Okay.

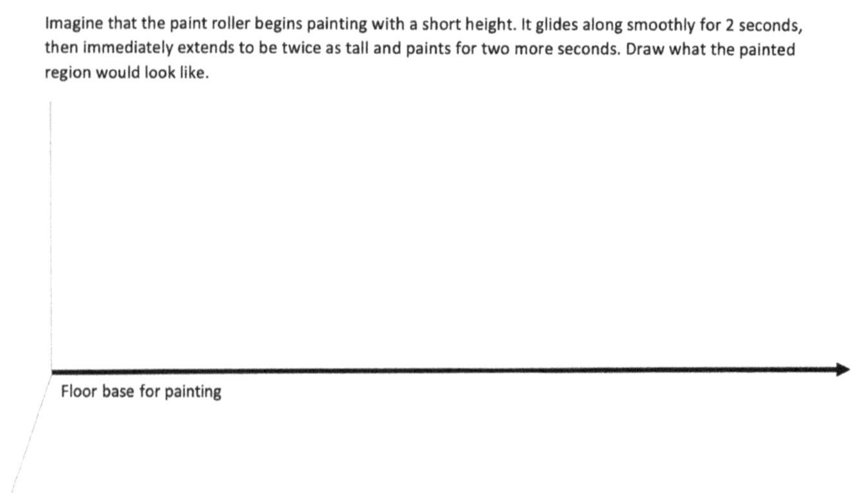

Fig. 6 Task introducing two rectangles with the second height twice as large as the first

Homer And then after that it starts rolling, it extends to be twice as tall and paints for two more seconds. So in that case, it would, I lost my train of thought. So, it would start for the next second it would go, it would get two extra m squared after rolling a meter, since it's twice as tall. It's 2 m tall now. So if you roll 1 m, you get times 2 m of height, you get 2 extra m squared. And then for the next second, it's the same.

By stating "it would go up to 4 m^2 at 2 ft", Barney doubled his composed unit of 2 m^2:1 m (his use of ft appeared to be a mis-speak). Similarly, Homer knew that after 2 s, the figure's area would be 2 m^2, and then once the height doubled, the area must also double, so "you get 2 extra m squared". He could also, therefore, double the area as a consequence of doubling the height.

Tasks that can encourage the development of equivalent ratios, the next step in quantifying covariation for linear growth, are ones that ask students to generate multiple area:length pairs. For instance, we provided the students with a task in which they could choose the rate of change of the area: "Draw a sweeping figure in which the rate of change of area per amount of length swept is **constant**. The figure accumulates area at _____ cm^2 per _____ cm." Olivia and Wesley chose a rectangle with a height of 4 cm, and Olivia wrote that the figure accumulates area at 4 cm^2 per cm. We asked her to generate other values that she could place in the blanks, and she iterated the composed unit 4 cm^2:1 cm to create other equivalent ratios such as 8 cm^2:2 cm and 16 cm^2:4 cm. Wesley also created the ratios 12 cm^2:3 cm and 2 cm^2:0.5 cm. Both students then worked together to develop more ratios, including 0.4 cm^2:0.1 cm, 0.8 cm^2:0.2 cm, and 1 cm^2:0.25 cm. In creating these ratios, the students moved beyond iterating and partitioning actions to create multiplicative comparisons. For instance, Olivia, relying on the composed unit 4 cm^2:1 cm, determined the 16 cm^2:4 cm ratio by multiplying: "I did four times four (cm^2) is 16, and so you have to do four times the number on the right (1 cm) to get four (cm)." Thus, Olivia mentally truncated the act of iterating 4 cm^2:1 cm four times by directly multiplying both the area and the length by 4. Wesley also began to rely on multiplication, but did so by multiplicatively comparing the area to the length. For instance, for the 2 cm^2:0.5 cm ratio, he took the length value, 0.5, and multiplied it by 4 to get 2 cm^2. Both students indicated that these ratios represented the same rate of change of area per length swept because, as Olivia explained, "Each of them could either be reduced to or bumped back up to four to one." Wesley agreed with Olivia and offered a geometric connection, stating that the rate of change of the area should not change because the height does not change: "The height doesn't, it's not a different shape, it's the same." Wesley also offered a generalized strategy to determine the area for any length swept: "If this (the length) is x, to get how much area accumulates by, you do x times four."

Wesley's use of an unspecified length x, combined with the creation of ratios with length values less than 1 cm, suggest that it may have been possible that he was imagining an equivalence class of ratios, but it was unclear whether this was the case. Wesley's understanding that all of the ratios were instantiated in the same rectangle height does suggest that he may have seen the height as a representation of the rate of change of area with respect to length swept, such that it does not depend

Fig. 7 Olivia's rectangle (**a**) and Wesley's vertical line (**b**)

on any specified length. Recall that creating a rate is the third part of quantifying covariation.

In order to help determine whether the students had developed a rate, we provided them with the following task: "Now imagine that you have a line, and it's about to start sweeping out an area at a rate of 3/2 cm^2 per cm swept. Draw that line." With this task, we explained to the students that the line was the height; we wanted to see whether they would simply draw a vertical line with a height of 3/2 cm, or whether they would need to produce a rectangle with a length. Olivia drew a vertical line segment with a height of 1.5 cm and an unspecified length amount, which she called x (Fig. 7a): "I did a line with thickness so that we could write down the height and then the length." We conjecture that Olivia needed to produce a length associated with the height, because she did not yet have a conception of an infinitesimal amount of length. Absent this conception, Olivia required a given (but unspecified) length x, along with the height of 3/2 cm, in order to conceive a ratio of change. In contrast, Wesley drew only a vertical line segment with a height of 1.5 cm (Fig. 7b), explaining, "It hasn't swept out any yet, but it's about to." He then lightly sketched both an extending horizontal line and an extending angled line, explaining that he could imagine the height sweeping out more than one possible figure. For instance, Wesley could imagine a rectangle with a constant rate of change of the area per length swept, but he could also imagine a triangle or a trapezoid, each which would have a constantly-changing rate of change.

As evidence of the final conceptual act, representing the constant rate of change in general form, all four students produced smooth graphs as representations of the area:length relationship; a typical example can be seen in Homer's graph of two different rectangles in Fig. 8. All four students could also express the area:length relationship for unspecified length amounts, by writing expressions such as "Area = $4x$" or "A = 2.5L". Developing general algebraic expressions for quadratic functions proved more intricate, as seen in the next section.

3.3.2 Quantifying Covariation for Quadratic Growth

In order to encourage the first conceptual act, quantifying growth across increments, we first asked the students to produce a graph comparing the area to the length of a sweeping triangle, such as the one in Fig. 3, but with a height to length ratio of 2:3. Olivia and Wesley calculated the area of different triangles with lengths of 1 in. through 5 in. and plotted points, such as (1, 1/2), (2, 4/3), (3, 3), and (4, 5 1/2), to create graphs (see Fig. 9 for Olivia's graph). We then asked Olivia to characterize

Fig. 8 Homer's graph of two area: length relationships

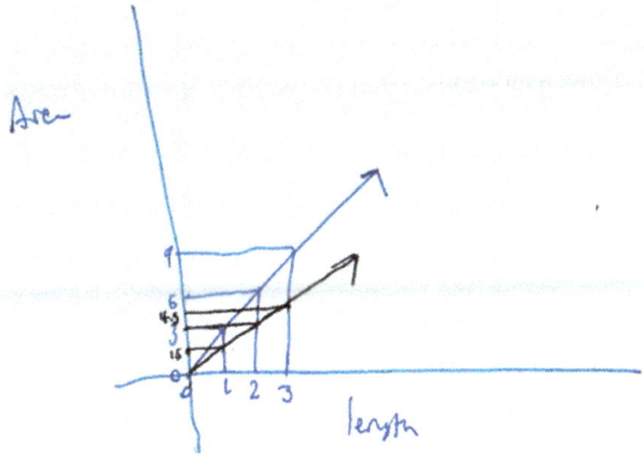

how steep her graph was. Olivia inspected her graph, comparing increases in area across same-size length increments of 1 in, and noticed that for each additional 1-in. increase in length, the area grew by a greater amount than for the previous inch. Olivia tried to quantify this amount by drawing in markers on her graph (Fig. 9): "I'm trying to visualize it, but, yeah. The farther you go along, the steeper it becomes than the first one. So it just keeps getting steeper. So this (pointing to the region between $x = 1$ and $x = 2$) is steeper than this (pointing to the region between $x = 0$ and $x = 1$). But this (pointing to the region between $x = 2$ and $x = 3$) is steeper than this (pointing to the region between $x = 1$ and $x = 2$) *and* this (pointing to the region between $x = 0$ and $x = 1$)." Olivia was unsure, however, how to quantify these area increases.

In order to follow up on this idea, we asked Olivia and Wesley to determine precisely how much new area would be added for each section of length added by providing the following task: "Yesterday Olivia explained why the graph of the area swept for the growing triangle would be a curve by describing how each new section or piece adds on more area than before. *How much* new area is added on for each section of length added? For example, let's think about the 1:1 triangle we thought about yesterday. For each additional increment in length, how much area is being added?" Both students chose 1-in. increments, and calculated area values for each 1-in. "column" on the triangle (see Fig. 10 for an example of a 1-in. "column" within the triangle, which was on a grid). For each column, the students calculated area values, ½ in^2, 1½ in^2, 2½ in^2, 3½ in^2, and so forth as the triangle swept out. (In Fig. 10, for instance, the highlighted column is between length $L = 3$ in. and $L = 4$ in., and its area is 3½ in^2.) The students then saw that the increase in area for each additional 1-in. "column" was one square inch greater than for the previous column, articulating an amount of area increase that grew linearly. Olivia stated, "Each time you increase by one column, you get an additional one unit squared in area for that column's area compared to the previous column."

Barney and Homer engaged in the same exercise, but for a triangle with a height to length ratio of 2:1. Barney decided to create a table of length and area values,

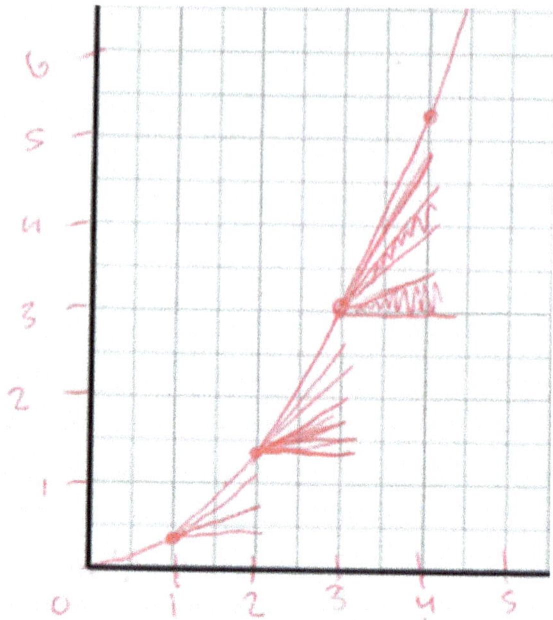

Fig. 9 Olivia's graph comparing length and area increases for a 2:3 triangle

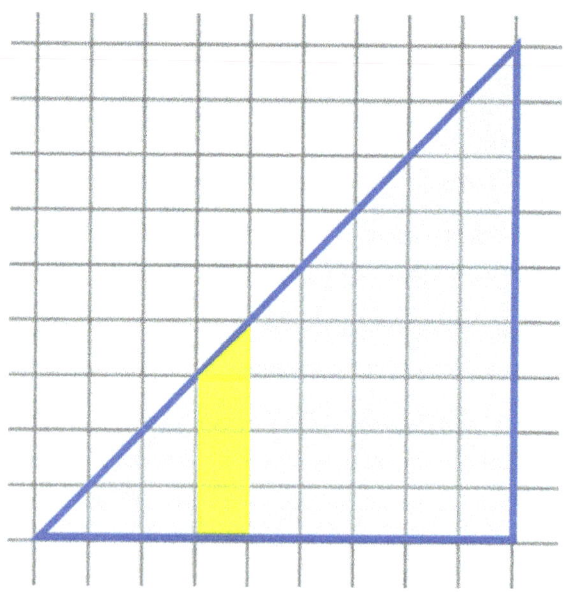

Fig. 10 The growing 1:1 triangle with 1-in. columns

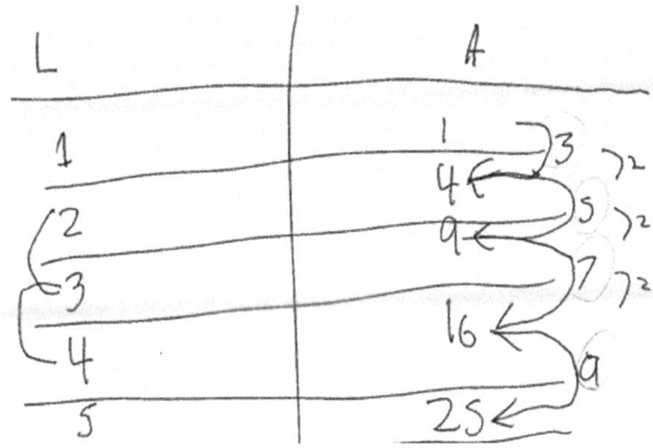

Fig. 11 Barney's table comparing area increases for 1-ft length increases

increasing by 1-ft increments, and then compare the amounts of increase in area for each 1-ft increase in length (Fig. 11). Barney likely focused on increases in area because he was attempting to determine how much area was added for each 1-ft increase in length. Barney quickly saw a pattern, noticing that the area grew by three, then five, then seven, then nine: "Yeah, it's, it's odd numbers. It's odd numbers but counting by two." The teacher-researcher asked Barney what those numbers represented in terms of the triangle, and Barney answered, "That here, in between lengths one and two, it (the area) increases three." Homer interjected to add, "And between two and three it'll be five, three and four it'll be seven."

After exploring the area increases for another right triangle with a height to length ratio of 2:3, Homer generalized the relationship of area increases for 1-unit increases in length for any triangle: "What you add more each time is twice the, twice whatever the area is when L equals one." He justified this geometrically, explaining with a 1:1 triangle that he used as a generic example (Fig. 12), "This (the first shaded triangle) is half of the (blue) rectangle and this is the whole rectangle. And then you add an extra whole rectangle at a time." Note that Homer has articulated what we can interpret as constantly-changing unit rate of change, but it is not clear whether he understood that the area's rate of change would be constantly changing for any increment size.

Recognizing that the rate of change is constantly changing regardless of increment size involves the next conceptual act, that of generalizing across increment sizes. In order to generalize across increment sizes, it is useful for students to be able to imagine changes between successive values of accumulated area and accumulated length. Early in the teaching experiment, Wesley and Olivia did not appear to have this image. For instance, when graphing the relationship between the area and length of a triangle, both students produced piecewise linear graphs and did not believe the increment size would affect the graph:

TR Would the graph look the same or different if you weren't choosing 1 in., 2 in., 3 in., but say you were choosing half an in., 1 in., 1 and a half in., 2 in.?

Olivia It should look the same shape, I think.

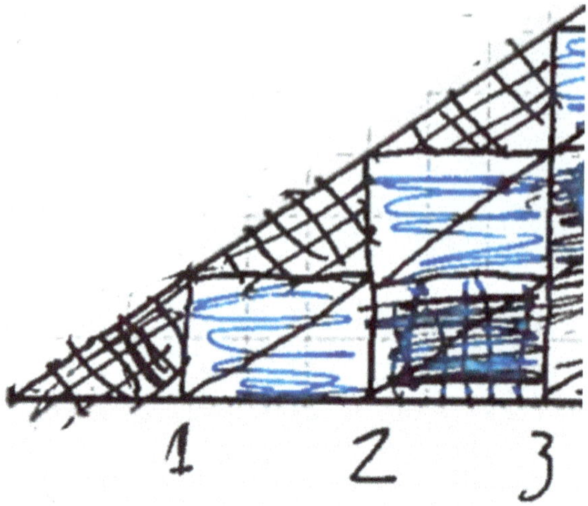

Fig. 12 Homer's geometric justification for the rate of change of the area

TR Okay. What about if you were choosing, like, 0.1 in., 0.2 in., 0.3 in., 0.4 in.? In other words, what if in between 0 and 1 you graphed 10 points? Would this portion of the graph look the same or different?

Olivia I think it'd look the same.

Olivia did not appear to envision length swept and area swept in chunks that tacitly contained intermediate values. If she had, she would have been able to re-size the length of her chunk, imagining a different graph with more line segments. However, after having developed chunky-continuous covariation (Thompson & Carlson, 2017), the students' beliefs about the effect of changing the increment sizes changed. For instance, when investigating a growing triangle with a height to length ratio of 2:5, Olivia again produced a graph with 1-in. increments for length, stating that "each line between each increment is just getting steeper." This time, though, Olivia now believed that changing the increment size would change the graph: "If you made the increments even smaller, like into 0.1 as your first point, then I think it'd be, all the little lines together I think they'd make a very subtle curve, but relatively straight. So when I did it with the increments as 1, I see them as straight, but if they were smaller they might look as if they were curved to make one big curve."

When asked to consider other increment sizes for comparing amounts of accumulated area, Olivia checked by calculating amounts of increase for 2-in. "columns" and discovered that although the amount of increase was different than for a 1-in. column, it still increased at a constantly-changing rate. Given the importance of encouraging reasoning about very small increment sizes, the teacher-researcher also asked whether they thought this phenomenon would hold true for a different increment size, such as a tenth of an in., or a thousandth of an inch. Both Wesley and Olivia said yes. Olivia explained, "It's going up at a consistent rate", gesturing to the triangle's area. Similarly, Homer and Barney also decided that the area would grow by a constantly-changing amount regardless of the increment size, with Homer declaring, "It would be the same". Homer explained that what that value would be would depend on the size of the first "section" of length: "For the first, whatever the first section is…if you were doing it every half inch, every half inch would be twice what the area is at the first half inch."

When creating a constantly-changing rate of change, the next conceptual act, we mentioned earlier that students must understand that the rate of the area's rate of change, or the second rate of change, remains invariant across all increments. That rate is dependent on the slope of the triangle's hypotenuse, specifically, the ratio of the triangle's height to length. Recall the task related to Fig. 10, in which students had to determine how much new area is added for each additional unit of length. We provided additional tasks of the same nature, but for triangles with different height to length ratios. The students determined the constantly-changing rate of change of the area per unit of length for triangles with height to length ratios of 2:3, 3:2, 5:4, and 2:5. Engaging in this form of repeated reasoning (Harel, 2007) across multiple triangles supported the students' noticing that the slope was a convenient way to express a unit ratio. When describing the rate of change of area, Wesley stated that for any triangle, "It increased by the slope." Barney related the slope to a triangle's height and length: "It's just the ratio (of the height to the length)." Further relating the slope to the triangle's area, Barney further explained, "So we're trying to get from slope to area. So you'd take the slope, you'd multiply times ½, or multiply the denominator by 2, then multiply that by L squared."

In order to encourage the final conceptual act of representing the constantly-changing rate of change in a general form, we provided the students with a triangle with a height to length ratio of 2:3, and asked them how they could determine the triangle's area for any given length swept, L. Barney indicated that first, he would need to know the triangle's height, which could be determined by considering the slope: "If you multiply L by two-thirds, that's going to be what the height is. Because height is two out of three of L." To determine the area, then, Barney knew that he had to use the formula for the area of a triangle, A = 1/2 H, which he expressed as "A = (2/2 L^2)/2". Barney also generalized this idea to any growing right triangle: "So if R stands for ratio (of height to length), it would be R L-squared over two equals area for triangles." Wesley and Olivia also expressed this idea generally, considering a generic right triangle with a height to length ratio of a:b, which Wesley expressed with the equation "$\left(\frac{a}{b} \cdot l \div 2\right) \cdot l = A$". He explained, "L would be the length of, you know, anything. Like L would be the length of any…it'd always be the length." Wesley understood the literal symbols "*a*" and "*b*" to represent specific but unknown height and length values, but "*l*" as a changing, unknown length for a triangle in the process of sweeping out. Olivia also explained that for an equation A = ml^2, the constant m would be half the slope: "It would always be the slope divided by two times L times L." In this manner the students were able to use the triangle's height to length ratio to determine a general equation to find its accumulated area based on the amount of length swept.

4 Task Design Principles for Supporting Function Reasoning Through Covariation

We have developed a set of task design principles to support the conceptual activities we want students to undergo in order to develop an understanding of functional growth. These design principles are intended for an introduction to functional reasoning, rather than for subsequent more advanced investigation into functional relationships. However, many of the principles are relevant for students' exploration of functional relationships at all levels. We propose five principles: (a) leverage contexts with continuously covarying quantities; (b) develop covariation before allowing calculation; (c) choose exact relationships; (d) choose genuine contexts; and (e) provide opportunities for visualization, manipulation, and justification. We discuss each in turn.

4.1 Leverage Contexts with Continuously Covarying Quantities

Thompson and Carlson (2017) argued that reasoning about the values of continuously covarying quantities played a key role in the development of the concepts that led to how we conceive of function relationships today. Several of the conceptual acts we outlined in the previous section would be difficult to accomplish with exploration of quantities that did not covary continuously. For instance, consider a typical quadratic figural pattern, such as seen in Fig. 13 (Chua & Hoyles, 2010):

Tasks such as these can support students' generalizing activities, particularly in terms of viewing each diagram in terms of its components in a manner that relates the shape number to the number of squares. One can express this relationship as $2n^2 + 2n$

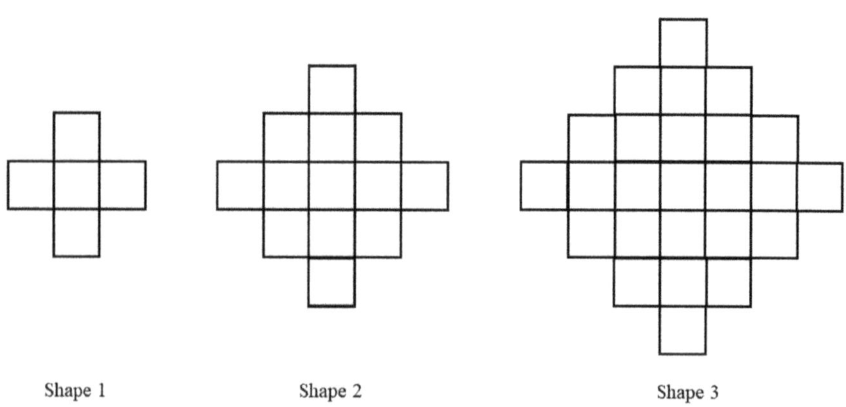

Fig. 13 Quadratic figural pattern (adapted from Chua & Hoyles, 2010, p. 71)

$+ 1$, or as $n^2 + (n + 1)^2$, or as $(2n + 1)(n + 1) - n$, among other options. However, when investigating change in a discrete situation, students do not have access to an image of what occurs between stages. What occurs at Shape 1.5, or Shape 1.001? At best, a student could reason with coordination of values (Thompson & Carlson, 2017), creating a discrete collection of pairs. When reasoning within discrete contexts, students cannot change the increment size, much less make a generalization about a ratio of change for a given increment size.

The form of covariational reasoning one might engage in with discrete contexts such as the above pattern is reminiscent of Confrey and Smith's (1994) discussion of examining a function in terms of coordinated changes of x- and y-values. This allows a student to attend to successive states of variation, but not to imagine what happens in between these states, imagine change in y for any given increment of x, or imagine change in y as happening simultaneously with change in x as both variables vary. For this reason, we urge the use of contexts with continuous covariation, such as growing rectangles and triangles, that enable students to identify relevant attributes, determine ways to measure those attributes, consider change across different increment sizes, and ultimately construct constant and changing ratios and rates.

4.2 Develop Covariation Before Allowing Calculation

Reasoning covariationally can be difficult, particularly for students who have not had many opportunities to do so. Students may attempt to avoid covariational reasoning when they have access to other means of assessing situations, such as leveraging calculation and measurement strategies. For instance, when investigating a growing rectangle with a provided height value, such as 1 cm, a student could simply calculate successive area values and plot those points, without ever needing to reason about constant rates of change. One way to avoid this pitfall is to provide tasks that require covariation without allowing students to fall back on calculational strategies. Tasks that require students to attend to two simultaneously covarying quantities that cannot be easily measured can support this way of thinking. For the growing rectangle and triangle tasks, one way to encourage covariation is to provide a growing figure without measurements, and ask students to create a sketch of a graph comparing area to length before making any area calculations. In our teaching sessions, we had students create graphs simultaneously while watching videos of the growing figures, and then afterwards consider more carefully the accuracy of their graphs.

4.3 Choose Exact Relationships

Students who are already familiar with functional relationships and are ready to pursue their investigation at more advanced levels can benefit from engaging in tasks that explore imprecise functional situations, even if the data are not exact.

When introducing function reasoning, however, approximate data may interfere with students' abilities to construct constant and changing rates of change. In order to necessitate the creation of ratios and rates of change, contexts that are exactly linear (or quadratic, cubic, exponential, etc.) are useful for affording opportunities to appropriately quantify and represent change. This is one reason why we chose the growing rectangle and triangle tasks: the comparison of area to length is exactly linear (in the case of the rectangle) and quadratic (in the case of the triangle), and do not require approximation or estimation.

4.4 Choose Genuine Contexts

Scenarios that are contrived to be linear or quadratic can be difficult for students to reason with when beginning to explore functional relationships. We do not mean to suggest that students should not encounter real-world problems. As Freudenthal (1983) argued, meaningful mathematical activity starts with phenomena that are experientially real to students. A key distinction here is that the situation must be real *to students*, by which we mean, within their current experiential space (Jurdak, 2006). This does not have to mean that a situation would occur in real life, if the students can engage with it in a manner that enables them to understand, visualize, and mathematize the relevant phenomena (Webb et al., 2011). In fact, following Webb et al. (2011), contexts such as the growing figure tasks, though not real-world contexts, are nevertheless realizable in that they leverage situations that are imaginable and can be idealized to motivate powerful mathematical strategies.

Students' natural sense-making should not have to be a barrier to engaging with a situation mathematically. A classic example of this phenomenon can be found in Taylor's (1989) study, which compared students' responses to two questions on fractions, one asking the fraction of a cake, and the other asking the fraction of a loaf. Taylor found that some students made different sense of these questions based on the word, "cake" or "loaf". The cake was regarded as a single entity that could be divided into any number of pieces, whereas the loaf was regarded as something that would have to be divided based on a pre-existing number of slices. For students who are beginning to make sense of functional relationships, we advocate the use of realizable contexts that are authentically linear (or quadratic, or exponential, etc.). The growing figure tasks are not real-world tasks, but within the premise of the tasks, the functional relationships are authentically linear and quadratic.

4.5 Provide Opportunities for Visualization, Manipulation, and Justification

Realizable contexts that are visualizable, imaginable, and manipulable will support students' abilities to (a) identify relevant attributes and the quantities affecting them; (b) anticipate, imagine, and represent covariation; and (c) quantify covariation by constructing constant or changing rates of change. Middle grades students, in particular, may need the ability to rely on a representation that enables them to manipulate relevant quantities and observe the outcome of such manipulation. In the growing rectangle and triangle contexts, the students' abilities to change the rectangle's height and the triangle's height-to-length ratio supported generalizations about constant and changing rates of change.

Another key element of these contexts is that they enable mathematical justification and proof. Because the growing rectangle and triangle contexts are authentically and exactly linear, they are able to justify why, for instance, any rectangle must have a constant rate of change of area to length. This enabled Wesley to generalize an equivalence class of area:length ratios of $4x$ with a rectangle of height of 4 cm, and it provided the students with a way to justify that any given ratio for any amount of length swept would be equivalent to the unit ratio, 4:1, because the height does not change. We consider it critical to situate students' explorations in contexts that support justification, not only because justification is a key mathematical practice that must be emphasized at all levels of schooling (National Governors Association, 2010; Turkish Ministry of National Education [MEB], 2018), but also because it supports a habit of mind that mathematics should make sense.

5 Conclusion

In this chapter we have argued that students' early encounters with functional relationships should leverage situations with continuously covarying quantities. This does not mean, however, that a correspondence perspective should be abandoned. Instead, we argue for a dual view of function that addresses both covariation and correspondence relationships. The covariation approach changes the emphasis on what it means for relations to be linear or quadratic functions. Rather than defining functional relationships through their algebraic or graphical forms, the covariation approach highlights a function's rate of change as its defining feature. Once students have constructed constant or constantly-changing rates of change, they can then begin to identify how these rates of change are represented algebraically and graphically. Key features, such as the degree of a function's polynomial, the role of parameters, and the characteristics of a function's graph, can all be established as a consequence of its rate of change.

We identified a number of task design recommendations to support students' visualization and manipulation of continuously covarying quantities. However, visualizing, imagining, and representing changes in these quantities may not be easy for students. In particular, understanding graphs as representing a continuum of states of covarying quantities is nontrivial (Saldanha & Thompson, 1998), and the ability to reason about and represent covariation can be difficult to foster (e.g., Best & Bikner-Ahsbahs, 2017). This may be particularly true for quadratic relationships, in which the construction of a constantly-changing rate of change demands significant conceptual effort (Ellis et al., 2020; Fonger et al., 2020). Dynamic software can be a powerful tool for supporting exploration with covariation, but it is important to keep in mind that what experts interpret as representations of continuous change may not be interpreted in the same way by students. Continuity and smooth continuous covariation are mental operations, and as such they are developed through interaction between a learner and a task, simulation, or representation.

Instructors play a key role in designing and implementing tasks, as well as in supporting students' function understanding as they reason through task situations. The design principles we described de-emphasize calculation in the initial stages of engagement in order to foster covariational reasoning. Once students are ready to transition to representing amounts of change, beginning with an increment size of 1 in. the independent variable and then extending it to other increment sizes, particularly to increments smaller than 1, can be pivotal in supporting students' shift from discrete reasoning to dynamic reasoning. Considering change across different increment sizes, particularly very small increments, is also useful for generalizing constant or constantly-changing rates of change. Teachers can also direct students' attention to covariation through their questioning and focusing moves, encouraging students to imagine, describe, and ultimately quantify joint variation. As Carlson and Oehrtman (2005) described, teachers can engage in targeted questioning to ask students what values change and what variables influence the quantity of interest, to encourage clarification about specific rates of change, and to require attention to both variables' values and the relationships between the changes in both quantities. It is through both the design and careful implementation of continuous covariation contexts that instruction can support a meaningful understanding of functional relationships, establishing a critical foundation for key ideas in higher mathematics.

References

Akkoc, H., & Tall, D. (2005). A mismatch between curriculum design and student learning: The case of the function concept. In D. Hewitt & A. Noyes (Eds.), *Proceedings of the Sixth British Congress of Mathematics Education University of Warwick* (pp. 1–8).

Australian Curriculum, Assessment and Reporting Authority [ACARA]. (2020). *Statements of learning for mathematics*. Curriculum Corporation.

Ayalon, M., Watson, A., & Lerman, S. (2017). Students' conceptualisations of function revealed through definitions and examples. *Research in Mathematics Education, 19*(1), 1–19. https://doi.org/10.1080/14794802.2016.1249397

Best, M., & Bikner-Ahsbahs, A. (2017). The function concept at the transition to upper secondary school level: Tasks for a situation of change. *ZDM Mathematics Education, 48*, 865–880. https://doi.org/10.1007/s11858-017-0880-6

Blanton, M. L., & Kaput, J. J. (2011). Functional thinking as a route into algebra in the elementary grades. In *Early algebraization* (pp. 5–23). Springer.

Carlson, M., Jacobs, S., Coe, E., Larsen, S., & Hsu, E. (2002). Applying covariational reasoning while modeling dynamic events: A framework and a study. *Journal for Research in Mathematics Education, 33*(5), 352–378.

Carlson, M., & Oehrtman, M. (2005). *Key aspects of knowing and learning the concept of function. Research sampler series*. Mathematical Association of America.

Carlson, M. P., Smith, N., & Persson, J. (2003). Developing and connecting calculus students' notions of rate-of change and accumulation: The fundamental theorem of calculus. *International Group for the Psychology of Mathematics Education, 2*, 165–172.

Castillo-Garsow, C., Johnson, H. L., & Moore, K. C. (2013). Chunky and smooth images of change. *For the Learning of Mathematics, 33*(3), 31–37.

Chua, B. L., & Hoyles, C. (2010). Generalisation and perceptual agility: How did teachers fare in a quadratic generalising problem? *Research in Mathematics Education, 12*(1), 71–72.

Clement, L. L. (2001). What do students really know about functions? *Mathematics Teacher, 94*(9), 745–748.

Confrey, J., & Smith, E. (1994). Exponential functions, rates of change, and the multiplicative unit. *Educational Studies in Mathematics, 26*, 135–164. https://doi.org/10.2307/749228

Confrey, J., & Smith, E. (1995). Splitting, covariation, and their role in the development of exponential functions. *Journal for Research in Mathematics Education, 26*(1), 66–86.

Cooney, T. J., Beckman, S., & Lloyd, G. M. (2010). *Developing essential understanding of functions for teaching mathematics in grades 9–12*. National Council of Teachers of Mathematics.

Ellis, A. B., Ely, R., Singleton, B., & Tasova, H. (2020). Scaling-continuous variation: Supporting students' algebraic reasoning. *Educational Studies in Mathematics, 104*(1), 87–103. https://doi.org/10.1007/s10649-020-09951-6

Fonger, N., Ellis, A. B., & Dogan, M. F. (2020). A quadratic growth learning trajectory. *Journal of Mathematical Behavior, 59*, 1–22.

Freudenthal, H. (1983). *Didactical phenomenology of mathematical structures*. Reidel Publishing.

Greeno, J. G. (1988, November). The situated activities of learning and knowing mathematics. In *Proceedings of the 10th Annual Meeting of the PME-NA* (pp. 481–521). International Group for the Psychology of Mathematics Education, North American Chapter.

Harel, G. (2007). The DNR system as a conceptual framework for curriculumdevelopment and instruction. In R. Lesh, J. Kaput, & E. Hamilton (Eds.), *Foundations for the future in mathematics education* (pp. 1–21). Erlbaum.

Herbert, S., & Pierce, R. (2012). Revealing educationally critical aspects of rate. *Educational Studies in Mathematics, 81*(1), 85–101.

Jones, M. (2006). Demystifying functions: The historical and pedagogical difficulties of the concept of the function. *Rose-Hulman Undergraduate Math Journal, 7*(2), 1–20.

Jurdak, M. E. (2006). Contrasting perspectives and performance of high school students on problem solving in real world, situated, and school contexts. *Educational Studies in Mathematics, 63*(3), 283–301.

Kaput, J. J. (1994). Democratizing access to calculus: New routes to old roots. In A. H. Schoenfeld (Ed.), *Mathematical thinking and problem solving* (pp. 77–156). Erlbaum.

Krüger, K. (2019). Functional thinking: The history of a didactical principle. In H.-G. Weigand, W. McCallum, M. Menghini, M. Neubrand, & G. Schubring (Eds.), *The legacy of Felix Klein*, ICME-13 Monographs (pp. 35–53). Springer. https://doi.org/10.1007/978-3-319-99386-7_3

Lamon, S. (1994). Ratio and proportion: Cognitive foundations in unitizing and norming. In G. Harel & J. Confrey (Eds.), *The development of multiplicative reasoning in the learning of mathematics* (pp. 89–122). State University of New York Press.

Leinhardt, G., Zaslavsky, O., & Stein, M. K. (1990). Functions, graphs, and graphing: Tasks, learning, and teaching. *Review of Educational Research, 60*(1), 1–64.

Lobato, J., & Ellis, A.B. (2010). *Essential understandings project: Ratios, proportions, and proportional reasoning (Gr. 6–8)*. National Council of the Teachers of Mathematics.

Madison, B. L., Carlson, M., Oerthman, M., & Tallman, M. (2015). Conceptual precalculus: Strengthening students' quantitative and covariational reasoning. *Mathematics Teacher, 109*(1), 54–61.

McCulloch, A. W., Lovett, J. N., Meagher, M. S., & Sherman, M. F. (2020). Challenging preservice secondary mathematics teachers' conceptions of function. *Mathematics Education Research Journal*, 1–26.

McDougal Little. (2008). *Holt McDougal Larson algebra 2: Student edition 2008* (2nd ed.). McDougal Little.

McGowen, M., DeMarois, P., & Tall, D. (2000). Using the function machine as a cognitive root. In M. L. Fernandez (Ed.), *Proceedings of the 22nd Annual Meeting of the North American Chapter of the International Group for the Psychology of Mathematics Education* (Vol. 1, pp. 247–254).

McGraw-Hill Education. (2012). *Algebra 1, student edition*. McGraw-Hill.

Milli Eğitim Bakanlığı (MEB). (2018). Matematik dersi öğretim programı 1–8. Sınıflar. Retrieved on October 15, 2020 from http://mufredat.meb.gov.tr/ProgramDetay.aspx?PID=329

National Governors Association & Council of Chief State School Officers. (2010). *Common core state standards for mathematics*. Authors.

Oehrtman, M., Carlson, M., & Thompson, P. W. (2008). Foundational reasoning abilities that promote coherence in students' function understanding. In M. P. Carlson & C. Rasmussen (Eds.), *Making the connection: Research and teaching in undergraduate mathematics education* (pp. 27–42). The Mathematical Association of America.

Panaoura, A., Michael-Chrysanthou, P., Gagatsis, A., Elia, I., & Philippou, A. (2017). A structural model related to the understanding of the concept of function: Definition and problem solving. *International Journal of Science and Mathematics Education, 15*, 723–740. https://doi.org/10.1007/s10763-016-9714-1

Rasmussen, C. L. (2000). New directions in differential equations: A framework for interpreting students' understandings and difficulties. *Journal of Mathematical Behavior, 20*, 55–87.

Sajka, M. (2003). A secondary school student's understanding of the concept of function: A case study. *Educational Studies in Mathematics, 53*(3), 229–254.

Saldanha, L. A., & Thompson, P. W. (1998). Re-thinking covariation from a quantitative perspective: Simultaneous continuous variation. In S. Berenson, K. Dawkins, M. Blanton, W. Coulombe, J. Kolb, K. Norwood, & L. Stiff (Eds.), *Proceedings of the Twentieth Annual Meeting of the North American Chapter of the International Group for the Psychology of Mathematics Education* (Vol. 1, pp. 298–304). ERIC Clearinghouse for Science, Mathematics, and Environmental Education.

Steele, M. D., Hillen, A. F., & Smith, M. S. (2013). Developing mathematical knowledge for teaching in a methods course: The case of function. *Journal of Mathematics Teacher Education, 16*(6), 451–482.

Stephens, A. C., Ellis, A. B., Blanton, M., & Brizuela, B. M. (2017). Algebraic thinking in the elementary and middle grades. In *Compendium for research in mathematics education* (pp. 386–420). National Council of Teachers of Mathematics.

Taylor, N. (1989). "Let them eat cake": Desire, cognition and culture in mathematics learning. In C. Keitel, P. Damerow, A. Bishop, & P. Gerdes (Eds.), *Mathematics, education, and society*. United Nations Educational Scientific.

Thompson, P. W. (1994). The development of the concept of speed and its relationship to concepts of rate. In G. Harel & J. Confrey (Eds.), *The development of multiplicative reasoning in the learning of mathematics* (pp. 179–234). SUNY Press.

Thompson, P. W., & Carlson, M. P. (2017). Variation, covariation, and functions: Foundational ways of thinking mathematically. In J. Cai (Ed.), *Compendium for research in mathematics education* (pp. 421–456). National Council of Teachers of Mathematics.

Thompson, P. W., Hatfield, N., Joshua, S., Yoon, H., & Byerley, C. (2017). Covariational reasoning among U.S. and South Korean secondary mathematics teachers. *The Journal of Mathematical Behavior, 48*, 95–111.

Thompson, P. W., & Thompson, A. G. (1992, April). *Images of rate*. Paper presented at the annual meeting of the American Educational Research Association, San Francisco, CA, USA.

UK Department for Education. (2009). The *national strategies: The framework for secondary mathematics*. Crown.

Vinner, S. (1983). Concept definition, concept image and the notion of function. *International Journal of Mathematical Education in Science and Technology, 14*(3), 293–305.

Vinner, S., & Dreyfus, T. (1989). Images and definitions for the concept of function. *Journal for Research in Mathematics Education, 20*(4), 356–366.

Webb, D. C., Van Der Kooij, H., & Geist, M. R. (2011). Design research in the Netherlands: Introducing logarithms using realistic mathematics education. *Journal of Mathematics Education at Teachers College, 2*, 47–52.

Yılmaz, Z., Dede, H. G., & Akkoç, H. (2020). Ders kitaplarında fonksiyon kavramına nasıl giriş yapılıyor?: Matematik öğretmen adaylarının değerlendirmeleri. *Necatibey Eğitim Fakültesi Elektronik Fen Ve Matematik Eğitimi Dergisi, 14*(1), 240–270.

Yoon, H., & Thompson, P. W. (2020). Secondary teachers' meanings for function notation in the United States and South Korea. *The Journal of Mathematical Behavior, 60*, 100804. https://doi.org/10.1016/j.jmathb.2020.100804

Zandieh, M. (2000). A theoretical framework for analyzing student understanding of the concept of derivative. *CBMS Issues in Mathematics Education, 8*, 103–127.

Geometric Transformations Through Quantitative Reasoning

Gülseren Karagöz Akar, İsmail Özgür Zembat, Selahattin Arslan, and Mervenur Belin

1 Geometric Transformations Through Quantitative Reasoning

In this chapter, we provide a conceptual analysis of the concept of isometry (translation, rotations, reflections) based on quantitative reasoning (QR). Curricular standards for mathematics in different countries (e.g., CCSSM, 2010; MEB-TTKB, 2018) highlight the importance of learning isometries as functions by the end of high school. However, the common practice for teaching geometric transformations, especially isometries, start with treating them as rigid motions only. Although such an approach seems sound, previous research reveals the difficulty of conceptualizing isometries as functions. We believe that treating isometries as motions only is problematic as geometric transformations are functions, and learners should go beyond thinking about motions by mathematizing them as functions. In this chapter, we provide an analysis of the end goal of what learners should understand or learn in order to conceptualize isometries quantitatively. Our analysis does not involve an articulation of a teaching sequence. Instead, what we propose is about the kind of conceptualizations learners should have to learn isometries that go beyond the motions. In so doing, we draw on the components of Quantitative Reasoning (QR) Theory. Investigating isometries via the use of QR Theory, as we noticed, is missing in the current literature. As future work, we intend to expand this analysis to provide an instructional sequence to help learners develop those conceptualizations of isometries.

G. Karagöz Akar (✉) · M. Belin
Bogazici University, İstanbul, Turkey
e-mail: gulseren.akar@boun.edu.tr

İ. Ö. Zembat
University of Glasgow, Glasgow, UK

S. Arslan
Trabzon University, Trabzon, Turkey

G. Karagöz Akar et al. (eds.), *Quantitative Reasoning in Mathematics and Science Education*, Mathematics Education in the Digital Era 21,
https://doi.org/10.1007/978-3-031-14553-7_8

In the following sections, we first define geometric transformations from a mathematical standpoint and then build on the relevant literature to explain our rationale for investigating isometries by focusing on the QR theory. Finally, we detail how isometries can be conceptualized based on QR notions (e.g., multiplicative object, continuous covariational reasoning) and discuss curricular and instructional implications.

1.1 Isometries from a Purely Mathematical Standpoint

As part of the geometric transformations set, isometries are distance-preserving one-to-one and onto functions that map plane to plane ($f : \mathbb{R}^2 \rightarrow \mathbb{R}^2, f$ is 1:1 and onto) (Fife et al., 2019; Martin, 1982). Isometry lexically consists of *iso* (meaning, same) and *metric* (meaning, measure). In this sense, an isometry is a distance preserving transformation. This chapter will focus on isometries such as translations, rotations and reflections as detailed below.

Parameters are crucial in defining isometries. For example, the parameter for *translations* is a *vector*, say $\|\vec{u}\|$. Consider translation of a specific quadrilateral *ABCD* (pre-image) on the plane that has the vector $\vec{u} = (2, 0)$ as a parameter. This pre-image quadrilateral corresponds to an identical image quadrilateral, $A'B'C'D'$, on the plane. The image quadrilateral consists of points that are two units (same as $\vec{u}$) away from the corresponding pre-image points (see Fig. 1a). In other words, all pre-image points in the form of $(a, b) \in ABCD$ are mapped to image points in the form of $(a + 2, b + 0) \in A'B'C'D'$ under such translation. If we consider the whole plane, any $(x, y) \in \mathbb{R}^2$ is mapped to $(x', y') = (x, y) + \vec{u}$ under such translation.

For *rotation*, *angle* and *center* of rotation are the parameters. Consider rotation of the same quadrilateral (pre-image) which has an angle, $\theta = -45°$, and a center point, $P(0, 0)$, as parameters. This quadrilateral corresponds to another image quadrilateral that is identical in shape to the pre-image rotated 45° clockwise about the center of rotation, the origin (see Fig. 1b). Such correspondence suggests that the distance of pre-image points to the center of rotation stays the same as the distance of the corresponding image points to the center. In contrast, the orientations of the pre-image and the image may change. *Reflection*, on the other hand, may have a *line* as a parameter. Reflection of the same quadrilateral *ABCD* by considering the x-axis as a parameter corresponds to the same size image quadrilateral (see Fig. 1c). The perpendicular distance of each pre-image point to the line of reflection is the same as those of corresponding image points.

The examples shown in Fig. 1a–c are just samples representing transformations of parts of planes. Several interconnected key ideas make isometries special (e.g., Argün et al., 2014; Yılmaz, 2020) as summarized below:

- Isometries preserve distances,
- Isometries are one-to-one and onto functions,
- The domain and range of isometries are $\mathbb{R}^2$,

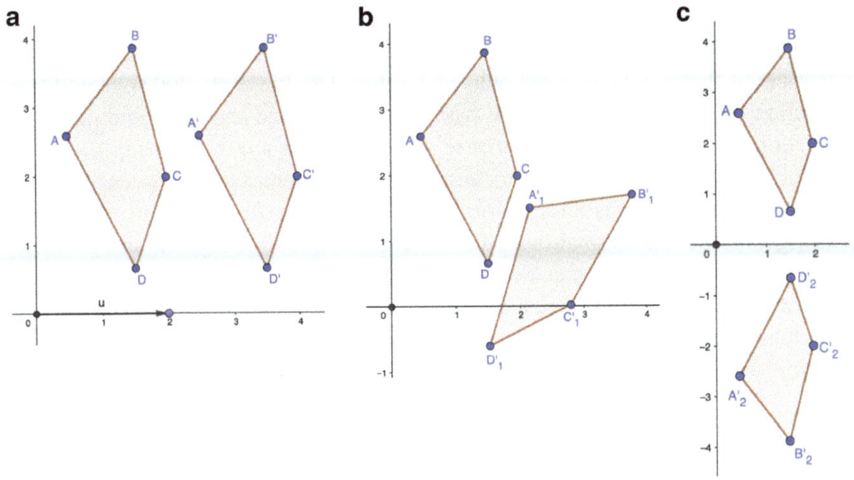

Fig. 1 **a** Translation of *ABCD* using $\vec{u}$ as a parameter. **b** Rotation of *ABCD* using origin and $-45°$ as parameters. **c** Reflection of *ABCD* using *x*-axis as a parameter

- Isometries map the whole $\mathbb{R}^2$ to the whole $\mathbb{R}^2$, and this mapping can be characterized as $(x, y) \to (x', y')$,
- Isometries have parameters (*vector* for translations, *angle* and *center point* for rotations, and *line* for reflections).

We will revisit these mathematical ideas as needed using QR throughout the chapter.

1.2 Importance of Geometric Transformations

Geometric transformations play a vital role in conceptualizing different mathematical concepts. Understanding geometric transformations as functions on the plane, for instance, is crucial (Fife et al., 2019; Hollebrands, 2003; Yanik & Flores, 2009) because it provides learners with opportunities to deepen their understanding of functions and to conceptualize transformations better (CCSSM, 2010; Fife et al., 2019). In particular, studying geometric transformations assist learners in moving from $\mathbb{R} \to \mathbb{R}$ settings to $\mathbb{R}^2 \to \mathbb{R}^2$ (Steketee & Scher, 2011) and engaging them in higher-level mathematical reasoning activities (Hollebrands, 2003). In this sense, learners extend their thinking of a function as a relationship between two variables, each referring to *a real number*, to thinking of function as a relationship between two variables, each of which is a point consisting of a combination of two real numbers (Steketee & Scher, 2011).

Geometric transformations also permit learners to conceptualize congruence and similarity and apply them to planar figures and families of functions (Jones, 2002).

For instance, studying transformations might allow learners to realize that the graphs of $y = ax^n$ and $y = ax^n + b$ ($n = 1, 2, 3, \ldots$ respectively and $a \in \mathbb{R}$, $b \in \mathbb{R}$) are congruent as they can be mapped onto each other (Usiskin et al., 2003). In addition, a function can only have an inverse when it is one-to-one and onto, and graphs of a function and its inverse are symmetric with respect to $y = x$. Utilizing geometric transformations might allow learners to understand the connection between a function and its inverse as there is a reflection that maps the function to its inverse. In this sense, geometric transformations allow learners to think of algebra and geometry interconnectedly rather than as isolated branches of mathematics.

Apart from such advantages, studying transformations provide learners with opportunities to develop conjectures and construct generalizations through critically thinking and imagining the relationships between variables (Yanik, 2014).

1.3 Difficulties in Understanding Transformations

Previous research on geometric transformations guides us in understanding learners' difficulties and conceptions that make isometries problematic. Research revealed that learners think about transformations as motions rather than functions (Edwards, 1991; Mhlolo & Schafer, 2013), as detailed below.

One of the factors preventing learners from conceiving geometric transformations as functions is conceiving the plane as a background for geometric figures resulting in considering planar figures isolated from the plane. Such understanding hinders learners' conceptualizations of domain and range as $\mathbb{R}^2$ (Yanik & Flores, 2009). Overcoming this difficulty is possible via understanding points as subsets of the plane (Edwards, 2003). This is important because when the plane is mapped onto the plane, everything, including the parameters of the geometric transformations (e.g., vectors, line of reflection, center of rotation), are also mapped as they are part of the domain, which is another difficulty for students to overcome (e.g., Sunker & Zembat, 2012; Yanik, 2006). Transformation of the whole plane is another problematic issue for both students and prospective teachers (Hollebrands, 2003; Yanik, 2011). Research in this area (e.g., Yanik, 2011) suggests that considering points on the plane as unique locations (Lakoff & Núñez, 2000) that are part of the domain and range and thinking about covariation between pre-image and image points (Steketee & Scher, 2016), may help learners understand isometries as functions. In this chapter, we will pay particular attention to conceptualizing points as representations of locations and the covariation of pre-image and image points.

Research also highlights problematic areas for learners in thinking about transformations. For example, learners do not consider that a vector with its components (both direction and magnitude) defines translations (Yanik, 2011). They only consider the angle of rotation by ignoring the center as a parameter for rotations (Flanagan, 2001). Learners do not pay attention to vertical distances between pre-images and the line of reflection when operating on reflections (Hollebrands, 2004; Zembat, 2007).

When we think about all these difficulties, an important question comes to mind: What would be an effective way of preventing learners from having these difficulties or overcoming them? We believe that engaging learners in quantitative reasoning based on the conceptualizations we highlight in this chapter may be a solution. In the following sections, we first define quantitative and covariational reasoning, reconceptualize the plane with respect to QR Theory, and then analyze isometries (translations, rotations, and reflections, respectively) in the context of QR without providing a teaching sequence.

1.4 Quantitative Reasoning and Covariational Reasoning

In his framework, Thompson (1990) describes ***quantity*** as a quality of an object that an individual conceives. Conceiving a quantity entails a ***measurement process*** based on an appropriate ***unit***, whether implicit or explicit. More specifically, to comprehend a quantity, an individual must have a mental image of an object and its measurable ***attributes*** (qualities) (Thompson, 1994). The comprehension of the object with its measurable attributes is called quantification. Thompson (2011, p. 37) defined ***quantification*** as a "process of conceptualizing an object and an attribute of it so that the attribute has a unit of measure, and the attribute's measure entails a proportional relationship (linear, bilinear, or multilinear) with its unit". According to Moore et al. (2009), "Conceiving of situations and measurable quantities of a situation" (p. 5) through ones' mental actions as well as "both developing and reasoning about relationships between these constructed quantities" (p. 3) is called ***quantitative reasoning***.

Covariation is another important construct in Thompson's theory of quantitative reasoning because conceptualizing a situation quantitatively and taking it as dynamic are important aspects of student reasoning (Thompson & Carlson, 2017). Covariation, here, is defined as one's "holding in mind a sustained image of two quantities' values (magnitudes) simultaneously" (Saldanha & Thompson, 1998, p. 299). Covariational reasoning encompasses "the cognitive activities involved in coordinating varying quantities while attending to the ways in which they change in relation to each other" (Carlson et al., 2002, p. 354).

1.5 Understanding Points as Multiplicative Objects and Conceptualizing $\mathbb{R}^2$ Quantitatively

We have previously mentioned that we will pay particular attention to points representing unique locations. We now describe what it means to think about a point on the plane quantitatively.

A *multiplicative object* is an object that is formed from two quantities by mentally uniting "their attributes to make a new attribute that is, simultaneously, one and the other" (Thompson et al., 2017, p. 96). It is possible to think about a point on the plane, an ordered pair in the form of (a, b), as a multiplicative object in the following sense. The point $A = (a, b)$ can be thought of as a cognitively uniting of two quantities' magnitudes (Saldanha & Thompson, 1998; Stevens & Moore, 2017). These quantities are directed lengths between the origin and $(a, 0)$, and the origin and $(0, b)$ [in other words, these quantities are the vectors $\begin{pmatrix} a \\ 0 \end{pmatrix}$ and $\begin{pmatrix} 0 \\ b \end{pmatrix}$]. An individual needs to unite their magnitudes in mind to conceive (a, b) as a multiplicative object. However, one does not need to physically measure these quantities but think about the possibility of measuring them for these to be conceived as quantities. Determining, for example, the magnitude of the quantity $(a, 0)$ requires one to think of the unit of measurement, as $1/m(a)$ of $\|a\|$, where $m(a)$ means measure of a, and $\|a\|$ means magnitude of a (Thompson, 2011) (Fig. 2).

For example, let us evaluate the point $(-5, 3)$ as a multiplicative object. One needs to unite the magnitudes of two quantities here, namely, magnitudes of the directed lengths between the origin and $(-5, 0)$, and the origin and $(0, 3)$, respectively [or magnitudes of the vectors $\vec{u} = \begin{pmatrix} -5 \\ 0 \end{pmatrix}$ and $\vec{v} = \begin{pmatrix} 0 \\ 3 \end{pmatrix}$]. Magnitude of the directed length between the origin and $(-5, 0)$ is $m(\vec{u}) \times \|\text{unit}\| = 5 \times 1 = 5$ since the unit here is 1, whereas the magnitude of $(0, 3)$ is $m(\vec{v}) \times \|\text{unit}\| = 3 \times 1 = 3$. Hence, mentally uniting these two quantities' magnitudes, 5 and 3, allows one to think of the point $(-5, 3)$ as a multiplicative object. In other words, understanding the location of a point (e.g., $(-5, 3)$) in two directions simultaneously from a reference point (e.g., origin) is an example of the multiplicative object (Thompson, 2011).

Conceptualizing a point on the plane, (a, b), as a multiplicative object, allows an individual to assimilate the following two measurable attributes:

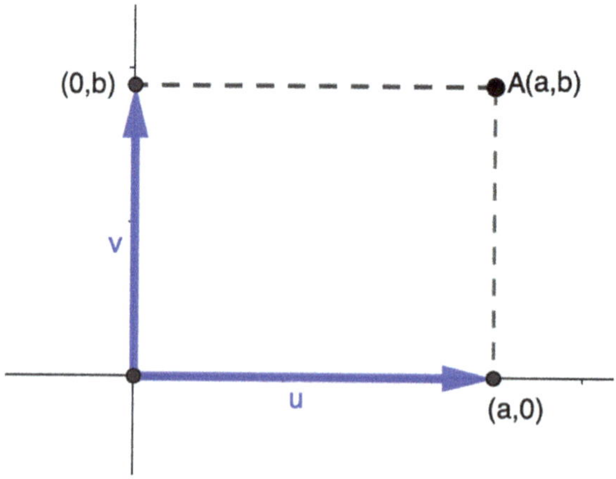

Fig. 2 Graph of a point $A(a, b)$ and its constituent vectors

(i) the distance of (a, b) to the origin—its measure can be determined via Pythagorean Theorem using the quantities $(a, 0)$ and $(0, b)$,
(ii) the slope[1] of the line passing through the origin and (a, b).

We claim that one's construction of any ordered pair having these attributes can potentially trigger the understanding that $\mathbb{R}^2$ is made up of all such points.[2] In other words, the individual's understanding of point should be generalized to all points that make up the plane, $\mathbb{R}^2$. We believe that learners should conceptualize $\mathbb{R}^2$ this way. Such generalization is possible by individual's thinking about the following three conceptualizations:

M#1. The learner conceptualizes a point (a, b) as the combination of two quantities' magnitudes (distances of $(a, 0)$ and $(0, b)$ to the origin) as opposed to conceptualizing a point (a, b) as a combination of two labels.
M#2. The learner conceptualizes every point on the plane, $\mathbb{R}^2$, as having a unique location that manifests itself as an element of $\mathbb{R}^2$—there is a unique combination of $(a, 0)$ and $(0, b)$.
M#3. The learner conceives that any point in $\mathbb{R}^2$ is part of $\mathbb{R}^2$ and similarly, any set of points (e.g., a geometric figure like a triangle, a line segment) is a subset of $\mathbb{R}^2$.

These conceptualizations are needed for learners to consider any set of points or geometric shapes as quantities relative to the whole, $\mathbb{R}^2$, rather than isolated and independent entities. We argue that thinking of any point on the Cartesian Plane as a multiplicative object as described above is vital for learning isometries.

1.6 *Conceptualizing Translations Through Quantitative Reasoning and Covariational Reasoning*

We first describe a geometric interpretation of translation. Assume we only focus on two preimage points, A and B, on the plane and a translation. These preimage points are mapped to corresponding image points, A' and B', under a translation T having a vector, $\vec{u}$, as the parameter.[3] Note that this mapping preserves the location

---

[1] Lobato and Thanheiser (2002) defined slope "as the rate of change in one quantity relative to the change of another quantity where the two quantities covary" (p. 163). Lines have slopes that have a constant rate of change. When two quantities covary under the condition that the accumulation of changes in a quantity is proportional to the accumulation of those of the other, one can understand the constant rate of change (Thompson, 1994). We adopted these definitions for slope throughout the paper.

[2] Note that we view the multiplicative object (a, b) in a dynamic sense in which "its coordinates represent a state of two quantities' covariation" (Thompson et al., 2017).

[3] An individual here may think about this translation as illustrated in Fig. 3a and consider T as a movement of preimage points A and B to A' and B'. In this case, we can say that the individual focuses on the figurative aspects of the translation and conceptualizes it as a movement or sliding rather than a function. Here, we are not focusing on such thinking.

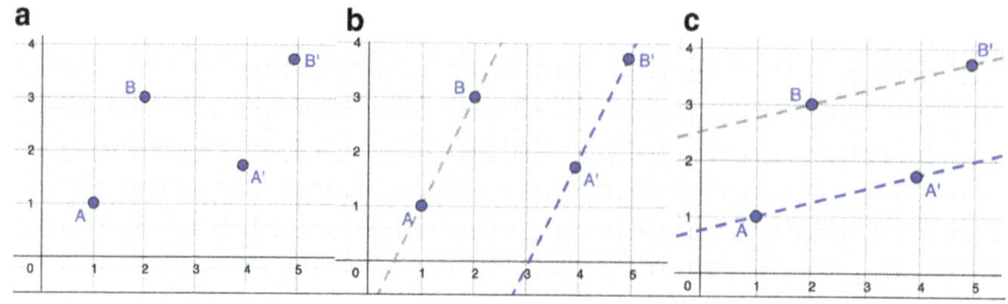

Fig. 3 **a** Translation of points A and B. **b** Comparing (A, B) to (A', B'). **c** Comparing (A, A') to (B, B')

of B' relative to the location of A' by considering the location of B relative to the location of A. However, this is a geometric interpretation of translation, and the literature (e.g., Portnoy et al., 2006; Sinclair et al., 2012) suggests that translations are geometrically treated in the textbooks or in mathematics classes. If this is the case, how can we quantify the preservation of relative locations of the preimage and the corresponding image points to allow our students to move beyond geometric interpretation of isometries? We believe that answering this question using QR is a key for changing curriculum design and established practices of teaching isometries. We delineate this issue using the key components of QR in what follows.

Once $\mathbb{R}^2$ is conceptualized as detailed previously, understanding a translation quantitatively requires coordination at two levels with their descriptive conditions (C_T*#1* and C_T*#2*), as detailed below and illustrated in Fig. 3a–c. In these conditions "m" refers to slope.[4]

C_T*#1*. $d(A, B) = d(A', B')$ and $m_{AB} = m_{A'B'}$.
C_T*#2*. $d(A, A') = d(B, B')$ and $m_{AA'} = m_{BB'}$.

In other words, an individual can think about preimage and image points and coordinate two sets of properties (the aforesaid measurable attributes of distance and slope) at two levels (as illustrated in Fig. 3b, c) and conclude that C_T*#1 and* C_T*#2* can only be true with a vector from the equivalence class of AA'.

More specifically, an individual needs to consider the condition that the magnitudes (as a result, the values) of the first set of quantities, namely the distance between A and B and the distance between A' and B', are the same. In addition, the magnitudes (as a result, the values) of the second set of quantities, namely the slopes of the lines passing from AB and $A'B'$ respectively, are also the same, $m_{AB} = m_{A'B'}$. In other words, the locations of A to B relative to the locations of A' to B' are the same. Such thinking is possible if the translation is considered for the distances among paired preimage points as quantities within the domain and the corresponding paired image points within the range. Figure 3b illustrates a comparison of the sets $\{A, B\}$ and

[4] One can argue that for a vector like $u = (0, b)$ the slope is undefined. However, by saying $m_{AB} = m_{A'B'}$ we mean that the locations of A to B relative to the locations of A' to B' are the same. The same logic applies to $m_{AA'} = m_{BB'}$.

$\{A', B'\}$—hereafter; we call such comparison as "***within comparison***", which is one level of analysis. Moreover, the magnitudes of the third set of quantities, namely the distance between A and A' and the distance between B and B', are also the same. In addition to this, the magnitudes of the fourth set of quantities, namely the slopes of the lines passing from AA' and BB' are correspondingly the same, $m_{AA'} = m_{BB'}$. Such thinking is also possible if the translation is considered for the distances as quantities in between preimage-image pairs from the domain and range. Figure 3c illustrates a comparison of the sets $\{A, A'\}$ to $\{B, B'\}$—hereafter; called as "***between comparison***", another level of analysis. Such two-level analysis is only possible by learners' coordinating these two sets of quantities (involving distance and slopes) and their interrelationships simultaneously. What makes the relationship between these sets of quantities unique is the vector from the equivalence class of AA', and this vector is the parameter that makes the coordination of these sets of quantities possible.

Such analysis can initially be expanded to a subset of the plane (e.g., a set of discrete points, a line segment, a line, a quadrilateral, etc.). The sample points we focused on in Fig. 3 can be considered the endpoints of a line segment, AB, and the same analysis can be executed for these two endpoints. However, there is a need for the learner to expand this thinking to all points that make up the line segment AB. In other words, the learner needs to be able to choose any two points on AB, say C and D, whether they are infinitely close to each other or not, and coordinate *within* and *between comparisons* for the quantities of distance and slopes—say $d(C, D) = d(C', D')$ and $m_{CD} = m_{C'D'}$ as well as $d(C, C') = d(D, D')$ and $m_{CC'} = m_{DD'}$.

In terms of *within comparison*, we can give the following articulation. The quantities involved in this coordination (e.g., $d(C, D)$, $d(C', D')$ as in Fig. 4) covary as such covariation is dependent on the covariation of preimage and corresponding image points (e.g., C and C' and D and D') and slopes (e.g., m_{CD}, $m_{C'D'}$) stay invariant. Indeed, any two chosen points on AB (e.g., C, D) can be infinitely close to each other, or one converges to the other, or one becomes the other ($C = D$ so $C' = D'$), which also suggests the continuous covariation of preimage and corresponding image points (in this case, $C = D$ and $C' = D'$). In other words, the learner needs to think of *continuous covariation* of the quantities included in the aforesaid coordination "one understands that if either quantity has different values at different times, it changed from one to another by assuming all intermediate values"[5] (Saldanha & Thompson, 1998, p. 299).

The result of this kind of reasoning is that quantity pairs used for *between comparison*, such as $d(C, C')$, $d(D, D')$ and $m_{CC'}$, $m_{DD'}$, are invariant depending on the parameter of the translation, the vector. Hence, the individual needs to conceptualize the vector as the invariant relationship between the covarying quantities (aforesaid distances and slopes based on points C, C', D, D'). "Under a translation, students should understand that a line and its image are parallel and that the distances from each pre-image point to its image point are equal to the magnitude of the translating

[5] Note that here "all intermediate values" for a student depends on their conceptions of the continuum as Sirotic and Zazkis (2007) pointed out that students may not conceptualize any two values as being connected by a continuum.

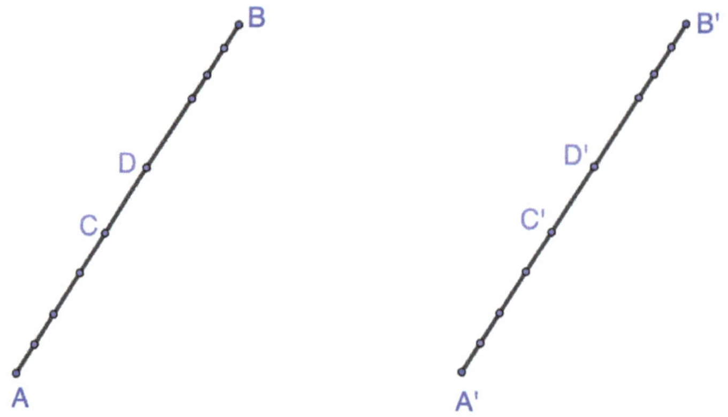

Fig. 4 Translation of a line segment AB

vector" (Hollebrands, 2003, pp. 65–66). Note that we have so far made our argument for the translation of a line segment only. The same reasoning can be expanded to a geometric figure or a polygon made up of several line segments (as in Fig. 5a), or to a planar surface (as in Fig. 5b), or to the whole plane (as in Fig. 5c).

At this point, the learners' assimilatory structure may be ready to assimilate translations as functions. Beyond the aforesaid coordination and the articulation of the role of vector, the learner also needs to comprehend that every preimage point of the plane is coupled with a unique image point of the plane so that translation is a function for the learner.

What about the domain and range of this function? Previously, we mentioned that the learner needs to conceptualize 'any set of points or geometric shape as quantities relative to the whole, $\mathbb{R}^{2}$' based on the conceptualisations of M#1, M#2 and M#3. Thus, an arbitrary point of $\mathbb{R}^2$, like F, here is coupled with a corresponding F' from the $\mathbb{R}^2$ with the vector $\vec{u}$ as a parameter [here, $F + \vec{u} = F'$]. In other words, for the learner, the translation function should consist of ordered pairs in the form of (F, F') as multiplicative objects, no two of which has the same first element. Note that both F and F', which are already multiplicative objects themselves, are elements of $\mathbb{R}^2$ and, therefore, $(F, F') \in \mathbb{R}^4$ (Fife et al., 2019). As a result, the domain of this function is the whole $\mathbb{R}^2$, and the range of this function is the whole $\mathbb{R}^2$, as illustrated for several representative points in Fig. 6.[6] Our argument up to this point suggests that such construction of translation as a function might allow learners to overcome their difficulties about translations (e.g., considering vector as an element of both domain and range, the translation of the vector itself and the translation of the whole plane (Hollebrands, 2003; Sunker & Zembat, 2012)).

---

[6] Note that the learner can assimilate at this point that the translation function is onto and 1:1. To be more specific, this function here is $T_{\vec{u}}: \mathbb{R}^2 \to \mathbb{R}^2$. For every $A'(x', y')$ there is a unique $A(x, y)$, which suggests that the function is onto. For every $A(x, y)$ and $B(x, y)$, if $A \neq B$, then $A' \neq B'$, which suggests that the function is 1:1.

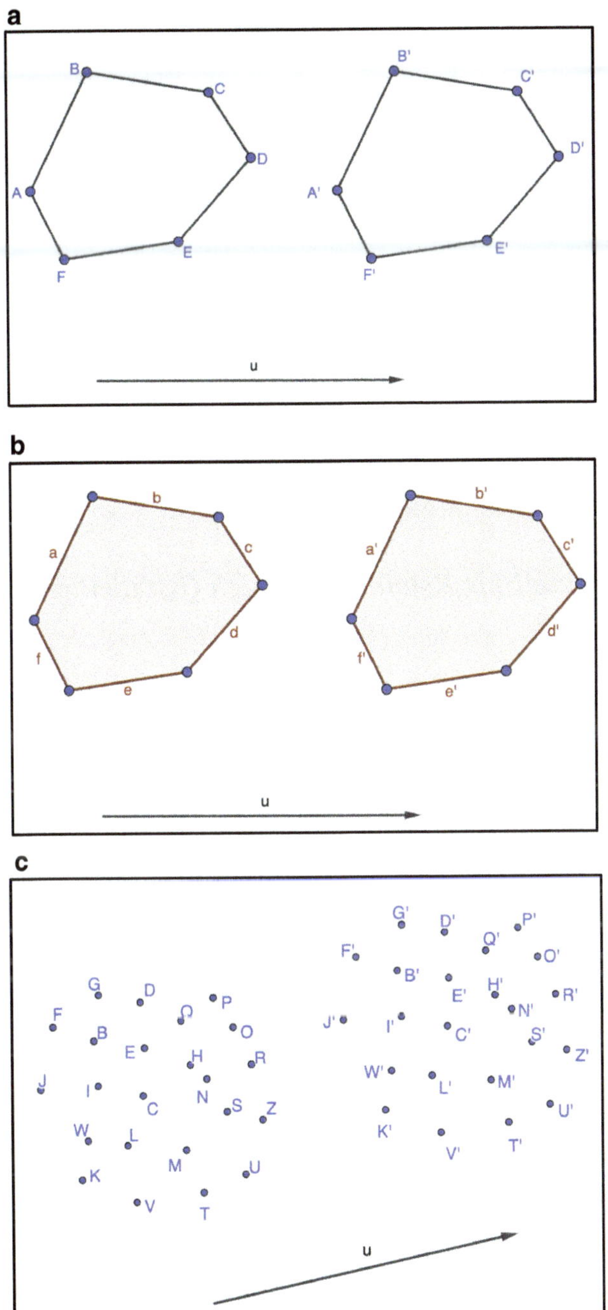

Fig. 5 **a** Translation of a geometric figure (or polygon) made up of several line segments with respect to a vector $\vec{u}$. **b** Translation of a planar surface with respect to a vector $\vec{u}$. **c** Translation of a representative set of points in the plane (to represent the whole plane) with respect to a vector $\vec{u}$

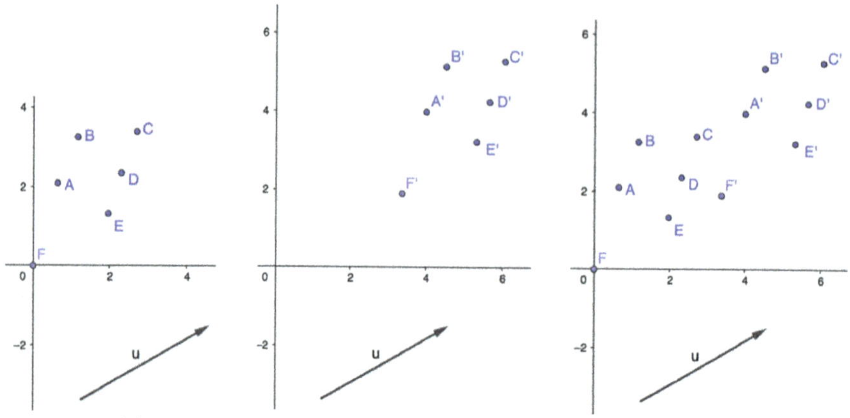

Fig. 6 A sample preimage and image points representing domain ($\mathbb{R}^2$), range ($\mathbb{R}^2$) and combination of both ($\mathbb{R}^2$) for a translation with respect to a vector $\vec{u}$

1.7 Conceptualizing Rotations Through Quantitative Reasoning

Assume we only focus on three preimage points, A, B, and D, on the plane and a rotation. These preimage points are mapped to corresponding image points, A', B', D', under a rotation R having a center, C, and an angle, θ, as the parameters. Note that this mapping preserves the location of, for example, B' relative to the location of A' by taking into account the location of B relative to the location of A. However, this is a geometric interpretation of rotation, and the literature (e.g., Sinclair et al., 2012) suggests that rotations, just like translations, are geometrically treated in the curricula or mathematics classes. The same question we raised in defining translation also arises here: How can we quantify the preservation of relative locations of the preimage and the corresponding image points to allow our students to go beyond geometric interpretation of rotations? We delineate this issue using the key components of QR in the following argument.

Assuming that $\mathbb{R}^2$ is conceptualized as detailed before, understanding a rotation requires a coordination of the following conditions involving different quantities:

C_R#1: The distance between any two preimage points, say A and B, and the distance between their corresponding image points, say A' and B', are equal. Namely, $d(A, B) = d(A', B')$.

C_R#2: An arbitrary preimage point, A, and its corresponding image point, A', on the plane are equidistant from a unique point, called center, C. Namely, $d(A, C) = d(A', C)$, etc. Note that the triangle formed by A, A', and C is isosceles.

C_R#3: Any arbitrary three colinear preimage points lying on a line having a particular slope have corresponding image points that are also colinear lying on

another line with another slope in the same way[7] as preimage points (as C_R#1 holds).

C_R#4: All angles formed by a preimage point (A), the center (C), and the corresponding image point (A'), $\widehat{ACA'}$ have the same measure θ, where θ is the parameter defining the angle of rotation.

These conditions are not necessarily hierarchical nor sequential. Given this, coordination of the conditions C_R#1, C_R#2 and C_R#3 necessitate that the angle mentioned in C_R#4 is unique. Note that we are not suggesting the combination of these conditions to give a formal mathematical definition of rotation. Instead, what we suggest is that the quantities given in these conditions are necessary to be conceptualized in order to think about rotations quantitatively as further detailed below. The conditions provided for the other isometries later in the chapter are also mentioned in the same manner.

We suggest that one way to think about rotations quantitatively requires one to understand and coordinate the quantities within these conditions in thinking about what is variant or invariant. As implied in the above conditions, these quantities are: the distances between preimage pairs, the distances between corresponding image pairs, the distances between preimage and the center and the distances between the corresponding image and the center, slopes of the lines passing through preimages and the corresponding images, the angle of center. An individual needs to think that the magnitudes (as a result, the values) of the distance between A and B and the distance between A' and B', are invariant—$d(A, B) = d(A', B')$. Such thinking is possible if rotation is considered for the distances among paired preimage points as quantities within the domain and the corresponding paired image points within the range. Figure 7a illustrates such comparison of the sets $\{A, B\}$ and $\{A', B'\}$, as previously called *within comparison*. Note that differently from translations, a comparison of the sets $\{A, A'\}$ to $\{B, B'\}$, *between comparison*, does not have the same invariances: it is not always true that $d(A, A') = d(B, B')$. However, the magnitudes of the distance between A and C and the distance between A' and C, *are* invariant: $d(A, C) = d(A', C)$ (see Fig. 7b).

For students to conceptualize rotations as functions from $\mathbb{R}^2$ to $\mathbb{R}^2$, the above reasoning needs to be expanded to all points that are colinear with A and B. In other words, the learner needs to be able to choose any point in between A and B that is colinear with A and B, say D, whether it is infinitely close to A, and coordinate within comparisons for the quantities of the distances, $d(A, D) = d(A', D')$ as well as $d(D, C) = d(D', C)$ and $d(A, C) = d(A', C)$ (see Fig. 7c). The following quantity pairs involved in this coordination covary:

- $d(A, D)$ and $d(A', D')$—Covariation #1
- $d(D, C)$ and $d(D', C)$ & $d(A, C)$ and $d(A', C)$—Covariation #2

Any chosen point, D, can be infinitely close to A, or it converges to A, or one becomes the other ($D = A$ and $D' = A'$), suggesting the continuous covariation of

[7] Note that the orientation may not be preserved under rotation.

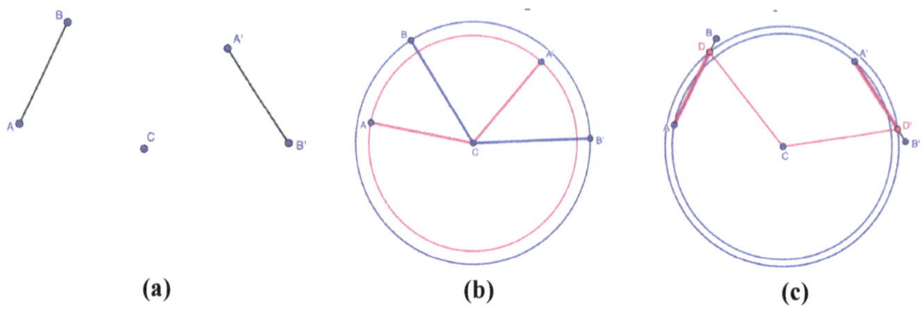

Fig. 7 Rotation of sample preimage points *A*, *B*, *D* with respect to center *C* and rotation angle of 120° clockwise

preimage and corresponding image points.[8] In other words, the learner needs to think of *continuous covariation* (Saldanha & Thompson, 1998) of the quantities included in Covariation #1 and Covariation #2 initially separately. Simultaneously thinking about (or coordinating) Covariation #1 and Covariation #2 (for infinitely many points like *D*) leads one to conserve collinearity in preimage and corresponding image points. Preservation of collinearity implies that there is an angle in between any 'preimage-center-image' point set. Therefore, previously stated Conditions C_R#1, C_R#2 and C_R#3 result in the uniqueness of this angle stated in C_R#4.

At this point, the learners' assimilatory structure may be ready to assimilate rotations as functions. Beyond the aforesaid coordination and articulations, the learner also needs to comprehend that every preimage point of the plane is coupled with a unique image point of the plane so that the rotation is a function for the learner. Previously, we mentioned that the learner needs to conceptualize 'any set of points or geometric shape as quantities relative to the whole, $\mathbb{R}^2$, based on the conditions of M#1, M#2 and M#3. Thus, an arbitrary point of $\mathbb{R}^2$, like *D*, is coupled with a corresponding D' from the $\mathbb{R}^2$; $\mathrm{R}_{C,\theta}\ (D) = D'$.[9] In other words, for the learner, the rotation function should consist of ordered pairs in the form of (D, D') as multiplicative objects, no two of which has the same first element. Note that both *D* and D', which are already multiplicative objects themselves, are elements of $\mathbb{R}^2$ and, therefore, $(D, D') \in \mathbb{R}^4$ (Fife et al., 2019). As a result, the domain of this function is the

[8] Note that the same argument can be made for points *B* and *D*. In that case, the coordination of following quantities is needed: $d(B, D) = d(B', D')$ as well as $d(D, C) = d(D', C)$ and $d(B, C) = d(B', C)$. The following quantity pairs involved in this coordination covary: $d(B, D)$ and $d(B', D')$—Covariation #1; $d(D, C)$ and $d(D', C)$ & $d(B, C)$ and $d(B', C)$—Covariation #2. Any chosen point, *D*, can be infinitely close to *B*, or converge to *B*, or same as B ($D = B$ and $D' = B'$), which also suggests the continuous covariation of preimage and corresponding image points.

[9] Note that the learner can assimilate at this point that the rotation function is onto and 1:1. To be more specific, this function here is $\mathrm{R}_{C,\theta}$: $\mathbb{R}^2 \rightarrow \mathbb{R}^2$. For every A' (x', y') there is a unique $A(x, y)$, which suggests that the function is onto. For every $A(x, y)$ and $B(x, y)$, if $A \neq B$, then $A' \neq B'$, which suggests that the function is 1:1.

whole $\mathbb{R}^2$, and the range of this function is the whole $\mathbb{R}^2$, as illustrated for several representative points in Fig. 7.[10]

Our argument up to this point suggests that such conceptualization of rotation as a function might allow learners to overcome their difficulties about rotations (e.g., considering center as an element of both domain and range, the rotation of the center itself and the rotation of the whole plane (Flanagan, 2001)).

1.8 Conceptualizing Reflections Through Quantitative Reasoning

Let us focus on two preimage points, A and B, on the plane and a reflection. These preimage points are mapped to corresponding image points, A' and B', under a reflection r having a line of reflection, l, as the parameter. Note that this mapping preserves the distances (e.g., $d(A, B) = d(A', B')$) but not relative locations in contrast to translations. When teaching reflections in schools, teachers mostly use paper-folding activities to talk about reflections as flipping-over-line activity—a movement rather than a special mapping (Zembat, 2007, 2010). Such treatment might not pave the way to thinking about reflections quantitatively or as functions. How can one quantify reflections keeping the covariation of preimage and image points in mind to go beyond such limited interpretation of reflections? We delineate this issue using the key components of QR in the following argument.

Assuming that $\mathbb{R}^2$ is conceptualized as detailed before, understanding a reflection requires a coordination of the following conditions involving different quantities:

C_r#1: The distance between any two preimage points, say A and B, and the distance between their corresponding image points, say A' and B', are equal. Namely, $d(A, B) = d(A', B')$.

C_r#2: An arbitrary preimage point, A, and its corresponding image point, A', on the plane are equidistant from every point falling on a line in between preimage and corresponding images, called the line of reflection. Namely, for an arbitrary A and C_i in the plane, $d(A, C_i) = d(A', C_i)$ where all such $C_i \in l$, as illustrated in Fig. 8.

C_r#3: Arbitrary three colinear preimage points lying on a line with a particular slope have corresponding image points that are also colinear lying on another line with another slope with a changing orientation that is determined by the line of reflection (as C_r#1 holds true).

Coordination of the conditions C_r#1, C_r#2 and C_r#3, not necessarily hierarchical or sequential, necessitate that the line mentioned in C_r#2 is unique.

We suggest that one way to think about reflections requires one to understand and coordinate the quantities within these conditions in thinking about what is variant or

[10] Note that, as is the case in translations, the center of rotation is also part of the preimage set as it is an element of $\mathbb{R}^2$—in fact, the image of the center in any rotation coincides with itself.

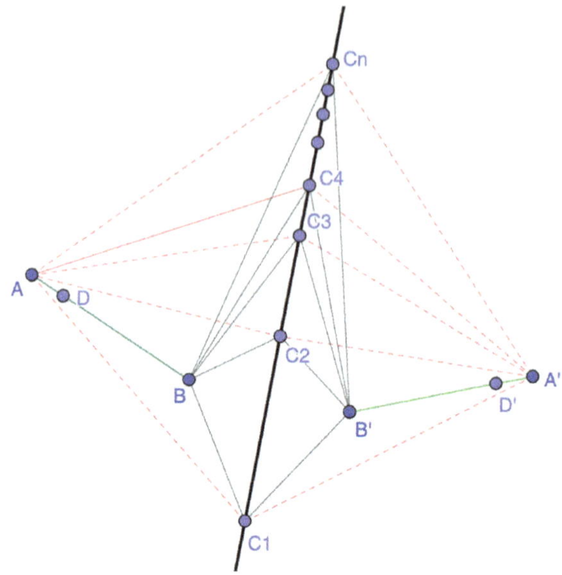

Fig. 8 Set of C_i equidistant from $\{A, A'\}$ and $\{B, B'\}$, and D that is colinear with A and B and D' that is colinear with A' and B'

invariant. An individual needs to think that the magnitudes (as a result, the values) of the distance between A and B and the distance between A' and B', are invariant—$d(A, B) = d(A', B')$. Such thinking is possible if reflection is considered for the distances among paired preimage points as quantities within the domain and the corresponding paired image points within the range. Figure 8 illustrates such comparison of the sets $\{A, B\}$ and $\{A', B'\}$, as previously called *within comparison*. Note that similar to rotations and different from translations, a comparison of the sets $\{A, A'\}$ to $\{B, B'\}$ does not yield to the same-size quantities, so $d(A, A') \neq d(B, B')$.[11] However, the magnitudes of the distance between A and C_i and the distance between A' and C_i, are also invariant—$d(A, C_i) = d(A', C_i)$ where $C_i \in l(i = 1, 2, \ldots)$. In the context of reflections, this second set of quantities is to be used for *between comparison*.

We have so far identified two sets of quantities that can be used for within and between comparisons: (1) '$d(A, B)$ and $d(A', B')$' for *within comparison*; (2) '$d(A, C_i)$ and $d(A', C_i)$' and '$d(B, C_i)$ and $d(B', C_i)$' for *between comparison*. The learner needs to coordinate these two sets of quantities. Articulating the role of C_i in this coordination is important.

The learner here needs to think about such C_i that are equidistant from a preimage point, A, and its image, A'. The same argument can be made for different preimage points (e.g., B as in Fig. 8) on the plane. In other words, there may be such C_i that another preimage point B and its corresponding image B' are equidistant from C_i (see Fig. 8). In order to have a unique line, C_i must have an invariant characteristic, which is that they must be geometrically the mid-points of the quantity pairs $\{A, A'\}$, $\{B, B'\}$,

[11] Note that this is not true for a preimage line parallel to the reflection line, as its image will also be parallel to the reflection line.

etc. This reasoning needs to be expanded to any point that is colinear with A and B. In other words, the learner needs to be able to choose any point in between A and B that is colinear with A and B, say D, whether it is infinitely close to A, or converges to A, or becomes same as A ($D = A$ and $D' = A'$), and coordinate *within comparisons* for the quantities of distances, $d(A, D) = d(A', D')$ as well as $d(D, C_i) = d(D', C_i)$ and $d(A, C_i) = d(A', C_i)$ (see Fig. 8). These quantity pairs involved in this coordination covary, suggesting the continuous covariation of preimage and corresponding image points.[12] In other words, the learner needs to think of *continuous covariation* (Saldanha & Thompson, 1998) of the two sets of quantities previously identified. Simultaneously thinking about (or coordinating) both sets lead one to conserve colinearity in both preimage and corresponding image points, respectively. Preservation of colinearity results in the uniqueness of the line of reflection. Hence, this line, called *line of symmetry*, consists of mid-points of all preimage and corresponding image points. Therefore, $d(A, C_i) = d(A', C_i)$, $d(B, C_i) = d(B', C_i)$, etc.

At this point, the learners' assimilatory structure may be ready to assimilate reflections as functions. Beyond the aforesaid coordination and articulations, the learner also needs to comprehend that every preimage point of the plane is coupled with a

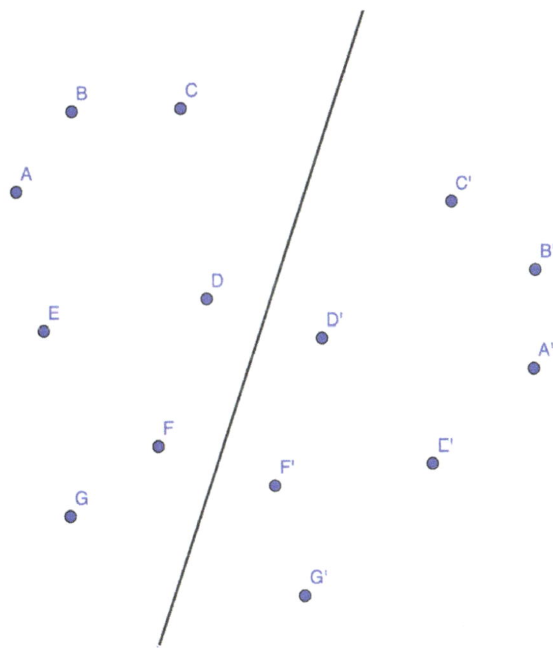

Fig. 9 Reflection of sample preimage points A, B, C, D, E, F, G with respect to line l

[12] Note that the same argument can be made for points B and D. In that case, the coordination of following quantities is needed: $d(B, D) = d(B', D')$ as well as $d(D, C) = d(D', C)$ and $d(B, C) = d(B', C)$. The following quantity pairs involved in this coordination covary: $d(B, D)$ and $d(B', D')$—Covariation #1; $d(D, C)$ and $d(D', C)$ & $d(B, C)$ and $d(B', C)$—Covariation #2. Any chosen point, D, can be infinitely close to B, or converge to B, or be same as B ($D = B$ and $D' = B'$), which also suggests the continuous covariation of preimage and corresponding image points.

unique image point of the plane so that the reflection is a function for the learner. Previously, we mentioned that the learner needs to conceptualize 'any set of points or geometric shapes as quantities relative to the whole, $\mathbb{R}^2$, based on the conditions of M#1, M#2 and M#3. Thus, an arbitrary point of $\mathbb{R}^2$, like D, is coupled with a corresponding D' from the $\mathbb{R}^2$; $r_l(D) = D'$. In other words, for the learner, the reflection function should consist of ordered pairs in the form of (D, D') as multiplicative objects, no two of which has the same first element. Note that both D and D', which are already multiplicative objects themselves, are elements of $\mathbb{R}^2$ and, therefore, $(D, D') \in \mathbb{R}^4$ (Fife et al., 2019).[13] As a result, the domain of this function is the whole $\mathbb{R}^2$, and the range of this function is the whole $\mathbb{R}^2$, as illustrated for a few representative points in Fig. 9.

2 Discussion

It is essential to view isometries as functions as it is dependent on

- conceiving of the whole plane (e.g., all points representing unique locations making up the plane) and its constituents (e.g., given preimages and corresponding images) as part of $\mathbb{R}^2 \times \mathbb{R}^2$,
- conceptualizations of domain and range as $\mathbb{R}^2$, consideration of the parameters as part of the transformations (e.g., vector of translation, the center of rotation, line of reflection), and
- thinking about covariation between pre-image and image points (Steketee & Scher, 2016) to understand the nature of the involved relations.

However, previous research informs us about the learner difficulties in these areas (Hollebrands, 2003; Steketee & Scher, 2011, 2016; Yanik, 2011), which seem to stem from the treatment of isometries as motions only. In this chapter, we have provided an analysis of a kind of thinking needed to understand isometries (translation, rotations, reflections) based on quantitative reasoning, which, we believe, lays a foundation for understanding isometries as functions.

We have argued in the chapter that the conceptualization of $\mathbb{R}^2$ is a key that may allow learners to understand both domain and range of the isometries as the whole plane, including all of its constituents (e.g., planar figures as preimages and their corresponding images, parameters). Our analysis suggests that such conceptualization of $\mathbb{R}^2$ is dependent on understanding any point, say (a, b), as a multiplicative object. This would allow learners to assimilate two measurable attributes: the distance of (a, b) to the origin and the slope of the line passing through the origin and (a, b).

---

[13] Note that, as is the case in translations, the line of reflection is also part of the preimage set as it is an element of $\mathbb{R}^2$—in fact, the image of line in any reflection coincides with itself. Here, by $(D, D') \in \mathbb{R}^4$ we mean $((x_D, y_D), (x_{D'}, y_{D'}))$ or $(x_D, y_D, x_{D'}, y_{D'})$, considering $D = (x, y)$ and $D' = (x', y')$.

Once $\mathbb{R}^2$ is conceptualized as such, understanding an isometry requires coordination of quantities at two levels with certain conditions that are specific to each isometry. These levels are *within comparison* (among distances between preimage couples and corresponding image couples as well as the slopes of the lines that connect these coupled points) and *between comparison* (among distances between preimage-image point couples and/or distances between preimage and image points to the parameters respectively). These comparisons involving the aforesaid quantities are to be coordinated for the learner to understand what varies or what stays invariant under the isometry at hand. The quantities (distances) involved in this coordination covary as such covariation is dependent on the covariation of preimage and corresponding image points while colinearity is preserved. This can trigger the learners' thinking of *continuous covariation* of the quantities included in the aforesaid coordination. Such coordination will be a precursor to assimilating the isometry at hand as a function as the learner comprehend that every preimage point of the plane is coupled with a unique image point of the plane. In other words, for the learner, the isometry function consists of ordered pairs in the form of (D, D') as multiplicative objects, no two of which has the same first element. This suggests that $(D, D') \in \mathbb{R}^4$ (Fife et al., 2019) as the domain and the range of this function is the whole $\mathbb{R}^2$.

Aforesaid analysis has some implications for curriculum design. We argue that the following conceptualizations would benefit students in understanding isometries based on quantitative reasoning:

1. Conceptualizing points on the plane as multiplicative objects,
2. Conceptualization of $\mathbb{R}^2$ consisting of such multiplicative objects by coming to understand two measurable attributes: (i) the distance between any point on the plane from a reference point (e.g., origin) (ii) the location of this point relative to the location of reference, which can be determined by using slopes,
3. Coordination of quantities (distances and slopes) at two levels (within and between comparisons) to understand what stays invariant under isometries,
4. Understanding that these coordinated quantities (distances) covary as colinearity is preserved,
5. Understanding of isometry as an invariant functional relationship (made possible by parameters) among multiplicative objects, ordered pairs in the form of (D, D'), no two of which has the same first element.

In their analysis of the learning progression for geometric transformations, Fife et al. (2019, p. 8) suggests that the "covariational approach to functions" is not applicable to transformations as their analysis is impacted by Confrey and Smith's (1994, p. 137) approach in which covariation "entails being able to move operationally from y_m to y_{m+1} coordinating with movement from x_m to x_{m+1}." However, the above list of conceptualizations in light of our analysis suggests the opposite as our approach adopts Thompson's theory of quantitative reasoning (Thompson & Carlson, 2017, p. 424), in which "a person reasons covariationally when she envisions two quantities' values varying and envisions them varying simultaneously." Though our analysis is a theoretical one and requires empirical validation with further research. A research-based teaching sequence that fosters these conceptualizations is also needed.

References

Argün, Z., Arikan, A., Bulut, S., & Halicioglu, S. (2014). *Temel matematik kavramların künyesi* (An index for fundamental mathematical concepts). Gazi Kitabevi.

Carlson, M., Jacobs, S., Coe, E., Larsen, S., & Hsu, E. (2002). Applying covariational reasoning while modeling dynamic events: A framework and a study. *Journal for Research in Mathematics Education, 33*(5), 352–378.

Confrey, J., & Smith, E. (1994). Exponential functions, rates of change, and the multiplicative unit. *Educational Studies in Mathematics, 26*, 135–164. https://doi.org/10.2307/749228

CCSSM. (2010). *Common core state standards for mathematics*. CCSSM.

Edwards, L. D. (1991). Children's learning in a computer microworld for transformation geometry. *Journal for Research in Mathematics Education, 22*(2), 122–137.

Edwards, L. (2003, February). The nature of mathematics as viewed from cognitive science. *Paper presented at 3rd Congress of the European Society for Research in Mathematics*, Italy.

Fife, J. H., James, K., & Bauer, M. (2019). A learning progression for geometric transformations. *ETS Research Report Series, 1*, 1–16.

Flanagan, K. (2001). *High school student' understandings of geometric transformations in the context of a technological environment*. Unpublished Ph.D. Dissertation, The Pennsylvania State University, University Park.

Hollebrands, K. F. (2003). High school students' understandings of geometric transformations in the context of a technological environment. *The Journal of Mathematical Behavior, 22*(1), 55–72.

Hollebrands, K. F. (2004). High school students' intuitive understandings of geometric transformations. *The Mathematics Teacher, 97*(3), 207–214.

Jones, K. (2002). Issues in the teaching and learning of geometry. In L. Haggarty (Ed.), *Aspects of teaching secondary mathematics: Perspectives on practice* (pp. 121–139). Routledge.

Lakoff, G., & Núñez, R. (2000). *Where mathematics comes from* (Vol. 6). Basic Books.

Lobato, J., & Thanheiser, E. (2002). Developing understanding of ratio as measure as a foundation for slope. In B. Litwiller & G. Bright (Eds.), *Making sense of fractions, ratios, and proportions* (pp. 162–175). National Council of Teachers of Mathematics.

Martin, G. (1982). *Transformation geometry: An introduction to symmetry*. Springer.

Mhlolo, M. K., & Schafer, M. (2013). Consistencies far beyond chance: An analysis of learner preconceptions of reflective symmetry. *South African Journal of Education, 33*(2), 1–17.

Milli Eğitim Bakanlığı Talim ve Terbiye Kurulu Başkanlığı (MEB-TTKB). (2018). *Ortaöğretim matematik dersi (9., 10., 11., 12. Sınıflar) öğretim programi* (Secondary mathematics curriculum for 9–12). MEB.

Moore, K. C., Carlson, M. P., & Oehrtman, M. (2009). The role of quantitative reasoning in solving applied precalculus problems. In *Proceedings of the Twelfth Conference on Research in Undergraduate Mathematics Education (CRUME)*. North Carolina State University.

Portnoy, N., Grundmeier, T. A., & Graham, K. J. (2006). Students' understanding of mathematical objects in the context of transformational geometry: Implications for constructing and understanding proofs. *Journal of Mathematical Behavior, 25*, 196–207.

Saldanha, L., & Thompson, P. W. (1998). Re-thinking co-variation from a quantitative perspective: Simultaneous continuous variation. In S. B. Berensah & W. N. Coulombe (Eds.), *Proceedings of the Annual Meeting of the Psychology of Mathematics Education—North America* (Vol. 1, pp. 298–306). North Carolina State University.

Sinclair, N., Pimm, D., & Skelin, M. (2012). *Developing essential understanding of geometry for teaching mathematics in grades 9–12* (pp. 71–82). National Council of Teachers of Mathematics.

Sirotic, N., & Zazkis, R. (2007). Irrational numbers: The gap between formal and intuitive knowledge. *Educational Studies in Mathematics, 65*, 49–76. https://doi.org/10.1007/s10649-006-9041-5
Steketee, S., & Scher, D. (2011). A geometric path to the concept of function. *Mathematics Teaching in the Middle School, 17*(1), 48–55.
Steketee, S., & Scher, D. (2016). Connecting functions in geometry and algebra. *The Mathematics Teacher, 109*(6), 448–455.
Stevens, I., & Moore, K. (2017). The intersection between quantification and an all-encompassing meaning for a graph. In E. Galindo & J. Newton (Eds.), *Proceedings of the 39th Annual Meeting of the North American Chapter of the International Group for the Psychology of Mathematics Education.* Hoosier Association of Mathematics Teacher Educators.
Sunker, S. F., & Zembat, I. O. (2012). Teaching of translations through use of vectors in Wingeom-tr environment. *Ilkogretim Online, 11*(1), 173–194.
Thompson, P. W. (1990). *A theoretical model of quantity-based reasoning in arithmetic and algebra (working paper).* Center for Research in Mathematics & Science Education—San Diego State University.
Thompson, P. W. (1994). The development of the concept of speed and its relationship to concepts of rate. In G. Vergnaud, G. Harel, & J. Coufrey (Eds.), *The development of multiplicative reasoning in the learning of mathematics* (pp. 179–234). SUNY Press.
Thompson, P. W. (2011). Quantitative reasoning and mathematical modeling. In L. L. Hatfiled, S. Chamberlain, & S. Belbase (Eds.), *New perspectives and directions for collaborative research in mathematics education—WISDOMe Monographs* (Vol. 1, pp. 33–57). University of Wyoming.
Thompson, P. W., & Carlson, M. (2017). Variation, covariation, and functions: Foundational ways of thinking mathematically. In J. Cai (Ed.), *Compendium for research in mathematics education* (pp. 421–456). National Council of Teachers of Mathematics.
Thompson, P. W., Hatfield, N. J., Yoon, H., Joshua, S., & Byerley, C. (2017). Covariational reasoning among US and South Korean secondary mathematics teachers. *The Journal of Mathematical Behavior, 48*, 95–111.
Usiskin, Z., Peressini, A. L., & Marchisotto, E. (2003). *Mathematics for high school teachers: An advanced perspective.* Prentice Hall.
Yanık, H. B. (2006). *Prospective elementary teachers' growth in knowledge and understanding of rigid geometric transformations.* Unpublished Ph.D. Dissertation, Arizona State University.
Yanik, H. B. (2011). Prospective middle school mathematics teachers' preconceptions of geometric translations. *Educational Studies in Mathematics, 78*(2), 231–260.
Yanik, H. B. (2014). Middle-school students' concept images of geometric translations. *The Journal of Mathematical Behavior, 36*, 33–50.
Yanik, H. B., & Flores, A. (2009). Understanding rigid geometric transformations: Jeff's learning path for translation. *The Journal of Mathematical Behavior, 28*(1), 41–57.
Yılmaz, R. (2020). Dönüşüm geometrisi (Transformation geometry). In A. Kaçar (Ed.), *Matematiğin temelleri* (pp. 335–365). Pegem Akademi.
Zembat, I. O. (2007). Yansıma dönüşümü, doğrudan öğretim ve yapılandırmacılığın temel bileşenleri (The main tenets of direct instruction and constructivism: The case of reflections). *Gazi University—Journal of Gazi Faculty of Education, 27*, 195–213.
Zembat, I. O. (2010). A micro-curricular analysis of unified mathematics curricula in Turkey. *ZDM—The International Journal on Mathematics Education, 42*(5), 443–455. https://doi.org/10.1007/s11858-010-0236-y

Instructional Conventions for Conceptualizing, Graphing and Symbolizing Quantitative Relationships

Marilyn P. Carlson, Alan O'Bryan, and Abby Rocha

1 Orienting to a Problem

We ask readers to think about the context in Table 1 before beginning this chapter and reflect on the reasoning you use to conceptualize and represent the relationships you consider.

What did you think about or imagine when reading the text? Make a drawing to represent the situation, then use that drawing to describe how pairs of quantities, whose values vary, are related and change together. How many distances appear in the description? Try describing each distance so that it is clear which distance you are referencing. What details did you need in your descriptions? Which distances have a value that varies, and which distances have a fixed value (or measure)? Can you verbalize how the distance between the Tortoise and Hare changes during the race? How might you support students in being able to conceptualize, verbalize, and represent this relationship?

M. P. Carlson (✉) · A. O'Bryan · A. Rocha
Arizona State University, Tempe, AZ, USA
e-mail: marilyn.carlson@asu.edu

A. O'Bryan
e-mail: avatar44@asu.edu

A. Rocha
e-mail: aerocha@asu.edu

G. Karagöz Akar et al. (eds.), *Quantitative Reasoning in Mathematics and Science Education*, Mathematics Education in the Digital Era 21,
https://doi.org/10.1007/978-3-031-14553-7_9

Table 1 One version of the tortoise and hare task (Carlson et al., 2020, p. 17)

A tortoise challenges a hare to a 100-m race and convinces the hare to give him a 60-m head start. They both are moving at a constant speed when the start gun is fired, with the hare running the entire race at a constant rate of 3.6 m/s and the tortoise moving at a constant rate of 0.4 m/s for the duration of the race

2 Introduction

Thompson (2008a) argued that "in the United States, the vast majority of school students rarely experience a significant mathematical idea and certainly rarely experience reasoning with ideas" (p. 31) due to "a systemic, cultural inattention to mathematical meaning and coherence" (2013, p. 57). Most U.S. students experience mathematics only as groups of procedures to memorize and employ (Boston & Wilhelm, 2017; Hiebert et al., 2005; Hill, 2021; Jackson et al., 2015; Laursen, 2019; Litke, 2020; Schmidt et al., 2005; Simon et al., 2000; Stigler et al., 1999; Stigler & Hiebert, 2009; Thompson, 2008a, 2013). In such a setting, students typically do not develop mathematical practices that lead to fluency in solving novel problems; nor do they construct strong meanings for key ideas necessary for success in calculus and STEM fields. For example, early studies of students' understanding of the function concept revealed weak meanings in students' function conception (e.g., Monk, 1992; Sierpinska, 1992; Vinner & Dreyfus, 1989). Studies report that students view (i) a function graph as a picture of an event (Bell & Janvier, 1981; Carlson, 1998; Leinhardt et al., 1990; Monk, 1992) or a static shape with specific properties (Carlson et al., 2002) and (ii) an algebraically defined function as a recipe for getting an answer (Breidenbach et al., 1992) or two expressions separated by an equal sign (Carlson, 1998).

We call for curriculum designers, professional development leaders, course coordinators, and instructors to foster learning experiences that support students in developing productive *ways of thinking* and coherent mathematical meanings essential for understanding calculus and continuing in STEM fields. It is our goal that students become confident and competent mathematical thinkers and problem solvers. Our data supports that students who understand ideas and acquire habits of reasoning meaningfully will be better equipped to spontaneously engage in productive reasoning and access their knowledge when learning new ideas and solving problems. This perspective is aligned with Harel and Thompson's call [as explained in Thompson et al. (2014)] for students to develop the "habitual anticipation of using specific meanings or ways of thinking in reasoning" (p. 13). For example, recent studies have identified ways of thinking that foster the emergence of function graphs as a record of a student's conception of how the value of one quantity varies with the value of another (Carlson et al., 2002; Moore & Thompson, 2015) and function formulas as a record of how pairs of quantities' values are related and vary together (Moore & Carlson, 2012; O'Bryan, 2020a; O'Bryan & Carlson, 2016).

Our call for mathematics instructors to support students in using and trusting their thinking when attempting novel problems and learning new ideas echoes the current recommendations from professional organizations (e.g., American Mathematical Association of Two-Year Colleges, 2018; National Governors Association Center for Best Practices, Council of Chief State School Officers, 2010; Winsløw, 2021). However, even when targeted and sustained professional development training is available, and instructors use research- and inquiry-based materials, researchers report only minor shifts in most teachers' instructional practices, with many instructors continuing to focus on lecture as a means of transmitting knowledge to students (e.g., Baş-Ader & Carlson, 2021; Jackson et al., 2015). Such teacher-centered instruction is not attentive to student thinking, nor does it reveal to teachers the variety of ways students are conceptualizing and reasoning about mathematical ideas. Understanding and assessing student progress in applying their reasoning has been shown to be valuable for informing teacher task selection, questions, and explanations. These *decentering* actions (Baş-Ader & Carlson, 2021; Piaget, 1955; Steffe & Thompson, 2000; Teuscher et al., 2012) are a critical component of responsive teaching whereby instructors act in the moment and adapt their instruction by leveraging students' thinking to successfully make progress toward a lesson's learning goals.

3 The Need for Conventions to Facilitate Changes in Pedagogy and Student Success

Thompson and Carlson (2017) call for introductory undergraduate courses in mathematics to be alert to the static images many students possess for variables, function formulas, and graphs. They call for curriculum and instruction to reengage precalculus level students in trusting and using their reasoning to conceptualize quantities in a problem context, and then illustrate how a student's image of how quantities are related can lead to constructing symbols and graphs that carry meaning for the one constructing them. However, during 15 years of working with and studying precalculus instructors in the Pathways Project, we rarely encountered an instructor who engaged in and valued pedagogical practices for supporting students' development of dynamic imagery related to function relationships and their representations. Most instructors have only experienced traditional curriculum in both their learning and teaching experiences. They need focused professional development to help them reconceptualize mathematical ideas they thought they understood and to reconceptualize effective teaching as focused on and affecting student thinking.

We developed Pathways curriculum materials (Carlson & Oehrtman, 2010) to support instructors in fostering productive reasoning patterns in their students that research has revealed to be essential for students' construction of meaningful function formulas (e.g. Moore & Carlson, 2012; Thompson, 1988, 1990, 1992) and graphs (e.g., Carlson et al., 2002; Moore & Thompson, 2015). These include specific support for: (i) conceptualizing and speaking about quantities and how their values vary together, (ii) representing how two quantities change together using a graph, and (iii) representing quantitative relationships with expressions and formulas. As we refined the Pathways course materials over time, we strove to design for coherence by emphasizing reasoning about and making connections between the three strands of school mathematics that prepare students for calculus: the mathematics of quantity, the mathematics of variation, and the mathematics of representational equivalence (Thompson, 2008a).[1] We included problems, teacher notes, and other resources that we believed would help instructors engage students in constructing strong meanings for ideas from these strands in each lesson and support students in making connections between ideas across multiple strands.

In scaling the use of the Pathways materials, however, we faced persistent challenges in shifting instructors' pedagogical actions and perspectives on student learning to achieve our intended learning goals. Many instructors routinely missed opportunities to conceptualize how the ideas in each lesson were related to one of these three strands or make connections between ideas in different strands. This led us to reexamine the focus of our professional development training and to develop pedagogical conventions that we believed would support both instructor and student learning if enacted. We were also guided by Thompson's (1990) elaboration of the role of quantitative reasoning in students' construction of meaningful algebraic expressions. This led to our introducing instructional conventions over the course of 15 years when working with instructors, including classroom observations, professional development workshops, preservice teacher courses, and training programs for graduate teaching assistants. For example, when students successfully reasoned about applied contexts, we noted instructor moves that supported students in orienting to the problem (Polya, 1957) and engaging in productive problem-solving behaviors (Carlson & Bloom, 2005). In contrast, when students struggled to make sense of ideas or problems, we noted pedagogical actions that were not taken that could have supported students in making sense of relationships in the problem. We also analyzed clinical interview data that suggested actions (such as drawing detailed diagrams) that led to students' successfully conceptualizing quantitative relationships (Moore & Carlson, 2012).

---

[1] Thompson (2008b) describes these strands as follows. The mathematics of quantity refers to how individuals conceptualize measurable attributes of a situation, create measurement schemes to quantify the attributes' magnitudes, represent the quantities in various ways, and generalize aspects of these attributes. The mathematics of variation refers to how individuals imagine quantities with magnitudes that can vary, how they represent this variation in different ways, and how they draw inferences from noticing what in a relationship remains invariant as two quantities change in tandem. The mathematics of representational equivalence refers to how individuals think of arithmetic and, eventually, algebraic expressions non-computationally as "segues into structural properties of numbers and quantitative relationship[s]" (p. 7). These three strands are interrelated, and opportunities always exist to discuss elements of one strand even within contexts emphasizing another strand. For example, while supporting students in conceptualizing quantitative relationships in some given scenario it can be very natural to explore how two or more of the conceptualized quantities co-vary in tandem. Thompson states that "The three strands in interaction, each receiving appropriate emphasis, and always with the other two in the background, builds a foundation for algebraic reasoning that simultaneously builds a foundation for schemes of meanings that are crucial for understanding the calculus" (pp. 8–9).

Our observations revealed that even when teachers were committed to making their instruction more engaging and meaningful for students, they continued to rely on familiar instructional practices of providing vague explanations and showing students how to find answers. For students to build personal meanings for mathematical ideas requires that instructors create opportunities for them to construct these meanings and engage in productive habits of reasoning as often as possible (Harel, 2008). Our attempts to support instructors in consistently engaging their students in meaningful mathematical activity that begins with their conceptualizing quantities and their relationships as a basis for their graphing and defining activities led to our introducing the instructional conventions described in this chapter. These conventions include underlining phrases in a problem statement that describes quantities, precisely referencing quantities when speaking, constructing a drawing that depicts how quantities are related, physically tracking quantities as their values vary, etc. We introduce them to focus and structure student mathematical activity in support of their meaningful engagement in conceptualizing problems and learning new ideas.

4 Elaborating Quantitative and Covariational Reasoning

We adopt Thompson's theory of quantitative reasoning to explicate the ways of thinking we desire students to construct. According to Thompson (1988, 1990, 1993, 1994, 2011, 2012) *quantitative reasoning* is rooted in a disposition to conceptualize situations in terms of measurable attributes of objects and relationships among them. Quantitative reasoning then is a way of thinking about situations whereby an individual conceptualizes measurable attributes of objects (*quantities*) and organizes relationships between these attributes to form a structured mental representation of the situation. Quantities are mental objects unique to an individual. The way an individual conceptualizes a quantity and the set of quantities deemed relevant provides the space of implications for the reasoning an individual can engage in relative to a given situation (Smith & Thompson, 2007; Thompson, 1994). A key element of quantitative reasoning is *quantification* whereby an individual develops a method for reliably representing a conceptualized quantity's magnitude with a numerical value (that is, its measure). The quantification process is critical for an individual to develop meaningful mathematical models because "[i]t is in the process of quantifying a quality that it [the quality] becomes truly analyzed" (Thompson, 1990, p. 5). The nature of the conceptualized quantity matters a great deal in this process, as does the sophistication of the individual's magnitude schemes (see Thompson et al. (2014) for a discussion of these ideas). For example, the quantity "distance between two people" can be quantified as a multiplicative comparison to some imagined unit (either a standard unit like "meter" or "inch", or a nonstandard unit such as "length of the measurer's foot" or "the length of the measurer's pace"). Some quantities, however, such as the speed of a baseball pitch or the impact force of an automobile striking another automobile arise via conceptualizing a *quantitative operation*, or a mental operation of comparison/coordination of other quantities the individual has already conceptualized (Thompson, 1990, 1994, 2011). The resulting quantity cannot be directly measured in the same way that "the distance between two people" can be directly measured. As a result, quantifying these more complex quantities requires a scheme dependent on the quantitative operations from which they arose (Johnson, 2015; Moore, 2010; Piaget, 1968; Schwartz, 1988; Simon & Placa, 2012; Thompson, 1990; Thompson et al., 2014).

It is also important that students distinguish between quantities with a fixed magnitude and those with a magnitude that can vary. This distinction is key in conceptualizing mathematical models of a dynamic situation as representing the simultaneous covariation of two quantities' values. An individual reasons covariationally when she envisions two quantities' values varying in tandem (Carlson et al., 2002; Saldanha & Thompson, 1998; Thompson & Carlson, 2017), and holds in mind a sustained image of the two quantities' values simultaneously (Saldanha & Thompson, 1998).

Thompson's (1988, 1990, 1993, 1994, 2011, 2012) theory of quantitative reasoning "is about a stratum of reasoning that lies beneath both applied arithmetic and applied algebra. It is about people using 'rigorously qualitative' reasoning, where rigor derives from the intention to attend to the quantification of a situation's qualities" (p. 3). The theory is most useful in considering how individuals come to understand quantifying qualities like heat, force, and torque—qualities that cannot be quantified via extensive measurement. However, for individuals to participate in quantifying and using this category of quantities in mathematical modeling they must begin by conceptualizing calculations, variables, and algebraic expressions as tools for representing the quantitative relationships they have conceptualized.

Despite the body of research pointing to the essential role of quantitative and covariational reasoning in students' mathematical development, there is a broad body of research in calculus learning that points to students' failure to conceptualize quantities, how they are related, and vary together as sources of challenges using calculus ideas to advance a problem's solution (Bressoud et al., 2016; Byerley, 2019; Carlson, 1998; Engelke, 2007; Mkhatshwa, 2020; Oehrtman, 2009; Thompson & Harel, 2021; Thompson, 1992; Zandieh, 2000). Our work in the Pathways Project focuses on operationalizing Thompson's theory of quantitative reasoning so that students develop the beliefs, expectations, and ways of thinking necessary to participate in meaningful mathematical modeling for success in calculus and STEM fields. In later sections, we elaborate how these constructs informed our thinking and design of Pathways conventions for supporting precalculus mathematics students' engagement in quantitative and covariational reasoning.

For now, we echo our claim that students' ability to construct meaningful formulas and graphs rely on their using quantitative reasoning to build a structured mental model of quantities (measurable attributes of objects) within a situation. To make this claim more transparent, we ask you to revisit the Tortoise and Hare context and follow-up questions we presented at the beginning of this chapter. Consider again how the following contributed to your ability to describe how the distance between the Tortoise and Hare changed during the race: (i) your conception of the quantities described in the text, (ii) your conception of the quantities to be related and how they vary together; (iii) the clarity with which you conceptualized the quantities (e.g., where a quantity's measurement begins, the direction of the measurement), and (iv) the clarity with which you represented the quantities and their relationships in a drawing.

Data from administering an item that presented the above context and asked students to define the Tortoise's distance (in meters) ahead of the Hare in terms of the number of seconds since the start of the race revealed that very few of over 1000 precalculus students were able to produce a correct answer. Further, data from administering the Mathematical Association of America's *Calculus Concept Readiness* (CCR) exam to 601 students from three different universities during their first week of calculus revealed widespread weaknesses in students' ability to define function formulas and interpret function graphs. In addition, only 28% of these 601 students selected the correct response (out of five multiple choice options) to an item that asked them to define the area A of a circle in terms of its circumference, C (Carlson et al., 2015). CCR data thus suggests widespread difficulties in students' ability to define function formulas to relate two quantities whose values vary as they begin calculus.

5 *Speaking with Meaning*: A Convention for Improving Instructors' Communication

Analysis of video data from instructors' professional learning communities (PLCs) (also reported in Clark et al., 2008) showed instructors being imprecise in referencing quantities and saying what a variable, expression, graph, and function formula represented when communicating with each other. These instructors regularly made vague references to a volume, height, time, etc. without making clear what volume, height, or time they were considering. We also noticed a pervasive use of pronouns that made it difficult for other instructors to understand what the speaker was imagining and conceptualizing when completing a problem. Their inability to be specific in describing and representing quantitative relationships appeared to reveal their weak conceptions and it was common for instructors to pretend to follow incoherent explanations resulting in meaningless exchanges among the instructors.

After our pointing out the difficulties we were experiencing in following their explanations, we collectively negotiated a specific goal to speak more meaningfully when discussing ideas and problems during the PLCs. This led to the project leaders negotiating with the PLC members patterns of speaking that we conjectured would improve communication about the mathematical ideas and how they are learned. We collectively decided to restrict the use of pronouns by requesting that all PLC members be precise in referencing the quantity they were imagining, including the direction of measurement, the starting point for the measurement, and the unit of measure. Our goal was to focus instructors' attention on the coherence of their speaking and how they might be interpreted by others. We were hopeful that reflecting on these issues would motivate them to expend the mental energy to improve in their ability to *speak with meaning*. Since retrospective analysis of the PLC videos revealed that the instructors' classroom explanations mirrored those they provided in the PLCs, we were hopeful that *speaking with meaning* would become normative within their classroom discussions as well.

Our subsequent analysis of the PLC videos, after agreeing on conventions for speaking, revealed instructors gradually becoming more fluent in referencing quantities and describing quantitative relationships. The instructors' language steadily shifted (with consistent reinforcing) to their describing the quantity they were conceptualizing by stating what was being measured, the unit of measurement, the starting point for the measurement, and the direction of the measure (e.g., Juan's distance in feet north of the stop sign). We further noticed that the instructors' ability to precisely reference the quantities and describe how pairs of quantities are related was accompanied by improvements in the instructors' ability to construct formulas and graphs that accurately represented the quantitative relationships described in a problem. When instructors expressed frustration while attempting to *speak with meaning* about a problem they were discussing, they typically had not taken time to conceptualize the quantities described in the problem context.

As our work to support instructors in communicating their thinking to their peers continued, we gradually introduced other conventions for communication, including: (i) the speaker verbalizing her thinking and the rationale for her choices, rather than describing what she did to get an answer; (ii) all members of the PLC attempting to make sense of the thinking of the speaker, instead of only listening to the words being spoken; and (iii) all members of the PLC asking a question if something was unclear, instead of pretending to understand when they were unable to follow. We formalized the convention *speaking with meaning* as a research construct for our continued study by saying,

> An individual who is *speaking with meaning* provides conceptually based descriptions when communicating with others about solution approaches. The quantities and relationships between quantities in the problem context are described rather than only stating procedures or numerical calculations used to obtain an answer to a problem. Solution approaches are justified with logical and coherent arguments that have a conceptual rather than procedural basis. (Clark et al., 2008, p. 297)

According to Yackel and Cobb (1996), considering how others might make sense of explanations requires a shift in perspective from only viewing explanations as something one gives or hears to making the explanations themselves an object of reflection. Thus, when one speaks, they concurrently imagine how their utterances might be interpreted. The capacity for an individual to anticipate how they might be interpreted has been termed *decentering* (Piaget, 1955; Steffe & Thompson, 2000). Our perspective on decentering is elaborated elsewhere (Baş-Ader & Carlson, 2021; Carlson et al., 2004; Teuscher et al., 2012), with these studies revealing that US secondary and university precalculus instructors are generally not oriented to making sense of students' thinking or considering how their explanations might be interpreted by others. This finding is consistent with findings reported in international studies (e.g., TIMSS) and research on teaching (Baş-Ader & Carlson, 2021; Teuscher et al., 2016; Thompson, 2013, 2016). As a result, our data and observations of the impact of instructors adopting the convention that a speaker *speak with meaning* and a listener's attempt to make sense of another's spoken words, revealed cognitive shifts in instructors' conceptions, and substantial shifts in instructors' conversation toward an improved understanding of each other's perspective (Clark et al., 2008).

We should note that in a few cases a PLC leader did not consistently model or reinforce *speaking with meaning*. The conversations and explanations in these PLCs did not shift to become more meaningful or coherent and instructors were observed agreeing with incorrect solutions and illogical explanations. According to Yackel and Cobb (1996), a sociomathematical norm refers to a normative behavior specific to mathematics, such as understanding what constitutes an acceptable mathematical solution or what counts as an acceptable mathematical behavior in a group setting. We reemphasize that the pattern of *speaking with meaning* as a new sociomathematical norm only became normative in settings where the PLC leader was consistent in modeling *speaking with meaning* and consistent in reinforcing speaking with meaning among the instructors.

6 Scaling the Convention of Speaking with Meaning Across the Pathways Project

After five years of research and development of Pathways interventions, the pre-post-gains of student learning using both the validated PCA (Carlson et al., 2010) and Calculus Concept Readiness (CCR) exams (Carlson et al., 2015) were highly significant. The mean PCA scores ranged from 13.5 to 18 (out of 25), representing pre- post-gains of 5–9 points on average. At this stage of the Pathways project, we made the curricular materials and professional development available for other universities, creating an opportunity for us to continue documenting speaking patterns among new communities of Pathways users.[2] The trends described previously among communities of new Pathways instructors, including using a single word to define a variable, providing calculational explanations, etc. were normative at the beginning of all 12 Pathways professional development workshops that proceeded a university deciding to adopt Pathways materials. In Table 2 we provide concrete examples of common speaking patterns and contrast vague speaking with what we considered to be speaking that is more meaningful.

[2] Since our initial introduction of the term *speaking with meaning*, 11 new colleges/universities have participated in Pathways professional development.

Table 2 Examples of *speaking with meaning* compared to statements that show an absence of *speaking with meaning*

Speaking with meaning	
Absence of speaking with meanings	Speaking with meaning
• The graph of the car's distance falls to the right	• The car's distance south of the stop sign (in feet) is decreasing as the number of seconds since the car started moving increases
• f (7) is 20 tells me that when I plug in 7 I get 20	• Since f (7) is equal to 20, the tank had 20 gallons of water 7 min after the tank started draining
• I multiplied 1.08 by $2000 to get my answer	• Since the amount I must pay is 8% more than the price, the amount I must pay is 1.08 times as large as $2000
• 24 divided by 5 is 4.8 because 5 goes into 24 4.8 times	• Since I need to determine how many times as long 24 in. is as compared to 5 in., I must divide 24 by 5. My answer tells me that 24 in. is 4.8 times as long as 5 in.
• Since the graph curves up the distance is getting larger from 0 to 5	• During the first 5 s of the race, the distance travelled by the runner over successive fixed amount of time increases. This also means that the runner is speeding up during the first 5 s or the race

6.1 Speaking with Meaning in Instruction and Curriculum

We leveraged the insights from our study of PLCs (Clark et al., 2008) in our initial workshops with instructors preparing to use Pathways materials at a particular university. In doing so, we consistently modeled *speaking with meaning* and asked for clarification when workshop participants were unclear about what quantity they were referencing. Imprecise quantity references that failed to describe the starting point of the quantity's measurement such as "time elapsed" or "hours passed" were called out for clarification (e.g., "the number of minutes since 9 am", the number of seconds since the car left home). When a workshop participant used pronouns or made imprecise statements like, "its distance is getting closer" the workshop leader might ask questions like, "What is getting closer?" "What distance?" "Closer to what?", hoping to raise the instructor's awareness of the need to be more specific in describing the distance the instructor was imagining. Our persistent probing typically generated a response like, "the car's distance north of the intersection is decreasing". The reinforcement of *speaking with meaning* in the initial workshop made the criteria for *speaking with meaning* public among all workshop participants. It is also noteworthy that workshop participants began to ask each other for clarification when speaking among themselves.[3] The Pathways curriculum materials further supported

[3] In universities where a commitment to *speaking with meaning* became a social mathematical norm and was highly valued by the local coordinator in our initial Pathways workshop, we have since documented the persistence of *speaking with meaning* in this local community of instructors.

an instructor's shift to *speak with meaning* by providing open-ended questions, opportunities for students to select and clearly define variables of interest, and requests for students to explain and justify their reasoning. The instructor materials also provide detailed solutions and explanations that model *speaking with meaning*, giving instructors clear examples of the preciseness in speaking that is needed when responding to specific problems and questions. Our observations of Pathways instructors' classrooms reveal gradual shifts in their effectiveness in using and reinforcing *speaking with meaning* with variation in their: (i) consistency in modeling *speaking with meaning*; (ii) commitment to making *speaking with meaning* a classroom convention all students adopt; and (iii) consistency in asking for clarification when a student produces vague descriptions and/or explanations.

6.2 *Emergent Shape Thinking* and Conventions for Meaningful Graphing Activity

Researchers have documented students' impoverished conceptions of graphs, including their conceiving of a graph as a picture of an actual event (e.g., Bell & Janvier, 1981; Carlson, 1998; Kaput, 1992; Leinhardt et al., 1990; Monk, 1992). For example, Carlson (1998) reported that high performing precalculus students interpreted the speed-time graphs of two cars as the paths on which the cars were driving and the intersection of the two graphs as a collision location. In the same study she reported that high performing second semester calculus students could say nothing more detailed about a graph's inflection point than it being the location on a graph where the graph changes concavity, and when pressed to explain what the concavity conveyed about the quantitative relationships they responded with comments about the curvature of the graph. In contrast, more recent studies have revealed five levels of student reasoning as they attempt to construct a graph of two quantities as their values varied together in a non-linear pattern (Carlson et al., 2002) and have demonstrated the utility of students' graphing activities as emerging from their conceptualizing two quantities' values varying in tandem while imagining how the two quantities' values are changing together (e.g., Carlson et al., 2002; Moore & Thompson, 2015).

Moore and Thompson (2015) classified and contrasted students' ways of thinking about graphs in terms of the thinking they used to construct the graph. They say that a student is engaging in *static shape thinking* if the student conceptualizes the "graph as an object in and of itself, and as having properties that the student associates with learned facts." For example, a student who constructs a graph by plotting points and applying algebraic methods to identify roots, inflection points, maximum/minimum values would be engaging in *static shape thinking*. In contrast, a student is said to be engaging in *emergent shape thinking* when the graph's trace emerges from the student considering two quantities' values as they vary together. A student engaging in the first three mental actions described in the Carlson et al. (2002) covariation framework would be engaged in *emergent shape thinking*. The mental actions as characterized

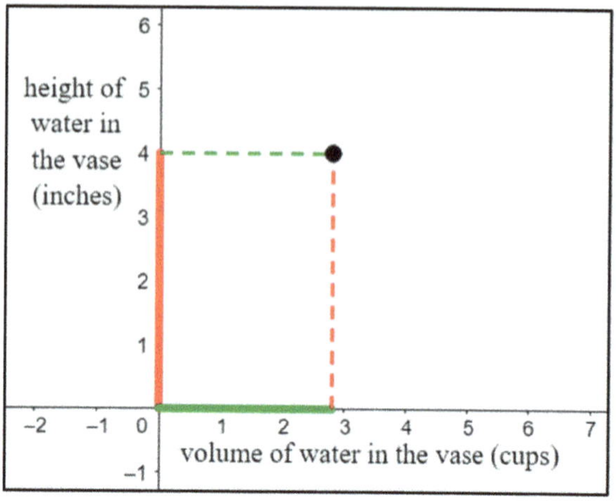

Fig. 1 A conception of a point as the simultaneous values of two quantities (a multiplicative object)

in the context of a student constructing a graph of the height of water in a vase in terms of the volume of water in the vase entails: (i) conceptualizing two quantities' values varying together (i.e., the volume of water in the vase and the height of the water in the vase) (MA1); (ii) conceptualizing the two quantities' values varying simultaneously and continuously, while considering the direction of the variation of each quantity's value (as the volume of water in the vase increases the height of the water in the vase increases) (MA2); (iii) conceptualizing how the values of the two quantities vary together by imagining a successive fixed amount of variation in one quantity while considering the amount of variation in the other quantity (considering how much the water's height varies while considering successive fixed increases in the volume of water) (MA3).

When the student has linked together two measurable attributes of the vase (the volume of water in the vase and the height of the water in the vase), they have conceptualized a *multiplicative object*, an object that simultaneously combines the attributes of two conceived quantities (Saldanha & Thompson, 1998; Thompson, 2011). A student who has conceptualized the two quantities as a multiplicative object may find it useful to sometimes consider the variation in one quantity only; however, when doing so the student will have a persistent awareness that the other quantity's value is also varying (Thompson & Carlson, 2017) (see Fig. 1). The volume of water in the vase (in cups) and the height of water in the vase (in inches) vary in tandem, and a point is used to represent the simultaneous correlated values of each quantity.

7 Pathways Conventions for Graphing

The Pathways Project uses specific conventions to support students in conceptualizing a graph as a record of how two quantities' values vary together. Prior to introducing graphing, we engage students in using what we call the *quantity tracking tool.*

While observing a dynamic event (e.g., someone walking across the room from one wall to another) or an applet that displays a dynamic event (e.g., a vase filling with water) in which at least two quantities' values are varying, students are prompted to move their index fingers to track the variation in two quantities' values. The *quantity tracking tool*, as first described in Thompson (2002), supports students in conceptualizing graphs as emergent traces produced from the coupling of values for two co-varying quantities. According to Thompson (2002), conceptualizing graphs as a record of simultaneous variation requires having students.

> internalize their perceptions of two quantities whose values vary, making that variation experientially concrete. To make covariation of quantities values experientially concrete, it is essential that they envision a single quantity's variation as itself having momentary states and therefore that the attribute whose value varies has momentary values. (p. 206)

When introducing the *quantity tracking tool*, the instructor negotiates a location for students' index fingers that represents a measurement of 0 units for each quantity either on (or beside if negative values are being tracked) each student's desk. The instructor also engages the students in deciding on a direction for moving each index finger when representing increasing and decreasing values.[4] The instructor then prompts students to use one index finger to track the variation in one quantity's value (e.g., the height of water in a vase in inches) as they observe the dynamic movement or applet. This is repeated several times, with the instructor prompting specific students to describe what they were imagining as they moved their finger. Students next pick another quantity (in the same problem context) whose value is varying in tandem with the first quantity (e.g., the volume of water in the vase in cups or the number of seconds elapsed since the waterspout was turned on). After the class has decided on a direction for the measurement, they track the quantity's value as they again observe the dynamic event or applet. The instructor may also stop the event or applet and prompt students to say what they are imagining as they move their index fingers. The instructor then prompts students to use both index fingers to simultaneously track the two quantities' values while observing the situation unfold. To ensure that students are engaging in quantitative reasoning, it is important that the instructor prompt students to explain what their finger movement represents, what the initial location of one or both of their index fingers represents, how they know what direction to move their fingers, etc. (see the next section for a detailed example of using the *quantity tracking tool*).

It is noteworthy that use of the *quantity tracking tool* requires students to both conceptualize and track a quantity's measurement, including where the measurement begins and the measurement's direction, prior to moving both index fingers to track the simultaneous value of the two quantities as their values vary. The convention that each finger be moved up or down (or right or left) from a common starting point keeps students' focus on the magnitude of the quantity's value in contrast to

[4] When initially using the *quantity tracking tool* students typically decide that a positive value of one quantity is represented by a distance upward from a starting point and that positive values of a second quantity are represented by a distance to the right of the same starting point, while negative values are downward and to the left respectively.

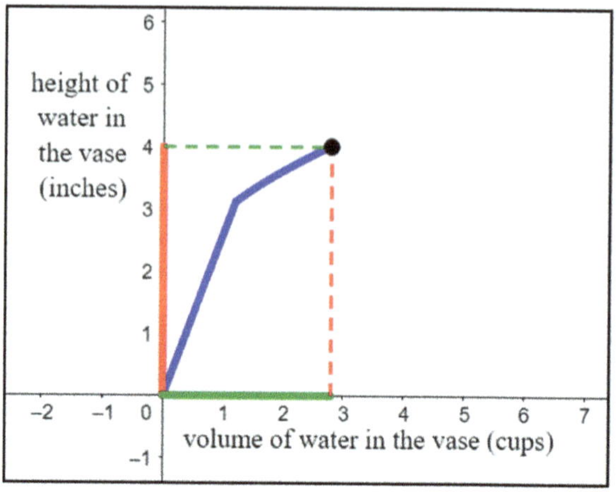

Fig. 2 Graphs emerge as traces

tracing a pre-imagined shape. We also see the *quantity tracking tool* as providing a meaningful foundation for conceptualizing coordinate axes as measurement tools for two quantities' values. Ours and others' studies (e.g., Frank, 2017a, 2017b) suggest that students typically lose sight of the fact that each axis is a measuring tool for one quantity, and that every point on a graph represents the simultaneous measurement of two quantities. The Pathways convention of drawing dashed lines from the point back to the axes (as in Fig. 1) are one attempt to reinforce the conception of a point as the simultaneous value of the two quantities at an instance during the unfolding of the dynamic event.

We encourage instructors to leverage the ways of thinking supported in the *quantity tracking tool* as a basis for constructing graphs in their lessons. As the unfolding of a dynamic event is displayed in an applet or animation, instructors support students in seeing the graph materialize with the leading point on the emergent trace projected back to the axes with dashed lines (Fig. 2) and the point representing the simultaneous values of two quantities (with values determined from the axes). In this way attributes of the relationship between the covarying quantities also emerge as "properties of covariation" (Moore & Thompson, 2015, p. 786), such as whether one quantity increases or decreases as the other quantity increases and where and why this behavior may change.

8 Implementing the *Quantity Tracking Tool*

We designed an applet to assist students in conceptualizing the quantities in the Tortoise and Hare context. Recall that the Hare agrees to give the Tortoise a 60-m head start (Fig. 3).

As the instructor moves a slider smoothly to vary the number of seconds since the start of the race (or hits "play" and allows the slider to move automatically), each

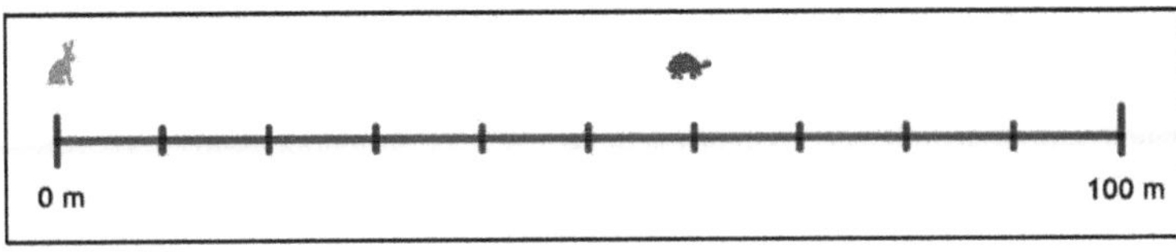

Fig. 3 The initial positions of the tortoise and hare

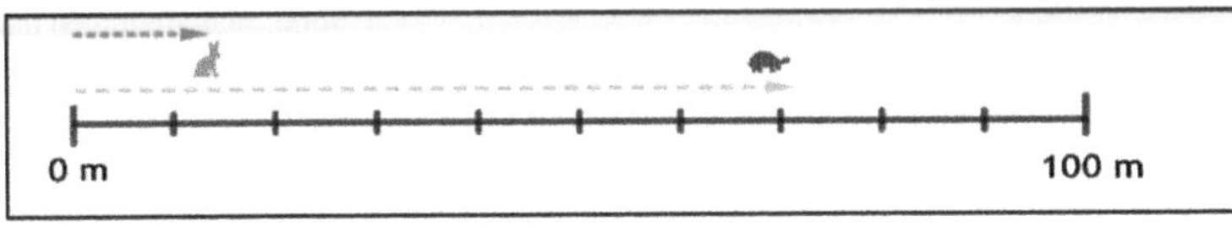

Fig. 4 Tracking each animal's distance from the starting line

animal's distance from the starting line is represented as a dashed line with an arrow pointing to the right (Fig. 4).[5]

As the applet plays, *the elapsed time since the start of the race* varies as does *the amount of time remaining until the Tortoise finishes the race* and *the amount of time remaining until the Hare finishes the race*. There are also five distances with values that vary as the race plays out. Let us assume that the instructor has begun exploring this context with her class and is now interested in exploring how *the distance (in meters) the Tortoise is ahead of the Hare* varies with the *elapsed time (in seconds) since the start of the race*. The conceptions and imagery students construct while the teacher engages them in a conversation using the applet and *quantity tracking tool* is impacted by the instructor's effectiveness in focusing students' attention on conceptualizing the relevant quantities and how they vary together. To illustrate this point, we provide a brief example of hypothetical instructor and student actions for promoting quantitative reasoning in students (see Table 3[6]). Note that the questions posed by the instructor in the context of the applet and *quantity tracking tool* continue to promote *speaking with meaning*.

Constructing a graph that carries meaning for the one constructing it relies on the individual having in mind a goal to represent the simultaneous variation of two quantities' values as they vary together. When enacting the *quantity tracking tool*, students' attention is focused on coordinating the magnitudes and directed measures of quantities with values that vary in tandem. As students make decisions about the starting position for their index fingers, they consider the starting point (0 value) for each measurement. As they begin to track a quantity's value by moving their

[5] The instructor selects a specific button to indicate which two quantities to isolate when exploring the concurrent variation in two quantities' values.

[6] The convention of *quantitative drawing*, as explained later in this chapter, can further support students in conceptualizing the quantities we want them to coordinate with the *quantity tracking tool*. The conventions in this chapter all support each other to maximize students' learning opportunities.

Table 3 Utilizing the *quantity tracking tool* in a class setting with students to help them conceptualize relevant quantities and relationships between quantities

	Instructor actions/statements to students	Student actions	Follow-up questions/observations
1	"Place your right index finger on your starting position for measuring the elapsed time since the race began."	Students choose a starting position for their fingers	Pose questions to specific students: "What does the position of your finger represent?" If student say "time," ask them, "What time?" Continue probing students until your weakest students can verbalize a precise description of the quantity (e.g., "The number of seconds elapsed since the start of the race.")
2	"As I move the slider on the applet, move your right index finger in a way to represent the number of seconds elapsed since the start of the race." (repeat if some students don't participate)	All students move their right index finger to the right from the designated starting point for measuring the number of seconds elapsed since the start of the race. The motion is smooth and continuous. See Fig. 5	Pose questions to specific students: "What were you imagining as you moved your finger? What does the starting position of your finger represent? What does the ending position of your finger represent? Should your finger be moved smoothly? Explain."
3	"To represent that the tortoise is a *distance of 0 m ahead of the hare*, position your left index finger at the same initial position as your right index finger. Positive values will be above this position."	Students position their left index fingers at the same initial position as their right index finger	Students often overlook exploring how the distance that the tortoise is ahead of the hare varies during the running of the race. Ask students to point out where they "see" this quantity's magnitude represented on the drawing
4	"Now, place your left index finger on a starting position that represents the approximate distance that the tortoise is ahead of the hare at the start of the race."	Students place their left finger some distance above the starting reference point	Note that some students will likely place their finger at the position that represents 0 m from the starting line. Ask these students to say what a 0 value represents. Ask questions like, "Where is the tortoise relative to the hare at the start of the race?"
5	"As I move the slider on the applet, move your left index finger in a way to represent the approximate distance between the tortoise and hare."	All students move their left index downward from their designated starting point. The motion is smooth and continuous. See Fig. 6	If some students do not participate or move their index finger in the wrong direction, prompt them to explain their thinking; then replay the race being run

(continued)

Table 3 (continued)

	Instructor actions/statements to students	Student actions	Follow-up questions/observations
6	Prompt students to model *the tortoise's distance ahead of the hare* in terms of *the number of seconds since the start of the race* **using both index fingers** at the same time	All students coordinate the motion of their right and left index fingers to represent the relationship between the two quantities' values. The motion is smooth and continuous. See Figs. 7 and 8	Since textbooks and instructors commonly make requests for students to represent one quantity "in terms of" another. It is important to introduce this language (and make sure students are clear on what it means) so students can attach this request to the mental operations of covarying *the tortoise's distance ahead of the hare* with the *amount of time that has elapsed since the start of the race*

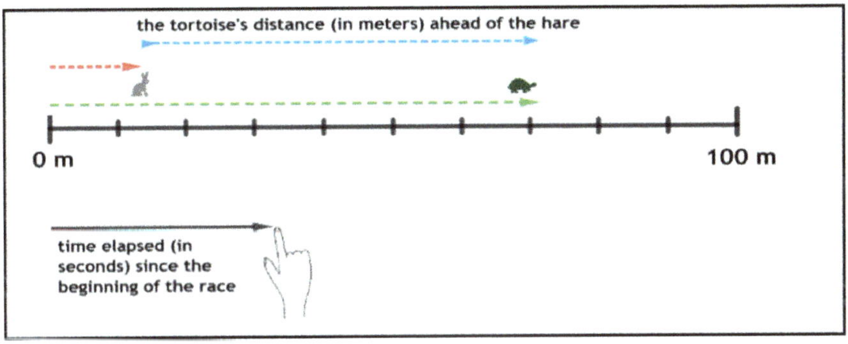

Fig. 5 A student's right hand as she uses her index finger to sweep out a duration of elapsed time while the animation plays

index finger, they associate a direction of movement for positive measurements and a direction for negative measurements (Figs. 5 and 6).

A point on a graph in the coordinate plane is then viewed as the co-occurring values of the two quantities (a multiplicative object) at an instance as the two quantities' values vary together (Figs. 7 and 8).

Given the broadly documented difficulties students encounter in creating and interpreting graphs (e.g., Carlson, 1998; Monk, 1992) and the commonly held view that a point on a graph is a result of a sequence of actions (count over 4 and down 5), shifting students to conceptualize graphs as an emergent trace of the values of two covarying quantities requires repeated reinforcement. Tracking the simultaneous variation in the two quantities' values by physically moving one's index fingers together promotes students' conceiving of the two quantities' values as coupled (a

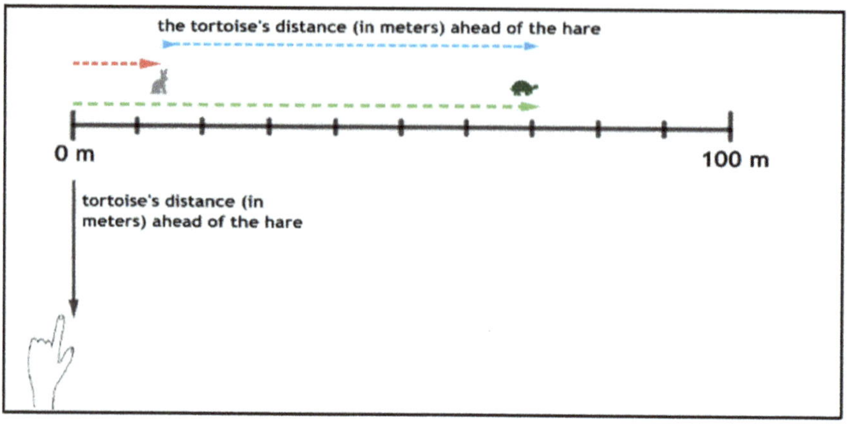

Fig. 6 A student's left hand as she uses her index finger to sweep out the distance the tortoise is ahead of the hare while the animation plays

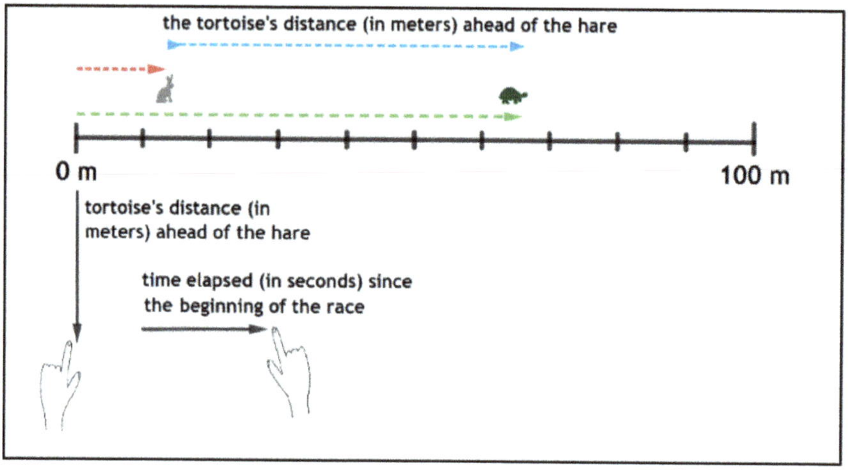

Fig. 7 A student using her index fingers to coordinate time elapsed (in seconds) since the beginning of the race with the Tortoise's distance (in meters) ahead of the Hare

multiplicative object) (Fig. 8), while considering how the values of the two quantities vary together.[7]

[7] The conceptualizations for enacting the *quantity tracking tool* parallels the thinking for constructing a graph of two quantities' values as they vary together.

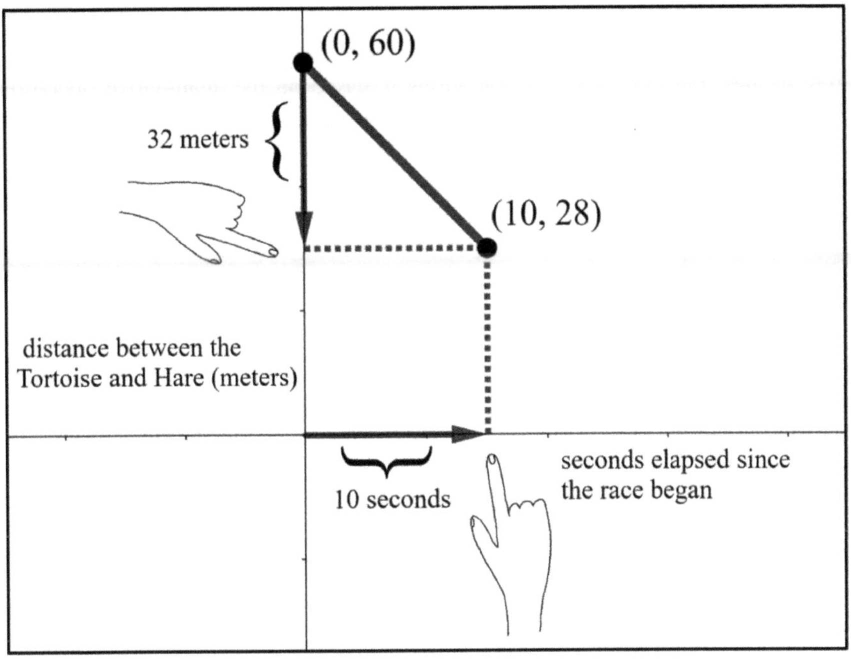

Fig. 8 The imagery we want students to develop for a graph as an emergent trace while using the *quantity tracking tool*

9 *Emergent Symbol Meaning* and Conventions for Meaningful Symbolization Activity

Generating a meaningful algebraic representation of a relationship between quantities' values relies on individuals organizing quantities they find relevant in a structured mental model. Constructing a structured mental model of the quantitative relationships depends on the individual's ability to conceptualize new quantities by relating two other quantities. As an example, one might conceptualize the relative size of quantity A with respect to quantity B and represent this new quantity as a quotient, (size of quantity A)/(size of quantity B), where the result, as well as any expression representing that result, is understood to be that relative size measurement. Representing a quantitative structure in a drawing can be useful for advancing an individual's image of a problem's quantitative structure while laying a foundation for producing a meaningful algebraic representation that relates two quantities' values as those values vary.[8]

[8] We repeat that the conventions and ideas described throughout this chapter support each other. *Speaking with meaning*, the *quantity tracking tool*, and *quantitative drawing* all support students in using symbols meaningfully and making connections between different representations of the same relationship.

Thompson (1990, 1993, 2011) draws a careful distinction between quantitative operations and arithmetic operations. A person uses arithmetic to calculate quantities' values, but the choice of operations is based on the quantitative operation conceptualized—the way the individual has formed a new quantity in their mind as a relation involving two other quantities. For example, "difference" as a quantitative operation is an additive comparison[9] between two quantities. However, a difference is not always evaluated using subtraction. Thompson uses the following problem to demonstrate this idea with a difference evaluated via division: "Jim is 15 cm taller than Sarah. This difference is five times greater than the difference between Abe and Sam's heights. What is the difference between Abe and Sam's heights?" (Thompson, 1990, p. 11). This example illustrates why the common pedagogical approach of training students to key on specific words like "difference" when determining the operation for combining two quantities to represent a new quantity can be problematic and sometimes leads to constructing symbols that do not represent the quantitative relationships described in the problem.

It is also our experience that instructors often focus on the "sameness" of the solutions produced by multiple solution paths instead of highlighting and emphasizing the unique reasoning, variety in quantities and structures conceptualized, different conceptualizations of problem goals, and so on. They also tend to ignore parallels between steps in a reasoning process (including the evaluation of various quantities) and the steps in a numerical or algebraic solution process (including highlighting which quantities are evaluated/represented at each step). Instead, they defer to algebraic "equivalence" even when the solutions represent unique ways of conceptualizing the context. The impact of this is that students receive the message that, even though there are multiple solution paths, there is one preferred path and algebraic form of a solution, which often does not foster students' confidence in their own mathematical reasoning and sense-making.

10 Quantitative Reasoning and Algebra

It is the attempt to generalize the quantification process for a quantitative relationship where algebraic expressions enter the picture and where distinctions between quantitative and arithmetic operations become critical. Thompson (1992, 1996) describes a study where fourth-grade students were free to develop their own notational methods for communicating their reasoning about decimal place value within activities he designed to encourage students to move back and forth between a situation and how the individual wanted to express that situation with notation. The result was that

[9] An "additive comparison" is the answer to the question, "By how much does the magnitude (or value) of one quantity exceed the magnitude (or value) of another quantity?" In contrast, one category of "multiplicative comparisons" is the answer to the question, "One quantity is how many times as large as a second quantity?".

students talked about their images of the situations presented "*as they spoke about notational actions* [italics in original]" (Thompson, 1996, p. 16).

As a precursor to seeing an algebraic expression as representing the value of a quantity (in underdetermined form), two understandings are paramount. (1) The individual has a productive conceptualization of the quantity for which he is attempting to represent the value. (2) The individual sees the expression that evaluates the quantity, as well as its numerical value, as representing the quantity's measurement. For example, imagine dropping an object from a tall building. Between two moments in time, the object's height above the ground changes from 42 to 29 ft. In order to evaluate the quantity, *the change in the object's height above the ground*, the individual must understand that the change in the object's height above the ground is a *difference*, and moreover there is a frame of reference (the 0 value is the ground and a positive value represents a distance above the ground) such that the difference from an initial value to a final value is a directed change that indicates the direction of the movement (upward or downward). With this conceptualization, he can understand that the expression "29–42 ft" represents the value of this change (directed difference), as does "– 13 ft". Individuals with both understandings are poised to understand how the variable expression "$h - 42$ ft" (where h represents the object's height above the ground in feet) represents the change in the object's height above the ground (from its initial height) at any moment during its fall. Thus, it is important that students see unevaluated expressions as representing a new quantity that is the result of performing a designated quantitative operation, in addition to the final numerical value (Thompson, 2011) (see Table 4). *Note that the table describes basic quantitative operations for sum, difference, product, and quotient. This list is not exhaustive. Again, we emphasize that student conceptualization of quantitative operations is what determines their choice of operation—not key words. And more complex quantities such as measurements of force, torque, work, and so on often require multiple levels of comparison and coordination.*

Within Thompson's theory, there is no assumption that proficiency with quantitative reasoning (or a disposition to reason quantitatively) is a natural outcome of participating in mathematics courses. In fact, research reports frequently point out how instruction in the United States produces students (and instructors) without this proficiency (e.g., Moore & Carlson, 2012; O'Bryan, 2020b; Thompson et al., 2017; Yoon et al., 2015). As we mentioned earlier, professional development training designed to introduce instructors to the usefulness of quantitative reasoning and provide support for implementing activities to support quantitative reasoning in the classroom often failed. Introducing Pathways course materials have helped make shifts in students' and instructors' meanings possible, but it remains an ongoing challenge to help instructors create lessons and consistently engage students in discourse focused on helping students strengthen their quantitative reasoning skills. O'Bryan and Carlson (2016) report on one instructor who did create such a learning environment for her students. The primary finding was that this instructor had internalized a set of expectations about engaging in mathematical reasoning (especially in how that reasoning connected to algebraic representations of mathematical relationships). It was these expectations that drove her use of quantitative reasoning and the

Table 4 Quantities to be evaluated along with the expressions that represent those quantities' values

Concept	Specific example	General example
Combining quantities additively	Mario is 6.5 years older than his sister Lexi. If x represents Lexi's age in years since she was born, Mario's age is $x +$ 6.5 years	Quantity A is measured in some unit and has a measurement of a in that unit. Quantity B is measured in a compatible unit and measures b in that unit. The sum $a + b$ represents the measurement of the quantity formed by combining Quantity A and Quantity B additively
Additive comparisons of two quantities	Michael is 62 in. tall and Maria is 55 in. tall. Michael is 62–55 in. taller than Maria	Quantity A is measured in some unit and has a measurement of a in that unit. Quantity B is measured in a compatible unit and measures b in that unit. The difference $a - b$ represents the amount by which the measurement of Quantity A exceeds the measurement of Quantity B
Combining quantities multiplicatively	The radius is 1.7 in. long and the arc length is 2.8 times as long as the radius. The length of the arc is (2.8)(1.7) times as long as the radius	Quantity A is measured in some unit and has a measurement of a in that unit. Making n copies of Quantity A produces a resulting quantity that is n times as large as Quantity A, with measurement $n \cdot a$ in the unit of Quantity A
Comparing quantities multiplicatively	The height of water in an empty pool increased 5 in. in 3 h as water flowed into the pool at a constant rate of change. The quotient 5/3 tells us the relative size of the height (in inches) of the water in the pool is, as compared to the number of hours since the water started flowing into the pool	Quantity A is measured in some unit and has measurement a in that unit. Quantity B is measured in some unit and has a measurement b in that unit. The quotient $a/b = $ c is the relative size of a and b, with the quotient c representing how many times as large a is than b (and is usually associated with conceptualizing average rates of change)

activities she designed for students. O'Bryan (2018, 2020a) called these beliefs and expectations *emergent symbol meaning* (see next section).

As students develop meanings for ideas like relative size, differences, change in a quantity's value, etc., it is critical that instruction and activities encourage a "dialogue" within the student about the quantities conceptualized and the numerical methods for calculating the quantitative relationship's value. It is also critical to emphasize how symbolic methods for representing this value as an unevaluated expression such that the order of operations parallels the calculations performed mirrors the steps in the reasoning process that motivated those calculations. The algebraic expressions then represent the same reasoning from which the quantities were conceptualized, but where the quantity's value is undetermined (Thompson, 1990, 2011).

10.1 Emergent Symbolization

As mentioned earlier, Moore and Thompson (2015) coined the term *emergent shape thinking* to describe reasoning about graphs so that (1) they emerge as traces of how two quantities change together in tandem, (2) individuals see already-complete graphs as having been generated as emergent traces and they can imagine the coordination that produced the graph, and (3) individuals conceptualize properties of the relationship between the co-varying quantities through this emergent trace. They argue that this way of thinking about graphs is useful because "students thinking about graphs emergently are positioned to reflect on their reasoning to form abstractions and generalizations from their reasoning… not constrained to a particular labeling and orientation" (pp. 787–788). We emphasize that *emergent shape thinking* describes both how someone could reason about generating a graph but also describes how that person could interpret a graph produced by someone else. In other words, the individual's graphing scheme contains an expectation about what it means to reason about graphs as well as actions related to producing graphs.

We argue for a similar idea related to productive reasoning about generating and interpreting algebraic formulas that relate two or more quantities' values in a situation. O'Bryan (2018, 2020a) described *emergent symbol meaning*[10] as a similarly productive set of expectations, beliefs, and meanings related to generating and interpreting quantitatively meaningful algebraic statements. Individuals may have any combination of these expectations to varying degrees of sophistication, and their expectations may be situation dependent. The following list is expanded from the original definition.

[10] Note that O'Bryan (2020a) uses *emergent symbol meaning* to describe a set of meanings and expectations that may guide an individual's algebraic symbolization activity and interpretation of the algebraic symbols that others generate. He uses *emergent symbolization* when describing the actions an individual engages in that are motivated by these meanings and expectations.

1. An expectation that performing calculations or generating expressions should reflect a quantification process for quantities that the individual conceptualizes.
2. An expectation that demonstrating calculations and producing expressions are attempts to communicate an individual's meanings. Thus, when given a set of calculations or an expression/formula, we can hypothesize how the individual conceptualized a situation based on analyzing the products of their reasoning.
3. An expectation that the order of operations used to perform calculations, evaluate expressions, and solve equations reflects the hierarchy of quantities within a conceptualized quantitative structure.[11]

Like emergent shape thinking, we emphasize that emergent symbol meaning describes the motivations and goals for how an individual could reason about the process of developing algebraic statements *and* how that person could reason about the algebraic statements someone else produces. As with emergent shape thinking, students with these expectations "are positioned to reflect on their reasoning to form abstractions and generalizations from their reasoning" (Moore & Thompson, 2015, p. 787).

O'Bryan (2018, 2020a) provides examples of how students with alternative sets of beliefs and expectations for their mathematical activity tended to be inattentive to quantities in choosing and justifying numerical operations and algebraic representations. It is worth mentioning that emergent symbol meaning is not just about framing a productive set of beliefs and expectations for students. We again point readers to O'Bryan and Carlson's (2016) report on how an instructor who internalized these beliefs and expectations was positioned to support productive mathematical discourse and to develop tasks that allowed her to decenter relative to her students' thinking. This is why we believe that introducing emergent symbol meaning as an explicit element of the theory of quantitative reasoning is useful. It can help orient researchers and curriculum designers to something important in attempts to foster quantitative reasoning—the beliefs and expectations students and instructors have regarding their mathematical activity. Without altering these beliefs and expectations we have found little success in shifting instructors or students to valuing and utilizing quantitative reasoning. See Table 5 for some examples of less and more productive beliefs about calculations and symbolization.

---

[11] Numbers 1 and 3 in this list might seem quite similar, but there is a different intent. The first item focuses on the expectation that all calculations or parts of expressions should represent an evaluation process for some quantity in the situation, and thus each calculation or part of an expression can be quantitatively justified (and, if the individual cannot justify it, then it provides a motivation to reconsider how she has conceptualized the situation). The third item is about how mathematically equivalent expressions with different orders of operations reflect different ways of understanding the situation and that manipulations, including "simplifying" or rewriting an equation to solve for a different variable, may require reconceptualizing the quantitative structure to make sense of the result.

Table 5 Some beliefs and expectations we and others have found to be unproductive for students along with a similar list of more productive beliefs and expectations grounded in Thompson's theory of quantitative reasoning

Some beliefs and expectations guiding mathematical activity: calculations and symbolization	
Examples of less productive beliefs and expectations	Examples of productive beliefs and expectations (emergent symbol meaning)
Variables always stand in for an unknown value to be solved for (Jacobs, 2002; Lozano, 1998)	A variable represents the value of a quantity when that value is not fixed. We also use variables to allow us to express the value of one quantity in terms of the value of another quantity when those values change in tandem
Numbers and letters are paired together by looking for key words like sum, difference, and product and should match the form of examples in the current textbook section or instructor demonstrations	Calculations and algebraic expressions reflect the relationships between quantities' values as conceptualized by the individual. The first steps in mathematical reasoning are to make sense of the problem context and identify quantities and their relationships
The equal sign in a statement is an indication that something must be "solved for" or calculated	An equal sign indicates that you have expressed the value of a quantity in two ways (and thus the expressions on each side of the equal sign represent the value of the same quantity), including the possibility that one of those is as a targeted constant value. Equal signs thus express equality in both value and meaning (or equality in value between two like quantities)
If the answer to a question is an algebraic expression, equation, or formula, then students should always and immediately simplify their answers as much as possible	The order of operations for evaluating an algebraic expression reflects the quantitative structure the individual conceptualized. Mathematically equivalent statements do not necessarily indicate equivalent reasoning, and much information can be gained from examining and discussing non-simplified expressions (practicing and understanding the purpose of simplification is a separate mathematical idea)

10.2 *Emergent Symbolization* in Instruction and Curriculum

Part of supporting the development of emergent symbol meaning, and thus a propensity to reason quantitatively and analyze others' reasoning quantitatively, involves instructors endorsing and highlighting key differences in students' reasoning. An instructor should look for opportunities to discuss the thinking that led to a student's choice of operations; she could also prompt students to explain the thinking that determined the order in which calculations were performed. As students produce algebraic expressions the instructor should ask them to explain what specific terms and expressions represent. The instructor might ask, "What quantity's value is being represented on each side of the equal sign?" An instructor's overarching goal is to support students in conceptualizing quantities and how they are related as a habitual way of acting for constructing symbols that carry meaning for the student. The instructor can foster students' habitual use of quantitative reasoning by consistently requiring students to make a drawing that represents the quantitative structure of a problem. Our challenge has been to get instructors to consistently adopt these practices in both their teaching and own reasoning. We are developing new approaches aimed at gradually shifting instructors' commitment to viewing symbols as emerging from their conceptions of quantities and how they are related (emergent symbol meaning) as a perspective for modeling dynamic situations in mathematics and science with function formulas.

11 An Example of Unproductive Beliefs in Action

We do not have space within this paper to provide multiple examples of how students without the expectations described by *emergent symbol meaning* operate when working in mathematical contexts. However, we present one brief example from O'Bryan (2020a). O'Bryan found that students tended to produce linear models for contexts where exponential models were expected. It appeared that students were trying to translate English goal statements using mathematical symbols without attending to the quantities involved or the meaning of the expressions they produced. For example, in trying to model the height of a plant that was 7 in. tall when first measured and that grew by 13% per week, a large majority of students provided the result shown in Fig. 9.

Students' explanations revealed a failure to notice that $0.13t$ did not represent a number of inches. One student, when trying to represent the height of a different plant that grew by 50% for two weeks and was four inches when first measured, produced the answer shown in Fig. 10. Even though his answer "8 in." is incorrect based on the conventional meaning of percent change, what is most interesting is that (1) the expression he wrote, $4 + 2 \cdot ½$, fits the pattern in Fig. 9, (2) he noticed and verbalized that the expression did not produce the value he claimed it did, and

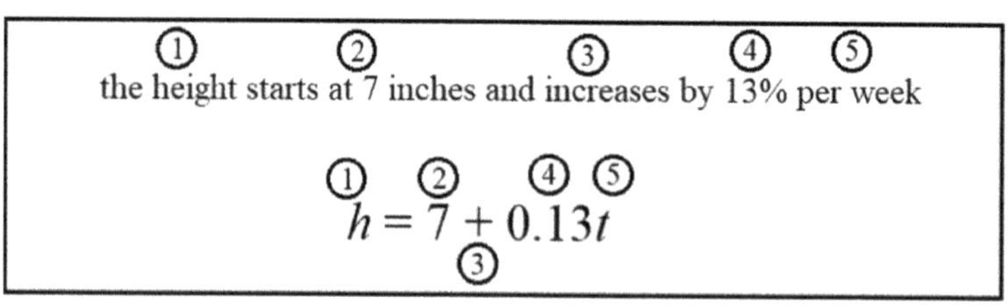

Fig. 9 Students appeared to produce answers as literal translations of English words into mathematical symbols (O'Bryan, 2020a, p. 453). The circled numbers are for emphasis only and were not written by students

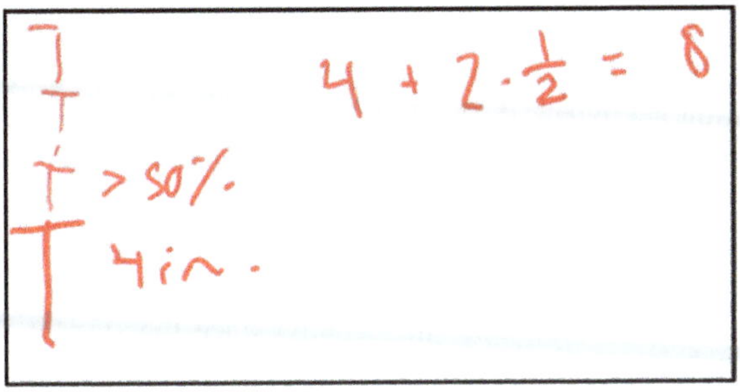

Fig. 10 A student's work justifying a plant's height after two weeks if its first measured height was four inches and it grew 50% per week (O'Bryan, 2020a, p. 451)

(3) he chose not to try to rewrite the incorrect statement since he was confident his final answer was correct.

According to Thompson (1996),

> the expression of an idea in notation provides [a student with] an occasion to reflect on what she said, an occasion to consider if what she said was what she intended to say and if what she intended to say is what she said. To act in this way unthinkingly is common among practicing mathematicians and mathematical scientists. Behind such a dialectic between understanding and expression is an image, most often unarticulated and unconsciously acted, of what one does when reasoning mathematically. This image entails an orientation to negotiations with oneself about meaning, something that is outside the experience of most school students. […] [T]he predominant image behind students' and instructors' notational actions seems to be more like 'put the right stuff on the paper.' (p. 12)

The effect of students' unreflective combining of symbols leads to students encountering significant barriers in their future math classes. For example, the students' work in Fig. 10 is likely the product of many years in mathematics classes without an emphasis on the quantitative significance of notational activity. Focusing instruction on promoting emergent symbol meaning and emphasizing important conventions such as speaking with meaning, emergent shape thinking, quantitative drawing, careful variable definitions, etc. helps students focus their thinking on conceptualizing quantitative relationships. This provides the necessary foundation to help shift students to viewing function formulas and graphs as two ways of representing how two quantities' values vary together.

11.1 *Quantitative Drawing* and Building Imagery for Quantitative Relationships

As noted earlier, a major initial challenge in supporting precalculus instructors to maximize the impact of the Pathways research-based materials has been advancing their understanding of the course's key ideas and how they are learned. This includes their acquiring productive conceptions of ideas of variable, function, function composition, constant rate of change, exponential growth, etc., and their viewing function graphs and formulas as ways of representing the constrained covariation of two quantities' values. A second major challenge has been to support instructors in shifting their teaching to have a primary focus on developing and leveraging student thinking toward the goal of supporting their students in relying on their reasoning as a foundation for emerging as confident and competent mathematical thinkers.

To provide a concrete example of an instructional shift we are trying to achieve, see Table 6. We contrast two Pathways instructors' approaches to helping students respond to a request to define the distance that the Tortoise is ahead of the Hare in terms of the time (in seconds) since the start of the race. Note that we would describe Instructor B as engaging in *quantitative drawing* because conceptualizations of quantities and their relationships to each other are foregrounded in the conversation and highlighted in representations. We once more emphasize that the convention of *quantitative drawing*, with practices that include writing clear variable definitions and using vectors to represent a quantity's magnitude when that magnitude can vary, becomes even more powerful when the instructor also emphasizes *speaking with meaning*, *emergent shape thinking*, and *emergent symbol meaning* (most of which are practices within these exchanges).

The contrasting approaches in Table 6 illuminate how an instructor's commitment to quantitative reasoning influences her instructional orientation, discussions, and questions. What we found to be surprising (and now predictable over time) is the strong commitment new Pathways instructors have to showing students steps for obtaining answers and how little value they place on helping their students use their own reasoning to make sense of a problem context prior to trying to write formulas and construct graphs. As examples of how these views surface during instruction, notice that Instructor A appears to be the one doing most of the thinking and her questions to students are almost exclusively focused on how to find the answer, what to write, or what to do (e.g., What is the formula for the Hare's distance? What did you do to get $60 - 3.2t$?). In contrast, Instructor B consistently makes requests and poses questions to engage students in identifying and conceptualizing the quantities in the problem context and considering how they are related (e.g., What are you imagining measuring? How should I represent the length of the race?) prior to making requests for students to represent quantities' values with symbols. It is also noteworthy that Instructor B consistently *speaks with meaning* when interacting with students and is careful to specify the quantity when defining variables.

Table 6 Comparing instructor actions when those instructors value quantitative reasoning or do not value quantitative reasoning

Instructor A: doesn't value QR	Instructor B: values QR
Focuses students' attention on finding an answer • Reads the problem statement once • Picks out values explicitly listed in the problem statement	The instructor focuses students' attention on reasoning about the situation in the problem statement • Reads the statement of the problem multiple times • Asks students what they can imagine measuring • Directs students to underline phrases that describe attributes that have or can assume a measurement value
Poses questions for the purpose of getting an answer • Any ideas on how to solve this problem?	Poses question for the purpose of gaining insight into how the students are thinking about the problem • How are you imagining this situation? • Can you make a drawing to represent how you are visualizing the situation?

(continued)

Table 6 (continued)

<table>
<tr><th>Instructor A: doesn't value QR</th><th>Instructor B: values QR</th></tr>
<tr><td>Makes a drawing and labels it with values stated in the problem context
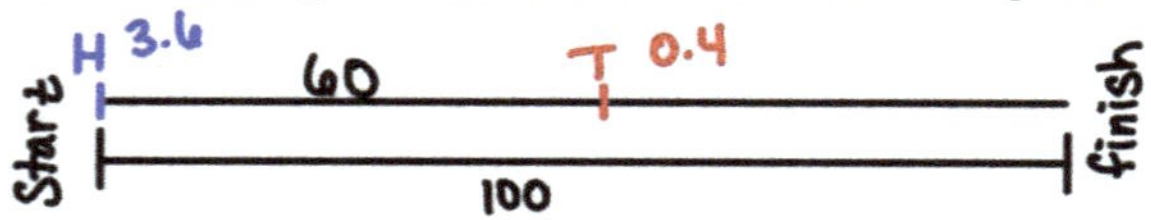
</td><td>Uses the students' interpretation of the problem statement to represent each quantity in the situation. Prompts students to consider the following:
• How should I represent the length of the race?
• Where should I place the hare? The tortoise?
• What values change during the race? How should I represent distances that vary?
• The instructor develops the drawing, one quantity at a time (labels picture carefully), with input from students
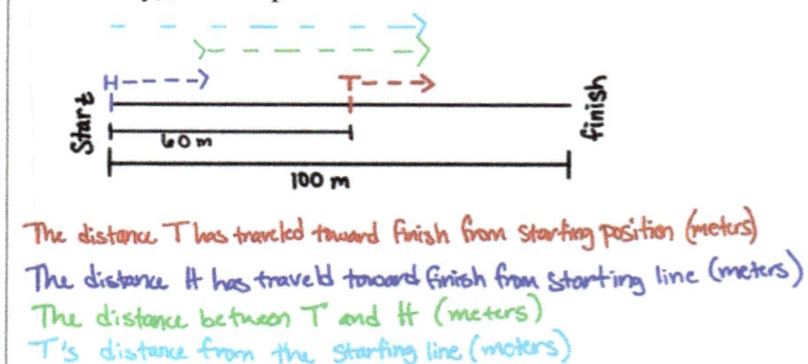
</td></tr>
<tr><td>Defines variables generically
• t = time
• d_T = Tortoise' distance
• d_H = Hare's distance
• d_b = Distance between</td><td>Defines variables carefully
• t = time in seconds (or number of seconds) since the start gun was fired
• d_T = the tortoise's distance traveled (in meters) since the starting gun was fired
• d_s = the tortoise's distance (in meters) from the starting line
• d_H = the hare's distance (in meters) from the starting line
• d_b = the distance (in meters) that the Tortoise is ahead of the Hare</td></tr>
</table>

(continued)

Table 6 (continued)

Instructor A: doesn't value QR	Instructor B: values QR
Updates drawing on the board 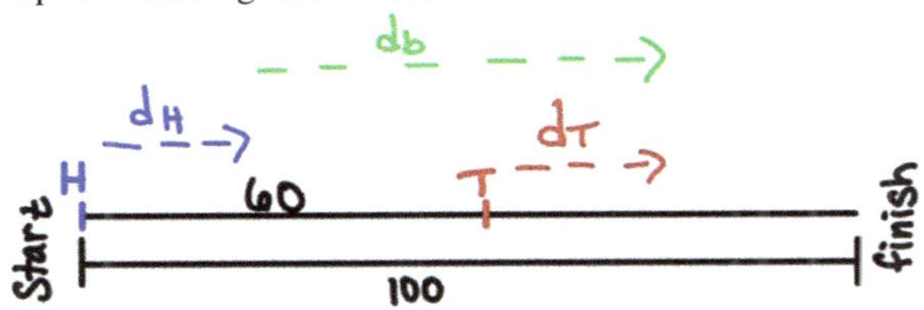 	Updates drawing on the board 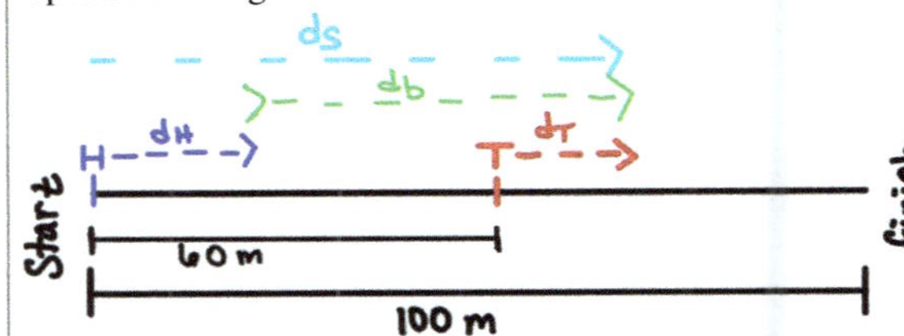
Poses questions for the purpose of getting an answer • What is the formula for the hare's distance? • What is the formula for the tortoise's distance? • What is the formula for the distance between the tortoise and the hare?	Poses question for the purpose of gaining insight into how the students are thinking about the problem • What is the tortoise's distance from the starting line at the beginning of the race? • As the time (in sec) since the start gun was fired increases, how is the Tortoise's distance (in meters) from the starting line changing? • As the time (in sec) since the start gun was fired increases, how is the Hare's distance (in meters) from the starting line changing? • As the time (in sec) since the start gun was fired increases, how is the distance between the Tortoise and the Hare (in meters) changing?

(continued)

Table 6 (continued)

Instructor A: doesn't value QR	Instructor B: values QR
Writes expressions on the board without asking students to explain what 60, – 3.2t, or 60 – 3.2t represent $d_b = =60 - 3.2t$	Written expressions emerge from students' responses to instructor questions. As students share their algebraic expressions the instructor prompts them to explain what each term and each expression represents. (e.g., What does 3.6t represent in this situation?) As the instructor writes expressions, she calls on a student to label their term/expression on a drawing displayed to the class on a whiteboard • $60 + 0.4t - 3.6t$ 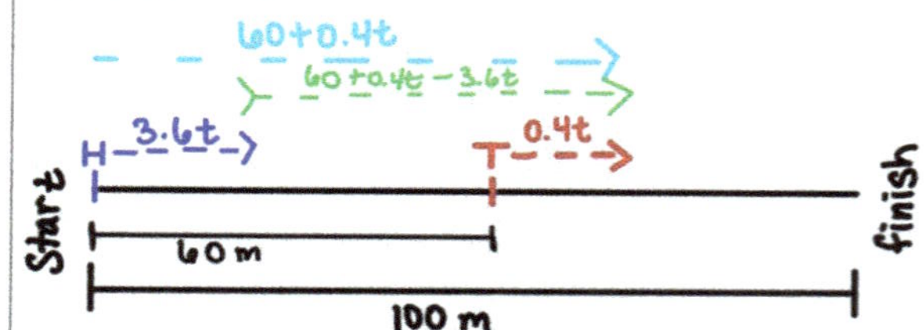
Poses questions that focus students' attention on the calculations that led to the answer • What did you do to get 60 – 3.2t? • Did anyone perform different calculations to get to the answer?	Poses questions that focus students' attention on the reasoning that led to the answer • What thinking led you to writing 60 – 3.2t? • Why did you subtract? • What thinking led you to writing 3.2t? • Did anyone think about the problem differently?

12 Discussion

Studies of precalculus instructors' pedagogical practices have revealed a predominant focus on instructor-led demonstrations of methods for obtaining answers, with instructors constructing incoherent drawings and doing the majority of the speaking in class (e.g., Carlson & Bas-Ader, 2019; Teuscher et al., 2016). Supporting precalculus instructors to commit to engaging students in quantitative reasoning (and all that this entails), and be equipped to do so, is more complex and challenging than we initially imagined 15 years ago. We have been successful in some contexts, and less successful in others, and continue to explore and investigate approaches (such as the conventions described in this chapter) for supporting precalculus faculty to make this shift.

Making the problem more difficult is (1) that many precalculus instructors possess relatively weak meanings of fundamental mathematical ideas (e.g., Baş-Ader & Carlson, 2021; Musgrave & Carlson, 2017; Tallman & Frank, 2018) and (2) instructors with weak conceptions of ideas they teach are unable to engage their students in conversations that leverage and advance their students' thinking (Carlson & Bas-Ader, 2019). For example, during one of our professional development workshops with 25 secondary precalculus instructors, the majority expressed that any statement with an equal sign was an equation that needed to be solved. In another context, we asked this same group of instructors to explain how solving an equation and evaluating a function formula differed. After a relatively long wait for a response, one instructor said she saw no difference since both are equations that need to be solved. As one more example, prior to intervention, many Pathways instructors conceive of a constant rate of change as a description of the "slantiness of a line" rather than the relative size of the changes in two quantities' values. Our observations are corroborated by Thompson's research group's studies (e.g., Byerley & Thompson, 2017; Yoon & Thompson, 2020) of U.S. instructors' mathematical meanings. Elaboration of the difficulties these impoverished conceptions create for students are discussed in Thompson and Carlson (2017).

Instructors who have only experienced traditional curricula in both their learning and teaching need sustained and focused support to reconceptualize mathematical ideas they thought they understood and to reconceptualize effective teaching as focused on and affecting student thinking. Our work with instructors continues to provide us with confidence that the culture of mathematics teaching in the U.S. can change and that such a change benefits students. We have observed that as instructors become more interested in understanding and affecting their students' thinking, and reflecting on their effectiveness in doing so, they (over time) acquire more robust images of diverse ways of thinking that students present and improved insights into what productive thinking entails (Carlson & Bas-Ader, 2019; Rocha & Carlson, 2020). As instructors' mathematical meanings, images of student thinking, and images of effective teaching develop, they are more able to effectively adjust their lessons and instruction to be more meaningful and coherent for students (O'Bryan & Carlson, 2019; Rocha & Carlson, 2020; Underwood & Carlson, 2012). Instructors who became committed to implementing the Pathways conventions for supporting quantitative reasoning are individuals who attended our workshops and shifted to value our focus on conceptualizing and relating quantities as a foundational way of thinking for generating meaningful formulas and graphs.

It is our goal to support all instructors in engaging in meaningful reflection about the impact of their teaching on students' learning. The Pathways research, development, and professional development teams are collectively committed to quantitative reasoning as providing a unifying lens for advancing and studying growth in instructor knowledge and instructional practices. Our commitment to this perspective emerged from many other attempts to improve precalculus and calculus students' learning, and consistently recognizing that students' difficulties in understanding ideas, constructing meaningful function formulas and graphs, etc. were rooted in their failure to conceptualize quantities in a problem context and then to consider how pairs of quantities are related and change together.

13 Concluding Remarks

We continue to study our effectiveness in supporting instructors' construction of strong conceptions of the key ideas taught in the Pathways curriculum. Silverman and Thompson (2008) argue that an instructor must become aware of the mental processes and operations that constitute coherent mathematical understandings for reorganizing their mathematical knowledge and engaging in effective teaching practices. Quantitative reasoning will not become a meaningful part of an instructor's teaching practices until she has an image of the conceptual affordances of this way of reasoning for students' learning. We have evidence that the Pathways conventions for representing quantitative relationships, if implemented consistently, lead to advances in students' mathematical thinking, including their expectation that symbols are useful for representing quantities and relationships between quantities and that graphs emerge as a trace of an individual's conception of how two quantities' values vary in tandem. Our work has revealed that an instructor's consistent implementation of the Pathways conventions results in both students and instructors constructing more robust meanings for specific mathematics ideas, and instructors showing greater interest in understanding student's reasoning. These shifts were frequently accompanied by an increased attention on quantitative reasoning in instructors' lesson design and delivery. The result is a profound shift in how students and instructors approach applied problems; in particular, their actions to conceptualize quantitative relationships, as a foreshadowing of their construction of function formula and graphs that are personally meaningful to the one constructing them.

Acknowledgements This work was supported by the National Science Foundation grant number 1323753.

References

American Mathematical Association of Two-Year Colleges (2018). *IMPACT: Improving mathematical prowess and college teaching*. Author.

Baş-Ader, S., & Carlson, M. (2021). Decentering framework: A characterization of graduate student instructors' actions to understand and act on student thinking. *Mathematical Thinking and Learning*. https://doi.org/10.1080/10986065.2020.1844608

Bell, A., & Janvier, C. (1981). The interpretation of graphs representing situations. *For the Learning of Mathematics, 2*, 34–42.

Boston, M. D., & Wilhelm, A. G. (2017). Middle school mathematics instruction in instructionally focused urban districts. *Urban Education, 52*(7), 829–861.

Breidenbach, D., Dubinsky, E., Hawks, J., & Nichols, D. (1992). Development of the process conception of function. *Educational Studies in Mathematics, 23*, 247–285.

Bressoud, D., Ghedamsi, I., Martinez-Luaces, V., & Törner, G. (2016). *Teaching and learning of calculus*. Springer Nature.

Byerley, C. (2019). Calculus students' fraction and measure schemes and implications for teaching rate of change functions conceptually. *The Journal of Mathematical Behavior, 55*, 100694.

Byerley, C., & Thompson, P. W. (2017). Instructors' meanings for measure, slope, and rate of change. *Journal of Mathematical Behavior, 48*, 168–193.

Carlson, M. (1998). A cross-sectional investigation of the development of the function concept. In E. Dubinsky, A. H. Schoenfeld, & J. J. Kaput (Eds.), *Research in collegiate mathematics education, III* (p. 7). Issues in Mathematics Education.

Carlson, M., & Bloom, I. (2005). The cyclic nature of problem solving: An emergent multidimensional problem solving framework. *Educational Studies in Mathematics, 58*, 45–75.

Carlson, M., Jacobs, S., Coe, E., Larsen, S., & Hsu, E. (2002). *Applying* covariational reasoning while modeling dynamic events: A framework and a study. *Journal for Research in Mathematics Education, 33*(5), 352. https://doi.org/10.2307/4149958

Carlson, M. P., & Ohertman, M. C. (2010). *Precalculus: Pathways to calculus*. Rational Reasoning, LLC.

Carlson, M. P., Ohertman, M. C., Moore, K. C., & O'Bryan, A. E. (2020). *Precalculus: Pathways to calculus* (8th ed.). Rational Reasoning, LLC.

Carlson, M. P., Oehrtman, M. C., & Engelke, N. (2010). The precalculus concept assessment (PCA) instrument: A tool for assessing students' reasoning patterns and understandings. *Cognition and Instruction, 28*(2), 113 145.

Carlson, M. P., Madison, B., & West, R. D. (2015). A study of students' readiness to learn calculus. *International Journal of Research in Undergraduate Mathematics Education, 1*, 209–233.

Carlson, M. P., & Bas-Ader, S. (2019). The interaction between a instructor's mathematical conceptions and instructional practices. In A. Weinberg, D. Moore-Russo, H. Soto, & M. Wawro (Eds.), *Proceedings of the 22nd annual conference on research in undergraduate mathematics* (pp. 101–110). Mathematical Association of America.

Clark, P. G., Moore, K. C., & Carlson, M. P. (2008). Documenting the emergence of "speaking with meaning" as a sociomathematical norm in professional learning community discourse. *The Journal of Mathematical Behavior, 27*(4), 297–310. https://doi.org/10.1016/j.jmathb.2009.01.001

David, E. J., Roh, K. H., & Sellers, M. E. (2019). Value-thinking and location-thinking: Two ways students visualize points and think about graphs. *The Journal of Mathematical Behavior, 54*, 100675. https://doi.org/10.1016/j.jmathb.2018.09.004

Engelke, N. (2007). *Students' understanding of related rate problems in calculus* (Unpublished doctoral dissertation). Arizona State University.

Frank, K. M. (2017a). *Examining the development of students' covariational reasoning in the context of graphing* (Ph.D. dissertation). Arizona State University.

Frank, K. M. (2017b). Tinker Bell's pixie dust: The role of differentiation in emergent shape thinking. In *Proceedings of the twenty-first annual conference on research in undergraduate mathematics education* (pp. 596–604).

Harel, G. (2008). DNR perspective on mathematics curriculum and instruction, part II: with reference to teacher's knowledge base. *ZDM—Mathematics Education, 40*, 893–907. https://doi.org/10.1007/s11858-008-0146-4

Hiebert, J., Stigler, J. W., Jacobs, J. K., Givvin, K. B., Garnier, H., Smith, M., Hollingsworth, H., Manaster, A., Wearne, D., & Gallimore, R. (2005). Mathematics teaching in the United States today (and tomorrow): Results from the TIMSS 1999 video study. *Educational Evaluation and Policy Analysis, 27*(2), 111–132.

Hill, J. A. T. (2021). *Using logistic regression to examine the relationship between early alert systems and success in mathematics* (Doctoral dissertation, Northern Illinois University).

Jackson, K., Cobb, P., Wilson, J., Webster, M., Dunlap, C., & Applegate, M. (2015). Investigating the development of mathematics leaders' capacity to support teachers' learning on a large scale. *ZDM Mathematics Education, 47*(1), 93–104.

Johnson, H. (2015). Secondary students' quantification of ratio and rate: A framework for reasoning about change in Covarying quantities. *Mathematical Thinking and Learning, 17*(1), 64–90. https://doi.org/10.1080/10986065.2015.981946

Kaput, J. J. (1992). Patterns in students' formalization of quantitative patterns. In G. Harel & E. Dubinsky (Eds.), *The concept of function: Aspects of epistemology and pedagogy* (pp. 290–318). Mathematical Association of America.

Laursen, S. L., & Rasmussen, C. (2019). I on the prize: Inquiry approaches in undergraduate mathematics. *International Journal of Research in Undergraduate Mathematics Education, 5*(1), 129–146.

Litke, E. (2020). Instructional practice in algebra: Building from existing practices to inform an incremental improvement approach. *Teaching and Teacher Education, 91*, 103030.

Leinhardt, G., Zaslavsky, O., & Stein, M. K. (1990). Functions, graphs, and graphing: Tasks, learning, and teaching. *Review of Educational Research, 60*(1), 1–64. https://doi.org/10.3102/00346543060001001

Lozano, D. (1998). *El Concepto de Variable: Evolución a lo largo de la Instrucción Matemática.* B.Sc. thesis, Instituto Tecnologico Autonomo de Mexico.

Mkhatshwa, T. P. (2020). Calculus students' quantitative reasoning in the context of solving related rates of change problems. *Mathematical Thinking and Learning, 22*(2), 139–161.

Monk, S. (1992). Students' understanding of a function given by a physical model. In G. Harel & E. Dubinsky (Eds.), *The concept of function: Aspects of epistemology and pedagogy, MAA notes* (Vol. 25, pp. 175–193). Mathematical Association of America.

Moore, K. C. (2010). *The role of quantitative reasoning in precalculus students learning central concepts of trigonometry* (Order no. 3425753). Available from Dissertations & Theses @Arizona State University; ProQuest Dissertations & Theses Global (757614251).

Moore, K. C., & Carlson, M. P. (2012). Students' images of problem contexts when solving applied problems. *The Journal of Mathematical Behavior, 31*(1), 48–59. https://doi.org/10.1016/j.jmathb.2011.09.001

Moore, K. C., & Thompson, P. W. (2015). Shape thinking and students' graphing activity. In T. Fukawa-Connelly, N. E. Infante, K. Keene, & M. Zandieh (Eds.), *Proceedings of the 18th meeting of the MAA special interest group on research in undergraduate mathematics education* (pp. 782–789). RUME.

Musgrave, S., & Carlson, M. P. (2017). Understanding and advancing graduate teaching assistants' mathematical knowledge for teaching. *The Journal of Mathematical Behavior, 45*, 137–149. https://doi.org/10.1016/j.jmathb.2016.12.011

National Governors Association Center for Best Practices, Council of Chief State School Officers (2010). *Common core state standards for mathematics.* National Governors Association Center for Best Practices, Council of Chief State School Officers.

Oehrtman, M. (2009). Collapsing dimensions physical limitation, and other student metaphors for limit concepts. *Journal for Research in Mathematics Education JRME, 40*(4), 396–426.
O'Bryan, A. E. (2018). Conceptual analysis in cognitive research: Purpose, uses, and the need for clarity. In *Proceedings of the 21st annual conference on research in undergraduate mathematics education.*
O'Bryan, A. (2019). *Exponential growth and related ideas: Examining students' meanings and learning in an online Precalculus course.*
O'Bryan, A. E. (2020a). Quantitative reasoning and symbolization activity: Do individuals expect calculations and expressions to have quantitative significance? In *Proceedings of the 23rd annual conference on research in undergraduate mathematics education.*
O'Bryan, A. E. (2020b). You can't use what you don't see: Quantitative reasoning in applied contexts. *OnCore: Journal of the Arizona Association of Instructors of Mathematics*, 66–74.
O'Bryan, A. E., & Carlson, M. P. (2016). Fostering instructor change through increased noticing: Creating authentic opportunities for instructors to reflect on student thinking. In T. Fukawa-Connelly, N. Infante, M. Wawro, & S. Brown (Eds.), *Proceedings of the 19th annual conference on research in undergraduate mathematics education* (pp. 1192–1200).
Piaget, J. (1955). *The language and thought of the child.* Meridian Books.
Piaget, J. (1968). Quantification, conservation, and nativism. *Science, 162*(3857), 976–979. https://doi.org/10.1126/science.162.3857.976
Polya, G. (1957). *How to solve it; a new aspect of mathematical method* (2nd ed.). Doubleday.
Rocha, A., & Carlson, M. (2020). The role of mathematical meanings for teaching and decentering actions in productive student-instructor interactions. In *Proceedings of the twenty-third annual special interest group of the mathematical association of America conference on research in undergraduate mathematics education.*
Saldanha, L., & Thompson, P. W. (1998). Re-thinking co-variation from a quantitative perspective: Simultaneous continuous variation. In S. B. Berenson & W. N. Coulombe (Eds.), *Proceedings of the annual meeting of the psychology of mathematics education.* North Carolina State University.
Schmidt, W. H., Wang, H. C., & McKnight, C. C. (2005). Curriculum coherence: An examination of US mathematics and science content standards from an international perspective. *Journal of Curriculum Studies, 37*(5), 525–559.
Schwartz, J. L. (1988). Intensive quantity and referent transforming arithmetic operations. In J. Hiebert & M. Behr (Eds.), *Number concepts and operations in the middle grades* (p. 41Y52). National Council of Instructors of Mathematics (NCTM) and LEA.
Sierpinska, A. (1992). On understanding the notion of function. In G. Harel & E. Dubinsky (Eds.), *The concept of function: Aspects of epistemology and pedagogy, MAA notes* (Vol. 25, pp. 59–84). Mathematical Association of America.
Silverman, J., & Thompson, P. W. (2008). Toward a framework for the development of mathematical knowledge for teaching. *Journal of Mathematics Instructor Education.*
Simon, M. A., Tzur, R., Heinz, K., Kinzel, M., & Smith, M. S. (2000). Characterizing a perspective underlying the practice of mathematics teachers in transition. *Journal for Research in Mathematics Education, 31*(5), 579–601.
Simon, M., & Placa, N. (2012). Reasoning about intensive quantities in whole-number multiplication? A possible basis for ratio understanding. *For the Leaning of Mathematics, 32*, 35–41.
Smith, J., & Thompson, P. W. (2007). Quantitative reasoning and the development of algebraic reasoning. In J. Kaput, D. Carraher, & M. Blanton (Eds.), *Algebra in the early grades* (pp. 95–132). Erlbaum.
Steffe, L. P., & Thompson, P. W. (2000). Teaching experiment methodology: Underlying principles and essential elements. In R. Lesh & A. E. Kelly (Eds.), *Research design in mathematics and science education.* Kluwer.
Stigler, J. W., Gonzales, P., Kwanaka, T., Knoll, S., & Serrano, A. (1999). *The TIMSS videotape classroom study: Methods and findings from an exploratory research project on eighth-grade*

mathematics instruction in Germany, Japan, and the United States. A research and development report.

Stigler, J. W., & Hiebert, J. (2009). *The teaching gap: Best ideas from the world's teachers for improving education in the classroom*. Simon and Schuster.

Tallman, M. A., & Frank, K. M. (2018). Angle measure, quantitative reasoning, and instructional coherence: An examination of the role of mathematical ways of thinking as a component of instructors' knowledge base. *Journal of Mathematics Instructor Education*. https://doi.org/10.1007/s10857-018-9409-3

Thompson, P. W. (1988). Quantitative concepts as a foundation for algebra. In M. Behr (Ed.), *Proceedings of the annual meeting of the North American chapter of the international group for the psychology of mathematics education* (Vol. 1, pp. 163–170). Dekalb.

Thompson, P. W. (1990). *A theoretical model of quantity-based reasoning in arithmetic and algebraic*. Center for Research in Mathematics & Science Education, San Diego State University.

Thompson, P. W. (1992). Notations, conventions, and constraints: Contributions to effective uses of concrete materials in elementary mathematics. *Journal for Research in Mathematics Education, 23*(2), 123–147.

Thompson, P. W. (1993). Quantitative reasoning, complexity, and additive structures. *Educational Studies in Mathematics, 25*(3), 165–208.

Thompson, P. W. (1994). The development of the concept of speed and its relationship to concepts of rate. In G. Harel & J. Confrey (Eds.), *The development of multiplicative reasoning in the learning of mathematics* (pp. 179–234). Albany.

Thompson, P. W. (1996). Imagery and the development of mathematical reasoning. In L. P. Steffe, P. Nesher, P. Cobb, G. Goldin, & B. Greer (Eds.), *Theories of mathematical learning* (pp. 267–283). Erlbaum.

Thompson, P. W. (2002). Didactic objects and didactic models in radical constructivism. In K. Gravemeijer, R. Lehrer, B. van Oers, & L. Verschaffel (Eds.), *Symbolizing and modeling in mathematics education*. Kluwer.

Thompson, P. W. (2008a). Conceptual analysis of mathematical ideas: Some spadework at the foundation of mathematics education. Plenary paper delivered at the 32nd annual meeting of the international group for the psychology of mathematics education. In O. Figueras, J. L. Cortina, S. Alatorre, T. Rojano, & A. Sèpulveda (Eds.), *Proceedings of the annual meeting of the international group for the psychology of mathematics education* (Vol. 1, pp. 45–64). PME.

Thompson, P. W. (2008b). *One approach to a coherent K-12 mathematics. Or, it takes 12 years to learn calculus*. Paper presented at the Pathways to Algebra Conference, June 22–25, Mayenne, France.

Thompson, P. W. (2011). Quantitative reasoning and mathematical modeling. In L. L. Hatfield, S. Chamberlain, & S. Belbase (Eds.), *New perspectives and directions for collaborative research in mathematics education WISDOMe monographs* (Vol. 1, pp. 33–57). University of Wyoming Press.

Thompson, P. W. (2012). Advances in research on quantitative reasoning. In R. Mayes, R. Bonillia, L. L. Hatfield, & S. Belbase (Eds.), *Quantitative reasoning: Current state of understanding WISDOMe Monographs* (Vol. 2, pp. 143–148). University of Wyoming Press.

Thompson, P. (2013). In the absence of meaning. In *Vital directions for mathematics education research* (pp. 57–93). https://doi.org/10.1007/978-1-4614-6977-3_4

Thompson, P. W. (2016). Researching mathematical meanings for teaching. In L. D. English & D. Kirshner (Eds.), *Handbook of international research in mathematics education* (pp. 435–461). Taylor & Francis.

Thompson, P. W., & Carlson, M. P. (2017). Variation, covariation, and functions: Foundational ways of thinking mathematically. In J. Cai (Ed.), *Compendium for research in mathematics education* (pp. 421–456). National Council of Instructors of Mathematics.

Thompson, P. W, Carlson, M. P., Byerley, C., & Hatfield, N. (2014). Schemes for thinking with magnitudes: A hypothesis about foundational reasoning abilities in algebra. In K. C. Moore, L.

P. Steffe, & L. L. Hatfield (Eds.), *Epistemic algebra students: Emerging models of students' algebraic knowing. WISDOMe monographs* (Vol. 4, pp. 1–24). University of Wyoming.
Thompson, P. W., & Harel, G. (2021). Ideas foundational to calculus learning and their links to students' difficulties. *ZDM–Mathematics Education*, 1–13.
Thompson, P. W., Hatfield, N. J., Yoon, H., Joshua, S., & Byereley, C. (2017). Covariational reasoning among U.S. and South Korean secondary mathematics instructors. *Journal of Mathematical Behavior, 48*, 95–111.
Teuscher, D., Moore, K. C., & Carlson, M. P. (2016). Decentering: A construct to analyze and explain instructor actions as they relate to student thinking. *Journal of Mathematics Instructor Education, 19*(5), 433–456. https://doi.org/10.1007/s10857-015-9304-0
Underwood, K., & Carlson, M. P. (2012). Understanding how precalculus instructors develop mathematic knowledge for teaching the idea of rate of change. In *Proceedings of the 15th annual conference on research in undergraduate mathematics education.*
Vinner, S., & Dreyfus, T. (1989). Images and definitions for the concept of function. *Journal for Research in Mathematics Education, 20*, 356.
Winsløw, C. (2021). Abell, Braddy, Ensley, Ludwig, Soto: MAA instructional practices guide. *International Journal of Research in Undergraduate Mathematics Education.* https://doi.org/10.1007/s40753-021-00141-0
Yackel, E., & Cobb, P. (1996). Sociomathematical norms, argumentation, and autonomy in mathematics. *Journal for Research in Mathematics Education, 27*, 458–477.
Yoon, H., Byerley, C., & Thompson, P. W. (2015). Instructors' meanings for average rate of change in U.S.A. and Korea. In T. Fukawa-Connelly, N. E. Infante, K. Keene, & M. Zandieh (Eds.), *Proceedings of the 18th meeting of the MAA special interest group on research in undergraduate mathematics education* (pp. 335–348). RUME.
Yoon, H., & Thompson, P. W. (2020). Secondary instructors' meanings for function notation in the United States and South Korea. *The Journal of Mathematical Behavior, 60*, 100804. https://doi.org/10.1016/j.jmathb.2020.100804
Zandieh, M. (2000). A theoretical framework for analyzing student understanding of the concept of derivative. In E. Dubinsky, A. Schoenfeld, & J. Kaput (Eds.), *Research in collegiate mathematics education, IV* (Vol. 8, pp. 103–127). American Mathematical Society.

Mathematization: A Crosscutting Theme to Enhance the Curricular Coherence

Hui Jin, Dante Cisterna, Hyo Jeong Shin, and Matthew Vonk

Quantitative reasoning played a crucial role in the development and revolution of scientific knowledge in the history of science (Crombie, 1961; Jin et al., 2019a; Kline, 1982). It has been emphasized as an important learning goal for K-12 students for many years (NGSS Lead States, 2013; National Research Council [NRC], 1996, 2000). In science education literature, the term mathematization of science, or mathematization in short, is often used to refer to the specialized ways that scientists use to quantify phenomena and construct knowledge; it emphasizes the relationship between quantitative reasoning and science disciplinary knowledge (e.g., Kline, 1982; Lehrer & Schauble, 1998). Therefore, in this chapter, we use this term to refer to quantitative reasoning in science.

Researchers describe scientists' specialized ways of using quantitative reasoning with different terms such as mathematical deduction (Kind & Osborne, 2017; Osborne et al., 2018), mathematization (Kline, 1982), postulation exemplified by the Greek mathematical sciences (Crombie, 1961; Hacking, 1994), and quantification (Crombie, 1961). Nevertheless, they all emphasize a process of quantification: Scientists generate mathematical descriptions of phenomena in the material world.

H. Jin (✉) · D. Cisterna · H. J. Shin
Educational Testing Service, Princeton, NJ 08541, USA
e-mail: hjin@georgiasouthern.edu

H. J. Shin
e-mail: hshinedu@sogang.ac.kr

H. Jin
Georgia Southern University, Savannah, USA

H. J. Shin
Sogang University, Seoul, Korea

M. Vonk
University of Wisconsin-River Falls, River Falls, WI 54022, USA
e-mail: matthew.vonk@uwrf.edu

G. Karagöz Akar et al. (eds.), *Quantitative Reasoning in Mathematics and Science Education*, Mathematics Education in the Digital Era 21,
https://doi.org/10.1007/978-3-031-14553-7_10

In those descriptions, entities (e.g., matter, energy, and force) are represented by algebraic symbols and numeric values; and the relationships among those entities are represented by mathematical equations, tables, and graphs. Scientists generate concepts, principles, and theories to conceptualize those mathematical relationships. The value of mathematical descriptions resides in their accuracy, universality, and deductive logic (Pereira de Ataide & Greca, 2013). Due to this value, mathematical descriptions allow precise predictions and generation of new concepts; they also provide an objective base for scientific argumentation and discussion (Holton & Brush, 2006; Kline, 1990; Osborne et al., 2018). Although existing literature of scientists' mathematization provides concrete ideas about the quantification process, additional effort is needed to identify key components that differentiate that quantification process from our everyday intuitive thinking. Such information will help teachers and researchers design more targeted instruction on quantitative reasoning.

Researchers have investigated how students use mathematization to solve problems and explain phenomena. These studies have documented the expert-novice differences across physics (Bing & Redish, 2009; Chi et al., 1981; Kuo et al., 2013; Niss, 2017; Schuchardt & Schunn, 2016; Sherin, 2001; Tuminaro & Redish, 2004, 2007), chemistry (Dori & Hameiri, 2003; Kozma & Russell, 1997; Schuchardt & Schunn, 2016; Taasoobshirazi & Glynn, 2009), and biology (Schuchardt & Schunn, 2016). While experts incorporate conceptual understanding of scientific knowledge with mathematical representations, novices tend to select mathematical equations based on surface features of the scenario and manipulate the mathematical symbols/equations without understanding their scientific meaning. These expert-novice differences are largely due to the different epistemological perspectives—while experts view mathematics and science as integrated, students tend to see mathematics as a mere instrument for calculation (Bing & Redish, 2009). Additionally, using graphs presents significant challenge for many students. Most existing studies on students' use of graphs were conducted in the context of kinematics. These studies show that students often misinterpret graphs as pictures (e.g., viewing a velocity–time graph as a picture of the object's trajectory) and do not use scientific ideas to interpret the relationships presented in the graphs such as slope, trends, and patterns (Beichner, 1994; Kozhevnikov et al., 2007; Planinic et al., 2012). Although empirical studies have generated significant findings about students' mathematization, more research is needed to investigate how students develop from their novice thinking to expert thinking is limited.

We addressed these two needs in a Mathematical Thinking in Science project. We used a learning progression (LP) approach to investigate student development of mathematization in physical and life sciences. LPs are "descriptions of successively more sophisticated ways of thinking about how learners develop key disciplinary concepts and practices within a grade level and across multiple grades" (Fortus & Krajcik, 2012, p. 784). It is well-recognized that coherence in science curriculum leads to high-quality instruction and student achievement (Fortus & Krajcik, 2012; Schmidt et al., 2005). Existing literature emphasizes two aspects of curricular coherence—logical coherence and cognitive coherence (Fortus & Krajcik, 2012; Schmidt et al., 2005; Shwartz, et al., 2008; Sikorski & Hammer, 2017). That is, curriculum, instruction, and assessment are aligned based not only on the logical structure and organization of the discipline but also on the cognitive theories about student learning of the disciplinary knowledge and practices. Science LPs are rooted in foundational theories about disciplinary knowledge and cognition. Therefore, they are powerful in enhancing curricular coherence (Jin et al., 2019b).

In the project, we defined mathematization based on a historical analysis and Thompson's theory of quantitative reasoning in mathematics (Thompson, 1993, 2011, Thompson, et al., 2014). We then used a learning progression (LP) approach to study student development in mathematization across several topics in physical sciences and life sciences (heat and temperature, kinetic and gravitational potential energy, and elastic energy in physical sciences; the carbon cycle and interdependent relationships in life sciences). Our study suggests that, by using an LP approach, mathematization can be used as a crosscutting theme to align curriculum, instruction, and assessment. In this chapter, we summarize the major findings of this work, including the definition of mathematization, the LP for mathematization, and preliminary evidence of mathematization as a crosscutting theme. Based on these results, we discuss the possibility and benefits of using mathematization as a crosscutting theme for building curricular coherence.

1 Defining Mathematization

We intended to develop a functional definition of mathematization that reflects how scientists used quantitative reasoning to construct scientific knowledge. To do so, we conducted a historical analysis. We identified and examined five events across physics, biology, astronomy, and chemistry. These five events include the development of the ideal gas law, Mendel's discovery of the laws of hybridization, Newton's derivation of universal gravitation from Kepler's law of planetary motion, the chemical revolution initiated by Lavoisier, and the paradigm shift from Aristotelian to Newtonian theories about forces and motion. They played a critical role in the knowledge development and revolution in the history of science. Our analysis focused on how measurement and quantification enabled the generation of fundamental ideas in science. Details of the analysis can be found in our previous publication (Jin et al., 2019a). In this chapter, we summarize one event that has led to the overthrown of the phlogiston theory and the establishment of modern chemistry—Antoine Lavoisier's chemical revolution.

Both phlogiston theorists and Lavoisier investigated phenomena of burning, calcination, and breathing. However, the ways of reasoning used in their investigations are vastly different. Take burning as an example. The phlogiston theorists observed that some materials were combustible, while other materials were not. To explain this observation, they conjectured those combustible materials must contain some type of essence. They named this essence phlogiston. The ashes after combustion often weigh less than the combustible material. To explain this phenomenon, phlogiston theorists supposed that phlogiston must escape into the air. These qualitative conjectures constitute the phlogiston theory: Materials that are rich in phlogiston can burn; when a material burns, its phlogiston is liberated into the air and only ashes are left.

Unlike the phlogiston theorists, Lavoisier used quantitative reasoning to analyze burning. He conducted experiments in closed systems and with accurate measurement. He studied burning of different materials in a closed vessel system (Holton & Brush, 2001). In the burning iron experiment, burning 100 grains [a unit of mass] of iron produced 135 or 136 grains oxide of iron. At the same time, the diminution of air was found to be exactly 70 cubical inches, which weighed 35 grains. Lavoisier analyzed the relationships among several quantities: The mass of iron, the volume of air, the mass of air, and the mass of oxide of iron. After many similar experiments, he found a mathematical pattern: the total mass of materials is conserved in burning. To explain this pattern, Lavoisier proposed a new theory of combustion, the oxygen theory: The total mass is conserved before and after the combustion because oxygen is involved in combustion and the mass of oxygen should be included in the calculation.

Thompson's theory of quantitative reasoning in mathematics (Thompson, 1993, 2011; Thompson & Carlson, 2017) offers unique insights for us to identify key components that differentiates Lavoisier's and other scientists' mathematization from the intuitive reasoning patterns that once appeared and then became obsolete in the history of science. Thompson (1993) defines quantitative reasoning as "the analysis of a situation into a quantitative structure—a network of quantities and quantitative relationships" (p. 165). In explaining this definition, Thompson emphasizes two ideas. First, a key characteristic of quantity is its measurability (Thompson, 1993, p. 165):

> Quantity is not the same as a number. A person constitutes a quantity by conceiving of a quality of an object in such a way that he or she understands the possibility of measuring it (Thompson, 1989, in press). Quantities, when measured, have numerical value, but we need not measure them or know their measures to reason about them.

This concept of **measurability**, or measurable quantities/variables, is one component that differentiates mathematization from intuitive reasoning. While phlogiston theorists focused the analysis on qualitative attributes (e.g., combustible materials turn into ashes; some materials are combustible, while others are not), Lavoisier analyzed measured variables (i.e., the mass of iron, the volume of air, the mass of air, and the mass of oxide of iron). As another example, consider two responses to the following question: "Does a person have more energy after a night's sleep?"

> Response A: After a night's sleep, a person will have *less* energy than the night before since a certain amount of energy stored in the person's body has been used to support body functions such as heart beating and breathing.
>
> Student B: The person has *more* energy because people normally feel more energetic after a good night's sleep.

In Response A, energy is treated as a measurable quantity because the response is about how the total amount of energy changes and where the reduced amount of energy goes. Response B does not treat energy as a measurable quantity because it uses a qualitative reason (i.e., feeling more energetic) to explain why the person has more energy after a night's sleep.

Second, understanding **relational complexity** is crucial for analyzing a network of quantities and quantitative relationships (Thompson, 1993; Thompson & Carlson, 2017). This understanding involves coordination of two aspects of quantitative difference: (1) difference as the amount left over after a comparison and (2) quantitative difference as an item in a relational structure. In this sense, understanding relational complexity is not just about obtaining the result of subtracting. It includes understanding the relationships among multiple differences in a structure. Thompson discusses relational complexity in contexts involving subtraction and addition. We modified and applied this component to fit scientific contexts. We define relational complexity as the complexity involved in the kinds of relationships that play an important role in scientific conceptualization. These relationships include quantitative conservation, *extensive* versus *intensive* variables, change versus the rate of change, proportionality, exponential growth, quadratic relationships, and so on. The historical analysis is about quantitative conservation. Lavoisier's notion of conservation is quantitative because it is based on calculation of numerical values measured in experiments. Phlogiston theorists hold a notion of 'qualitative conservation'. They recognize that the ashes cannot weigh more than the combustible material. Something must come out from the material and that something must go somewhere. They label that something as phlogiston. This type of conservation is qualitative because it is based on humans' perception of less or more. Unlike Lavoisier's quantitative conservation, this "qualitative conservation" does not involve numerical values measured in experiments or real-world situations.

A third component involved in Lavoisier's mathematization is **conceptualization**. Lavoisier identified a quantitative relationship in his experiments—quantitative conservation, that is, the mass of materials before and after combustion is conserved. He then conceptualized this relationship into the oxygen theory of combustion: Oxygen is involved in combustion. If we calculate all substances involved in combustion, we will find that mass is conserved before and after the combustion. In the history of science, many concepts, principles, and theories were conceptualized from quantitative relationships. They are usually counter-intuitive, and therefore present significant challenges to students. For example, a student who understands relational complexity will understand the scientific implication of the equation of kinetic energy ($E = ½ mv^2$) and explain that doubling the vehicle speed can lead to collision damage that is much larger than doubling (due to the quadratic relationship between energy and speed). However, a student who does not recognize the relational complexity involved in the same equation may know that a higher vehicle speed is associated with more damage, but the student would not recognize the scientific implication of the quadratic relationship between the speed and the energy.

The above discussion suggests three components of mathematization of science—**measurable variables, relational complexity, and scientific conceptualization**.

Subsequently, we define mathematization of science as *abstracting measurable variables from 'messy' phenomena, identifying mathematical relationships among the variables, and using scientific ideas to conceptualize the mathematical relationships.*

2 The Learning Progression for Mathematization of Science

In the Mathematical Thinking in Science project, we developed an LP for mathematization in across topics in physical and life sciences: heat and temperature, kinetic and gravitational potential energy, and elastic energy in physical sciences; the carbon cycle and interdependent relationships in life sciences. We first carried out an interview, where 44 students from suburban and urban high schools each completed a set of mathematization tasks. Based on the interview data, we developed an initial mathematization LP.

Next, we conducted a large-scale field study. In the study, 57 assessment items, including 36 physical science items and 21 life science items were assembled into multiple computer-delivered tests forms, based on the courses taught by the participating teachers. In addition, most students took 24 mathematics assessment items developed by Wylie et al. (2015). The mathematics items assess student understanding of linear functions and proportional reasoning. These two concepts are essential in middle school mathematics curriculum. They also constitute the foundational knowledge for students to learn and conceptualize a variety of mathematical relationships in high school science. Therefore, they are used as a proxy for students' mathematics baseline understanding. The test forms were administered to 5353 students from 22 high schools in 14 US states. Among these students, 34% were in 11th grade, 27% in 10th grade, and 24% in 12th grade. Urban, suburban, and rural schools participated in the pilot study. Approximately 65% of the students were White, 10% Asian or Asian American, 8% African American, 8% Hispanic or Latino. We used students' assessment responses to revise the LP. The assessment results also provide two pieces of evidence that the mathematization LP is applicable to topics in both physical sciences and life sciences.

In this section, we use students' assessment responses to illustrate the LP levels. Then, we provide the evidence for using mathematization as a crosscutting theme across science topics and disciplines.

3 The LP for Mathematization of Science

The learning progression contains four levels, with each level describing a characteristic way of reasoning that students use to solve scientific problems and to explain real-world phenomena (Fig. 1). These four levels are named holistic phenomenon,

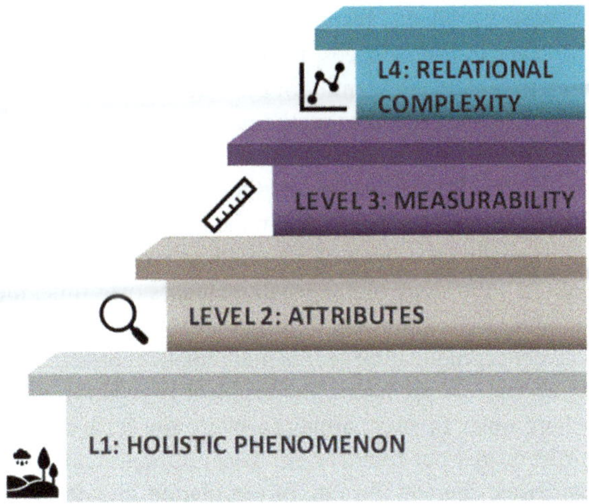

Fig. 1 The learning progression for mathematization for problem-solving

attributes, measurability, and relational complexity. Together, they present a developmental trend, where students progress from intuitive qualitative reasoning to scientific quantitative reasoning.

- *Level 1. Holistic Phenomenon*: At Level 1, students do not 'analyze', meaning that they do not identify any relevant attributes of the phenomena. Instead, they describe the phenomenon, or tell a story related to the phenomenon, or express personal feeling about the phenomenon.
- *Level 2. Attributes*: Students identify relevant attributes of a phenomenon considering their everyday concepts. However, they do not 'quantify', meaning that they treat those attributes as qualitative characteristics rather than measurable quantities/variables. The phlogiston theorists' analysis of burning is an example of reasoning about attributes.
- *Level 3. Measurability*: Students analyze phenomena in terms of measurable quantities/variables. They can abstract some relevant variables from the messy phenomena and identify some mathematical relationships. However, conceptualizing the mathematical relationships in terms of scientific ideas presents significant challenge for them.
- *Level 4. Relational Complexity*: Students distinguish among different types of quantities/variables and understand the complex relationships among those quantities/variables. The complex relationships include relationships between change and rate of change (e.g., velocity and acceleration), distinctions between extensive and intensive variables (e.g., thermal energy and temperature; mass and density), proportional relationship (e.g., gravitational potential energy is proportional to height), quadratic relationship (e.g., the relationship between kinetic energy and speed of a car), exponential relationship (e.g., the population size and the time), and so on.

We use students' responses in two assessment items, one in physical sciences and the other in life sciences, to illustrate this LP. In these responses, pseudonyms are used to protect the identity of the students. As presented below, the life science item (Fig. 2) asks students to mathematize the growth of reindeer population. In the item, the relevant variables are birth rate, the number of births, death rate, the number of deaths, the population size, and the population growth rate. The relationships among these variables are presented in the graph in Fig. 2. More specifically, the part about the exponential growth of the reindeer population shows an important mathematical pattern—the slope of the graph increases over time, meaning the population growth becomes more rapid over time, or in other words, the population growth rate increases. The conceptualization of this mathematical pattern is: Given adequate resources and an absence of predators, the reindeer population would increase exponentially for a long time. In such situation, while the birth rate and death rate (the number of births/deaths per reindeer per year) do not change, the total number of organisms increases, causing the rate of population growth (i.e., the absolute growth rate) to increase.

Table 1 presents students' responses that were scored at each level of the LP. The responses at Level 1 indicate that Diego did not identify any qualitative factors or attributes that explain the observed pattern—the population grew faster in timespan 2 than in timespan 1. Instead, he claimed that the observed pattern is due to reindeer's intention to increase their population. Diego treated the phenomenon holistically and did not analyze and abstract any variables or attributes. The responses at Level 2 suggests that Cindy identified two factors affecting the reindeer population: predation and starvation. She further explained how these qualitative factors affect the reindeer population. Cindy did not reason about any quantitative relationships or measurable variables. She only reasoned at a qualitative level. The responses at Level 3 suggest that Mike reasoned at a quantitative level. He explained that during the timespan 2, the reindeer has adapted to their surroundings; consequently, the reindeer were able to increase the breeding rate significantly, which caused the reindeer population to increase more rapidly. This explanation focuses on the relationship between two measurable variables—the breeding rate and the population growth rate; the increase of breeding rate caused the increase of population growth rate. Although Mike began to reason about the quantitative relationships between measurable variables, he was not successful in identifying and conceptualizing the relational complexity involved in the problem. The responses at Level 4 show that Amber was able to identify relevant measurable variables and conceptualize the complex relationships among the variables. Amber explained that, although individual reindeer produced offspring at the same rate, the total number of reindeers increased. As a result, the population size increased exponentially. Her explanation targets the complex relationship among three measurable variables—the reproduction rate per individual reindeer, the population size, and the population growth rate.

A physical science item is provided in Fig. 3. This item assesses how well students identify and differentiate between heat/energy and temperature. High school students are expected to understand the following distinctions among heat, energy, and temperature (Kesidou & Duit, 1993, p. 90):

In 1911, scientists released 25 reindeer on Saint Paul Island, a small Alaskan island. There were no predators of reindeer on the island. Scientists collected data on the reindeer population over many years. The graph below shows the scientists' data.

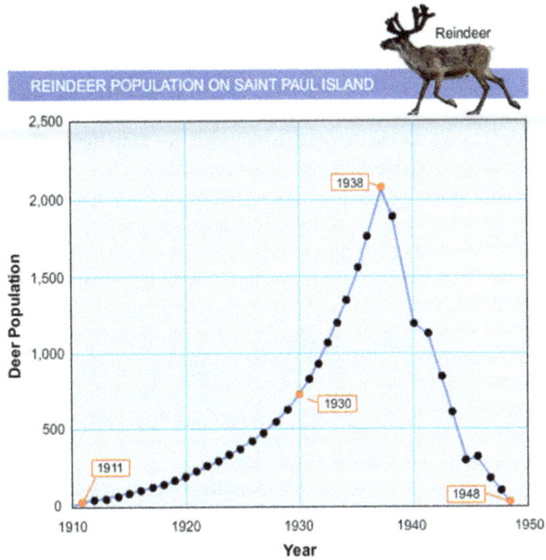

1. Please compare the population growth in these two timespans.
   Time span 1: 1911 to 1932
   Time span 2: 1932 to 1938
   Which of the three patterns below best describes the changes in reindeer population?

   A. The population grew faster in timespan 1 than in timespan 2.
   B. The population grew faster in timespan 2 than in timespan 1.
   C. The population grew at the same rate in these two timespans.

2. Why do you think the reindeers on Saint Paul Island exhibited this pattern?

Fig. 2 The life sciences item

Heat is the form of energy that is transported from one system to another due to temperature differences. From the physicist's point of view, heat is a process variable. Therefore, it is wrong to state that a body contains a certain amount of heat. But it makes sense to view heat as an extensive quantity. If a specific amount of heat (Q1) is transported and if this is followed by another amount of heat (Q2) the total amount of heat transported is Q1 + Q2. Temperature, on the other hand, is an intensive quantity. If two bodies at temperature T are brought into contact, then the temperature of the two bodies is still T.

Table 1 Middle and high school students' responses in the life science task

Learning progression levels	Responses
Level 4. Relational complexity	Responses from Amber Choice: B. The population grew faster in timespan 2 than in timespan 1 Explanation: the population increased exponentially because more individuals means that there is a greater number of animals capable of producing offspring. When they produce this amount of offspring, the population will increase, and then those offspring will go on to have offspring of their own, showing population growth
Level 3. Measurability	Responses from mike Choice: B. The population grew faster in timespan 2 than in timespan 1 Explanation: I believe the reindeer on Saint Paul grew faster of the course of time span 2 because the reindeer need to be adjusted to their environmet [environment]. The island only hosted 25 reindeer to start, but as the reindeer adapted themselves to their surroundings, they were able to utilize whatever that helped them breed at a significantly faster rate
Level 2. Attributes	Responses from Cindy Choice: A. The population grew faster in timespan 1 than in timespan 2 Explanation: they exhibit this pattern because there were no predators which says that they won't die, but there are other problems when there aren't predators, because they will they [then] die from starvation and etc.
Level 1. Holistic phenomena	Responses from Diego Choice: B. The population grew faster in timespan 2 than in timespan 1 Explanation: they exhibited this because they wanted to increase their population

Note that the distinction between heat as a process variable and energy as a status variable is not assessed in the item illustrated in Fig. 3. The item focuses on the distinction between heat/energy and temperature: the former are extensive variables, while the latter is an intensive variable. Successful mathematization involves identifying and distinguishing variables from three observations. The first observation is the oven setting (or the fire), which indicates the amount of heat transferred into the water in the pot. Since the oven is set at high heat in Situation 1 and at medium heat in Situation 2, less heat is transferred into the water in Situation 2. The second observation is that, in both situations, the water is at the boiling stage, indicating the water temperature as 100 °C. The third observation is the degree of vigorousness in boiling, which indicates how much evaporation is going on. In Situation 1, more energy/heat is used to evaporate the water, so the water boils more vigorously.

Table 2 provides students' responses at each LP level. The responses at Level 1 indicate that Zane did not identify any attributes or variables to support his claim

Paulo put a pot with water on the stove at high heat. After a few minutes, the water started to boil vigorously. Paulo turned down the heat setting to medium, and the water kept boiling but less vigorously.

Situation 1: The water boiled vigorously.

Situation 2: The water kept boiling but less vigorously.

1. Do you think the temperature of the water is the same in these two situations?
   A. The water temperature in Situation 1 and the water temperature in Situation 2 are the same.
   B. The water temperature in Situation 1 and the water temperature in Situation 2 are different.
2. [Different sets of questions are shown when the student chooses A or B.]

   Choosing A: If the water temperatures in these two situations are the same, why does the water in Situation 2 boil less vigorously than the water in Situation 1?

   Choosing B: What evidence can be used to support the claim that the water temperature in Situation 1 and the water temperature in Situation 2 are different? Please explain why this evidence can be used to support the claim.

Fig. 3 The physical sciences item

that the water temperatures in both situations are the same. Instead, he described macroscopic observations in an everyday activity—boiling water to cook pasta. The responses at Level 2 shows that Mia associated 'boiling more vigorously' with 'being hotter' and with higher temperature. As such, Mia treated temperature as hotness, which is a qualitative attribute and therefore is not measured and has no numerical values. The responses at Level 3 show that Lucy reasoned about the values of temperature, indicating that she recognized measurability as a key characteristic of variables. However, she does not differentiate between energy/heat and temperature in terms of extensive and intensive variables. Instead, she assumed that more heat input causes the water to boil more vigorously; and that water boiling more vigorously has a higher temperature. However, she did learn that boiling water has a temperature of 100 °C. To reconcile the discrepancy, she conceptualized a new theory—water begins to boil at 100 °C, and the temperature of the water will keep increasing when the water is boiling more vigorously. This way, the input energy/heat, degree of boiling, and temperature are equivalent. The exemplar responses at Level 4 suggest that Kai was able to identify and differentiate heat/energy and temperature. Although more energy goes to the water in Situation 1 than Situation 2, the water temperature stayed the same (100 °C) in the two situations. The reason is that the input energy is used to make water evaporate. Because more evaporation happens in Situation 1 than Situation 2, we observe that the water in Situation 1 boils more vigorously.

Table 2 Students' responses in a physical science task

Learning progression levels	Responses
Level 4. Relational complexity	Responses from Kai Choice: A. The water temperature in situation 1 and the water temperature in situation 2 are the same Explanation (if the water temperatures in these two situations are the same, why does the water in situation 2 boil less vigorously than the water in situation 1?): because with more heat the water is turning to steam more quickly, but at water's boiling point no matter how much heat is added it does not increase in temperature in this state
Level 3. Measurability	Responses from Lucy Choice: B. The water temperature in situation 1 and the water temperature in situation 2 are different Explanation (what evidence can be used to support the claim that the water temperature in situation 1 and the water temperature in situation 2 are different? please explain why this evidence can be used to support the claim.): the evidence that can be used to support this claim is that they tell you in situation 1 the water is boiling but in situation 2 it is not boiling as much. this evidence can be used because boiling starts to occur at 100 °C but that is not just where it stops
Level 2. Attributes	Responses from Mia Choice: B. The water temperature in situation 1 and the water temperature in situation 2 are different Explanation (what evidence can be used to support the claim that the water temperature in situation 1 and the water temperature in situation 2 are different? Please explain why this evidence can be used to support the claim.): the water in situation 1 was boiling much more vigoriously [vigorously] than the water in situation 2. This means that situation 1 had significantly more energy to use, meaning that it was hotter
Level 1. Holistic phenomena	Responses from Zane Choice: A. The water temperature in situation 1 and the water temperature in situation 2 are the same Explanation (if the water temperatures in these two situations are the same, why does the water in situation 2 boil less vigorously than the water in situation 1?): your pot of water is on the stove, you've turned on the maximum heat, and the wait for boiling begins. You are staring impatiently at the pot when the water looks like it's starting to swirl. You're anxious to see the bubbles that signify that you can put your pasta into that water

4 Evidence for Mathematization to Be Used as a Crosscutting Theme

For mathematization to be used as a crosscutting theme, a framework of mathematization must be developed to guide the development of curriculum, instruction, and assessment across topics and disciplines. In the project, we conducted quantitative analyses of the student assessment data. Our analyses provide two pieces of evidence that the mathematization LP is applicable to topics in both physical sciences and life sciences. Therefore, the mathematization LP is a potential framework to guide the development of curriculum, instruction, and assessment across science topics and disciplines. In this chapter, we describe these two pieces of evidence.

First, we scored the item responses in terms of the four levels of the LP (score 1 for Level 1 responses, etc.) and used the item response theory (IRT) models to analyze those scores. The results suggested that the mathematization LP *can* be used to evaluate student proficiency in both physical science topics and life science topics. More specifically, the Rasch model was used to fit dichotomous items; the Partial Credit model (Masters, 1982) was used for polytomous items. Results of the IRT analysis are presented in Wright maps (Fig. 4). The Wright maps provide quantified locations of item difficulties and students' performances on the same scale, called the logit scale. The left side of the Wright Map displays the distribution of students' performance estimates while the right side represents the distribution of the Thurstonian thresholds for each item. Each item has two to four threshold values. These values are 1, 2, 3, or 4, representing the transition between a zero score (responses such as "I don't know" or random letters) and Level 1, between Level 1 and Level 2, between Level 2 and Level 3, and between Level 3 and Level 4, respectively. For example, the location of the second threshold (labeled as 2) for a life science item, "LS18", is close to zero logit. This suggests that students located at zero logit value of performance have about 50% chance of transition from Level 2 to Level 3 for LS18.

Wright Maps allow a visual determination of whether the LP levels for mathematization of science were differentiated from each other. Undifferentiated levels in the Wright map would indicate that the scoring rubric or the LP is not empirically supported and should not be used to evaluate students' performance. The two Wright maps (Fig. 4) provide the following evidence. For all items, the data supports the hypothesized order of the four learning progression levels. For most items, the data shows that the learning progression levels are differentiated from each other. It is also important to note that the data of some items do not support the distinction between adjacent levels. For example, for a physical science item, "PS21", the threshold 1 and the threshold 2 are located remarkably close to each other. This evidence indicates that the transition between the zero score and Level 1 and the transition between Level 1 and Level 2 are not clearly distinguishable. One possible cause is that the small number of responses at those levels caused unreliable estimates. The distinction between levels is clear in the Wright map for the life science items, but not in the Wright map for the physical science items. This is probably because the physical

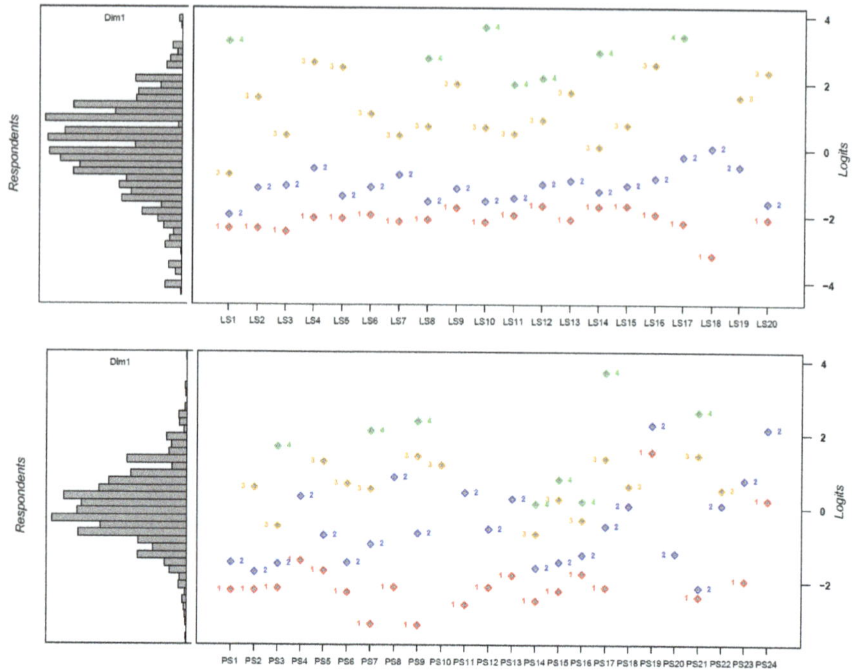

Fig. 4 Wright maps of mathematization of science in life science (upper panel) and in physical science (lower panel)

science assessments contain more multiple-choice items and multiple-choice items are not as effective as constructed response items. This may also be due to that more topics are involved in physical science assessments. In summary, the Wright Maps in both the life science and physical science domains support the internal structure of the assessment (Wilson, 2004) that the learning progression levels provide useful measures of mathematization of science.

Second, more advanced IRT modeling revealed the potential that the mathematization LP is applicable in both life sciences and physical sciences. More specifically, the same data was analyzed through a special type of IRT model (Shin et al., 2017) to investigate the relationships between students' mathematics ability and their mathematization proficiency in physical/life sciences. For this analysis, the same set of math item parameters were used to put mathematization in physical science and mathematization in life science on the same scale in relation to the mathematics ability measure. Next, thresholds between two adjacent learning progression levels were computed. Table 3 provides the estimated thresholds in the life science and the physical science referenced to the mathematics items.

As shown in Table 3, thresholds were estimated as the median values across items on the logit scale (Shin et al., 2012). Thus, the differences between thresholds are allowed to be varied (e.g., Level 1 could have a smaller range than Level 2). For Thresholds 1 and 2, the values in two different disciplines were estimated to be quite

Table 3 Estimated thresholds of the learning progression for mathematization

Thresholds	Life sciences	Physical sciences
Threshold 1	– 2.08	– 2.16
Threshold 2	– 1.09	– 0.96
Threshold 3	0.97	0.55
Threshold 4	2.92	2.09

similar, although not identical. This pattern is not observed for Thresholds 3 and 4. This may be due to that many physical science items were multiple-choice items and that the physical science items were divided into multiple topics to fit the courses taught by the participating teachers.

5 Conclusions

Quantitative reasoning played a crucial role in the development and revolution of scientific knowledge (Crombie, 1961; Kline, 1982). It is also an essential learning goal for K-12 students (NGSS Lead States, 2013; NRC, 1996). In this chapter, we use mathematization of science to refer to quantitative reasoning in science, because it was used to refer to the specialized ways of reasoning that scientists used to quantify phenomena and construct knowledge (Kline, 1982; Lehrer & Schauble, 1998). We conducted an analysis of five events that played critical role in the development and revolution of scientific knowledge. This analysis is inspired by Thompson's theory of quantitative reasoning in mathematics (Thompson, 1993, 2011). We found Thompson's ideas about measurability and relational complexity very useful for us to understand how mathematics and quantitative reasoning were used in the history of science. Our historical analysis suggests three components that differentiate mathematization from people's everyday intuitive reasoning. These three components are measurability, relational complexity, and scientific conceptualization. Together, they illustrate a quantification process, by which scientists abstract measurable variables from messy phenomena and observations; use mathematical operations to identify relationships among those variables; and conceptualize concepts, principles, and theories to explain the identified relationships. By using this quantification process, scientists have made significant breakthroughs in the history of science (Jin et al., 2019a). Mathematization is also one of the six styles of scientific reasoning embedded in all science disciplines (Crombie, 1994; Kind & Osborne, 2017; Osborne et al., 2018). Like other styles of reasoning, the value of mathematization includes "explaining the diversity to be found within the sciences, elegantly capturing the forms of reasoning, and helping to identify the intellectual achievement that the sciences represent" (Osborne & Rafanelli, 2019, p. 530). Therefore, mathematization, as well as other styles of reasoning, are good candidates for crosscutting themes to build curricular coherence.

Schmidt et al. (2005) compared the mathematics and science standards of the United States with those of top-achieving countries in the Third International Mathematics and Science Study (TIMSS). They found that, "coherence is one of the most critical, if not the single most important, defining elements of high-quality standards" (p. 554). They further point out that the U.S. has a 'mile-wide inch-deep' science curriculum that covers a wide range of science topics, but the topics are not organized in ways reflecting the logical nature of the disciplinary content. In other words, the U.S. curriculum is not coherent. The current science standards, NGSS, present a significant improvement in curricular coherence. NGSS were developed under the guidance of the NRC Framework. In the Framework, the three dimensions of science learning (two to four core ideas in each discipline, eight scientific and engineering practices, and seven crosscutting concepts) are integrated to achieve the logical coherence in science disciplines; learning progressions for the components in each dimension are used to ensure the cognitive coherence in science learning. Moreover, the seven crosscutting concepts (patterns; cause and effect; energy and matter, etc.) "provide students with connections and intellectual tools that are related across the differing areas of disciplinary content and can enrich their application of practices and their understanding of core ideas" (NRC, 2012, p. 233).

Osborne and colleagues (Kind & Osborne, 2017; Osborne & Rafanelli, 2019; Osborne et al., 2018) propose using the six styles of reasoning to replace the seven crosscutting concepts as crosscutting themes. While we agree with Saleh and colleagues (Saleh et al., 2019) that the crosscutting concepts have been proved effective when being used as a crosscutting theme, we also believe it is valuable to explore styles of reasoning as alternative crosscutting themes to build curricular coherence. After all, diversity drives innovation and advancements. A variety of approaches are needed to promote teaching and learning of science. For example, if mathematization is taught and assessed consistently across science topics and disciplines, students will learn to use mathematization more effectively. They will also develop deep understanding of the content knowledge in different topics and disciplines, because mathematization requires using disciplinary knowledge to explain mathematical relationships.

For mathematization to be used as a crosscutting theme, evidence in both logical coherence and cognitive coherence should be provided. In terms of logical coherence, researchers have conducted extensive and thorough analysis of scientific knowledge and found that mathematization is embedded in the knowledge across science disciplines (Crombie, 1961; Kline, 1982; Jin et al., 2019a). In terms of cognitive coherence, research of student understanding must be conducted to show that mathematization can be taught and assessed across topics and disciplines. Our research provides preliminary evidence for the cognitive coherence. We developed an LP that describes and evaluates student performance in terms of four levels of achievement—holistic phenomenon, attributes, measurability, and relational complexity. Moving up these levels, students demonstrate increasingly sophisticated mastery of mathematization. Our analyses of students' assessment data suggest that the mathematization LP can be used to assess mathematization across several topics in physical and life sciences.

Further research is needed to use the LP to guide assessment and instruction in more topics.

Acknowledgements This work was supported by Institute of Education Sciences under Grant R305A160219.

References

Ataíde, A. R., & Greca, I. M. (2013). Epistemic views of the relationship between physics and mathematics: Its influence on the approach of undergraduate students to problem solving. *Science & Education, 22*(6), 1405–1421. https://doi.org/10.1007/s11191-012-9492-2

Beichner, R. J. (1994). Testing student interpretation of kinematics graphs. *American Journal of Physics, 62*(8), 750–762. https://doi.org/10.1119/1.17449

Bing, T. J., & Redish, E. F. (2009). Analyzing problem solving using math in physics: Epistemological framing via warrants. *Physical Review Special Topics-Physics Education Research, 5*(2), 020108. https://doi.org/10.1103/PhysRevSTPER.5.020108

Chi, M. T. H., Feltovich, P. J., & Glaser, R. (1981). Categorization and representation of physics problems by experts and novices. *Cognitive Science, 5*(2), 121–152. https://doi.org/10.1207/s15516709cog0502_2

Crombie, A. C. (1994). *Styles of scientific thinking in the European tradition: The history of argument and explanation especially in the mathematical and biomedical sciences and arts*. Duckworth.

Crombie, A. C. (1961). Quantification in medieval physics. In H. Woolf (Ed.), Quantification: A history of the meaning of measurement in the natural and social sciences (pp. 13–30). New York: Bobbs-Merrill.

Dori, Y. J., & Hameiri, M. (2003). Multidimensional analysis system for quantitative chemistry problems: Symbol, macro, micro, and process aspects. *Journal of Research in Science Teaching, 40*(3), 278–302. https://doi.org/10.1002/tea.10077

Fortus, D., & Krajcik, J. (2012). Curriculum coherence and learning progressions. In *Second international handbook of science education* (pp. 783–798). Springer.

Hacking, I. (1994). Styles of scientific thinking or reasoning: A new analytical tool for historians and philosophers of the sciences. In *Trends in the Historiography of Science* (pp. 31–48). Springer.

Holton, G. J., & Brush, S. G. (2001). *Physics, the human adventure: From Copernicus to Einstein and beyond*. Rutgers University Press.

Jin, H., Delgado, C., Bauer, M. I., Wylie, E. C., Cisterna, D., & Llort, K. F. (2019a). A hypothetical learning progression for quantifying phenomena in science. *Science & Education, 28*(9), 1181–1208. https://doi.org/10.1007/s11191-019-00076-8

Jin, H., Mikeska, J. N., Hokayem, H., & Mavronikolas, E. (2019b). Toward coherence in curriculum, instruction, and assessment: A review of learning progression literature. *Science Education, 103*(5), 1206–1234. https://doi.org/10.1002/sce.21525

Kesidou, S., & Duit, R. (1993). Students' conceptions of the second law of thermodynamics—An interpretive study. *Journal of Research in Science Teaching, 30*(1), 85–106. https://doi.org/10.1002/tea.3660300107

Kind, P., & Osborne, J. (2017). Styles of scientific reasoning: A cultural rationale for science education? *Science Education, 101*(1), 8–31. https://doi.org/10.1002/sce.21251

Kline, M. (1982). *Mathematics: The loss of certainty*. Galaxy Books.

Kline, M. (1990). *Mathematical Thought from Ancient to Modern Times: Volume 2* (Vol. 2). Oxford university press.

Kozhevnikov, M., Motes, M. A., & Hegarty, M. (2007). Spatial visualization in physics problem solving. *Cognitive Science, 31*(4), 549–579. https://doi.org/10.1080/15326900701399897

Kozma, R. B., & Russell, J. (1997). Multimedia and understanding: Expert and novice responses to different representations of chemical phenomena. *Journal of Research in Science Teaching, 34*(9), 949–968. https://doi.org/10.1002/(SICI)1098-2736(199711)34:9%3c949::AID-TEA7%3e3.0.CO;2-U

Kuo, E., Hull, M. M., Gupta, A., & Elby, A. (2013). How students blend conceptual and formal mathematical reasoning in solving physics problems. *Science Education, 97*(1), 32–57. https://doi.org/10.1002/sce.21043

Lehrer, R., & Schauble, L. (1998). Reasoning about structure and function: Children's conceptions of gears. *Journal of Research in Science Teaching, 35*(1), 3–25. https://doi.org/10.1002/(SICI)1098-2736(199801)35:1%3c3::AID-TEA2%3e3.0.CO;2-X

Masters, G. N. (1982). A Rasch model for partial credit scoring. *Psychometrika, 47*(2), 149–174. https://doi.org/10.1007/BF02296272

NGSS Lead States (2013). *Next generation science standards: For states, by states.* National Academies Press.

Niss, M. (2017). Obstacles related to structuring for mathematization encountered by students when solving physics problems. *International Journal of Science and Mathematics Education, 15*(8), 1441–1462. https://doi.org/10.1007/s10763-016-9754-6

NRC (1996). *National science education standards: Observe, interact, change, learn.* National Academies Press.

NRC (2000). *Inquiry and the national science education standards: A guide for teaching and learning.* National Academies Press.

NRC (2012). *A framework for K-12 science education: Practices, crosscutting concepts, and core ideas.* National Academies Press.

Osborne, J., & Rafanelli, S. (2019). A response to Saleh et al.: The wrong call to action. *Journal of Research in Science Teaching, 56*(4), 529–531. https://doi.org/10.1002/tea.21536

Osborne, J., Rafanelli, S., & Kind, P. (2018). Toward a more coherent model for science education than the crosscutting concepts of the next generation science standards: The affordances of styles of reasoning. *Journal of Research in Science Teaching, 55*(7), 962–981. https://doi.org/10.1002/tea.21460

Planinic, M., Milin-Sipus, Z., Katic, H., Susac, A., & Ivanjek, L. (2012). Comparison of student understanding of line graph slope in physics and mathematics. *International Journal of Science and Mathematics Education, 10*(6), 1393–1414. https://doi.org/10.1007/s10763-012-9344-1

Saleh, A., Weiser, G., Rehmat, A. P., Housh, K., Cisterna, D., Liu, L., & Hmelo-Silver, C. (2019). A call to action: A response to Osborne, Rafanelli, and Kind (2018). Journal of Research in Science Teaching, 56(4), 526–528. https://doi.org/10.1002/tea.21537

Schmidt, W. H., Wang, H. C., & McKnight, C. C. (2005). Curriculum coherence: An examination of US mathematics and science content standards from an international perspective. *Journal of Curriculum Studies, 37*(5), 525–559. https://doi.org/10.1080/0022027042000294682

Schuchardt, A. M., & Schunn, C. D. (2016). Modeling scientific processes with mathematics equations enhances student qualitative conceptual understanding and quantitative problem solving. *Science Education, 100*(2), 290–320. https://doi.org/10.1002/sce.21198

Sherin, B. L. (2001). How students understand physics equations. *Cognition and Instruction, 19*(4), 479–541. https://doi.org/10.1207/S1532690XCI1904_3

Shin, H., Choi, J., & Draney, K. (2012). *Using item response theory models for classifying students onto levels of achievement.* Presented at the international objective measurement workshop (IOMW), Vancouver, BC, Canada.

Shin, H. J., Wilson, M., & Choi, I.-H. (2017). Structured constructs models based on change-point analysis. *Journal of Educational Measurement, 54*(3), 306–332. https://doi.org/10.1111/jedm.12146

Shwartz, Y., Weizman, A., Fortus, D., Krajcik, J., & Reiser, B. (2008). The IQWST experience: Using coherence as a design principle for a middle school science curriculum. *The Elementary School Journal, 109*(2), 199–219.

Sikorski, T. R., & Hammer, D. (2017). Looking for coherence in science curriculum. *Science Education, 101*(6), 929–943. https://doi.org/10.1002/sce.21299

Taasoobshirazi, G., & Glynn, S. M. (2009). College students solving chemistry problems: A theoretical model of expertise. *Journal of Research in Science Teaching, 46*(10), 1070–1089. https://doi.org/10.1002/tea.20301

Thompson, P. W. (2011). Quantitative reasoning and mathematical modeling. In L. L. Hatfield, S. Chamberlain, & S. Belbase (Eds.), *New perspectives and directions for collaborative research in mathematics education, WISDOMe monographs* (Vol. 1, pp. 33–57). University of Wyoming.

Thompson, P. W., Carlson, M. P., Byerley, C., & Hatfield, N. (2014). Schemes for thinking with magnitudes: A hypothesis about foundational reasoning abilities in algebra. In L. P. Steffe, L. L. Hatfield, & K. C. Moore (Eds.), *Epistemic algebra students: Emerging models of students' algebraic knowing, WISDOMe monographs* (Vol. 4, pp. 1–24). University of Wyoming.

Thompson, P. W., & Carlson, M. P. (2017). Variation, covariation, and functions: Foundational ways of thinking mathematically. *Compendium for research in mathematics education*, 421–456.

Thompson, P. W. (1993). Quantitative reasoning, complexity, and additive structures. *Educational Studies in Mathematics, 25*(3), 165–208. https://doi.org/10.1007/BF01273861

Tuminaro, J., & Redish, E. F. (2004). Understanding students' poor performance on mathematical problem solving in physics. In J. Marx, S. Franklin, & K. Cummings (Eds.), *AIP 2013 conference proceedings* (Vol. 720, pp. 113–116). https://doi.org/10.1063/1.1807267

Tuminaro, J., & Redish, E. F. (2007). Elements of a cognitive model of physics problem solving: Epistemic games. *Physical Review Special Topics-Physics Education Research.* https://doi.org/10.1103/PhysRevSTPER.3.020101

Wilson, M. (Ed.). (2004). *Towards coherence between classroom assessment and accountability.* University of Chicago Press.

Wylie, E. C., Bauer, M. I., Arieli-Attali, M. (2015, April). *Validating and using learning progressions to support mathematics formative assessment.* Paper presented at the annual meeting of the National Council on Measurement in Education, Chicago, IL, United States.

Applying Quantitative and Covariational Reasoning to Think About Systems: The Example of Climate Change

Darío A. González

1 Introduction

Climate change is a variation in the long-term average temperature and weather patterns of the planet. Human activities (e.g., generation of electricity, transportation, or food production) release large amounts of greenhouse gases into the atmosphere, which trap heat and increase the average temperature of the planet. The Intergovernmental Panel on Climate Change (IPCC) has warned us that exceeding a global warming of 1.5 ºC above the preindustrial era average could bring devastating and irreversible consequences to our social, economic, and natural systems. We have, at most, 30 years before passing that threshold (IPCC, 2018), and staying within that safe limit requires everyone's commitment to support and adopt mitigation strategies, which is more likely to happen when people possess knowledge about climate change (Sewell et al., 2017).

Unfortunately, climate change is not an easy phenomenon to understand. First, it is happening at a planetary scale which makes it difficult for a single individual to directly experience or grasp all of its consequences. This can make climate change feel like a distant problem that is too large or too abstract to be tackled. Second, the Earth's climate is a complex system that involves interactions between several components (Sun, oceans, land, atmosphere, among others) and even other systems (economies, societies, and ecosystems). Therefore, understanding climate change requires what are known as *systems thinking competences* (Ghosh, 2017; Roychoudhury et al., 2017; Schuler et al., 2018), which include the ability to think in terms

This work was supported by Centro de Modelamiento Matemático (CMM), ACE210010 and FB210005, BASAL funds for centers of excellence from ANID-Chile.

D. A. González (✉)
Centro de Modelamiento Matemático, Universidad de Chile, Beauchef 851, Santiago, Región Metropolitana, Chile
e-mail: dario.gonzalez@fulbrightmail.org

G. Karagöz Akar et al. (eds.), *Quantitative Reasoning in Mathematics and Science Education*, Mathematics Education in the Digital Era 21,
https://doi.org/10.1007/978-3-031-14553-7_11

of systems and the ability to model aspects of systems dynamics such as multiple interrelated variables (interconnectedness), causality loops between variables (feedback), and patterns of variation over time (dynamic relationships) (Riess & Mischo, 2010).

Mathematics can play a prominent role in helping students and teachers "experience" climate change as a tangible and real problem and develop the systems thinking competences needed to understand it (Barwell, 2013a, 2013b; Renert, 2011). However, studying the mathematics behind climate change rarely forms part of teacher education courses, and thus mathematics teachers may not be prepared to incorporate this topic in their instruction. As a result, mathematics teachers may not be familiar with two central concepts that are necessary to understand climate change: The Earth's energy budget and the link between carbon dioxide (CO_2) pollution and global warming (Lambert & Bleicher, 2013). A starting point, therefore, may involve helping preservice mathematics teachers (PSTs) understand these two constructs framed as mathematical situations.

This chapter discusses the role of quantitative reasoning in developing an understanding of the energy budget as a system formed by multiple interacting components in terms of quantities and the relationships between them. The chapter also discusses the role of covariational reasoning in developing an understanding of the energy budget's response to CO_2 pollution in terms of variation in the planet's surface temperature over time (dynamic relationships). This includes making sense of the greenhouse effect as the covariation of two quantities with respect to time that obeys a circular causality relationship (feedback loop). Thus, this chapter discusses how quantitative reasoning and covariational reasoning can support students and teachers' understanding of climate change. The chapter illustrates these claims with examples from two preservice mathematics teachers who participated in a study examining how they made sense of the mathematics involved in modeling climate change (González, 2017). The chapter also discusses how using climate change as a context for applying quantitative and covariational reasoning can attend to curriculum requirements (Common Core State Standards Initiative [CCSSI], 2010; National Research Council [NRC], 2013), as well as implications for teaching and research in both mathematics and science education.

2 The Earth's Energy Budget

The Earth's energy budget is a representation of how the energy flows between the main components of the climate system. Let us consider a simple energy budget model that includes three of these components: the Sun, the planet's surface, and the atmosphere. The energy flows S, R, L, B, and A (Fig. 1) are quantified in terms of *irradiance*, which is defined as the energy incident to a surface per unit of time per unit area so that it is measured in Joules per second per square meter ($Js^{-1}m^{-2}$). The Sun radiates energy, in the form of light and heat, towards the Earth, most of which passes through the atmosphere and is absorbed by the surface (S). As the

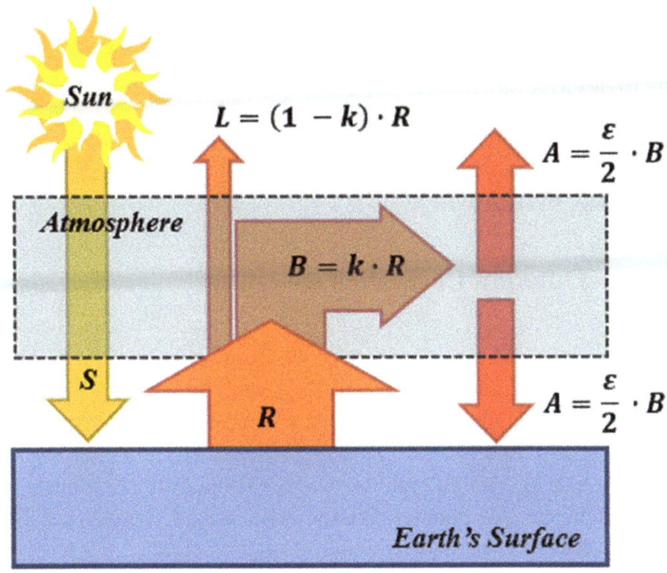

Fig. 1 Diagram of the Earth's energy budget

surface absorbs the solar radiation, it heats up and begins to radiate infrared energy, in the form of heat, upward towards the atmosphere (R). A small fraction of infrared energy (L) escapes to space, but the majority of it (B) is absorbed by *greenhouse gases* (GHG) in the atmosphere such as water vapor (H_2O), carbon dioxide (CO_2), or methane (CH_4). As the atmosphere absorbs surface radiation, it heats up and begins to radiate infrared energy in both directions, out to space (A) and back to the surface (A). The latter further increases the temperature of the surface, and the continuous and dynamic energy exchange between the surface and the atmosphere represents what is known as the *greenhouse effect*, which has an important impact on the planet's average temperature.

The notion of *radiative equilibrium* is key to understanding how the energy budget regulates the planet's average temperature. Let us consider the energy inflow into the surface ($S + A$) and the energy outflow from the surface (R). When $S + A = R$, it is said that the energy budget is in radiative equilibrium and the surface temperature remains constant. A system analogous to the energy budget would be a container with water flowing into it at rate r and water coming out of it at rate q. If $r = q$, then the volume of water in the container remains constant. The water flows and water volume are analogous to the energy flows and temperature in the energy budget, respectively. The radiative equilibrium can be disrupted due to different factors known as climate forcing agents, which result in changes in the planet's average temperature. A forcing agent of particular interest for this chapter involves quantifying changes in the energy flows and radiative equilibrium due to changes in the abundance of CO_2 in the atmosphere, which is quantified as concentration—the volume of CO_2 in the air[1] relative to the volume of air—and measured in parts per million (ppm). The parts-per notation is a set of pseudo-units used in science and engineering to describe small values of concentration of substances in solutions.

3 Conceptual Framework

3.1 *Systems Thinking Competencies*

Students and teachers need to make sense of the Earth's climate as a system in order to be able to understand climate change (Roychoudhury et al., 2017). Therefore, understanding climate change requires systems thinking competencies, or the ability to recognize, describe, model, and explain complex aspects of reality as systems (Riess & Mischo, 2010). Several authors have proposed different models of systems thinking competencies; this chapter utilizes Assaraf and Orion's (2005) Systems Thinking Hierarchical (STH) model, which describes eight competencies grouped into three hierarchical levels (Fig. 2).

According to the STH model, making sense of the Earth's energy budget as a system requires the *analysis* and then the *synthesis* of that system's components and processes (Levels 1 and 2 in Fig. 2). This can be accomplished by constructing quantities associated with those components and defining relationships between those quantities to model the processes relating the components, thus creating a quantitative structure that gives coherence to the energy budget as a whole. Quantitative reasoning can support the activation of systems thinking competencies because, by definition, quantitative reasoning is the set of "mental actions of a student who conceives of a mathematical situation, constructs quantities in that situation, and then relates, manipulates, and uses those quantities to make a problem situation coherent" (Weber et al., 2014, p. 24).

In particular, the ability to identify components and processes in the energy budget requires the conceptualization of quantities related to measurable attributes of those components and processes. For instance, one needs to conceptualize the abundance of CO_2 in the atmosphere and the energy flows between the energy budget's components in terms of quantities such as concentration and irradiance, respectively, which can be a very challenging task considering that these quantities make use of difficult concepts such as energy and peculiar units of measure such as ppm and $Js^{-1}m^{-2}$ (Aneye et al., 2019; de Berg, 2012; Liu & McKeough, 2005; Raviolo et al., 2021). Also, the ability to identify relationships between components requires the conceptualization of how the quantities associated with the components relate to each other. For instance, one needs to conceptualize that the energy emitted by the atmosphere, A, changes with changes in the energy emitted by the surface, R, or with changes in the atmospheric CO_2 concentration.

[1] The air in the atmosphere is a mixture of several gases such as nitrogen (78.09%), oxygen (20.95%), argon (0.93%), carbon dioxide (0.04%), and small amounts of other gases.

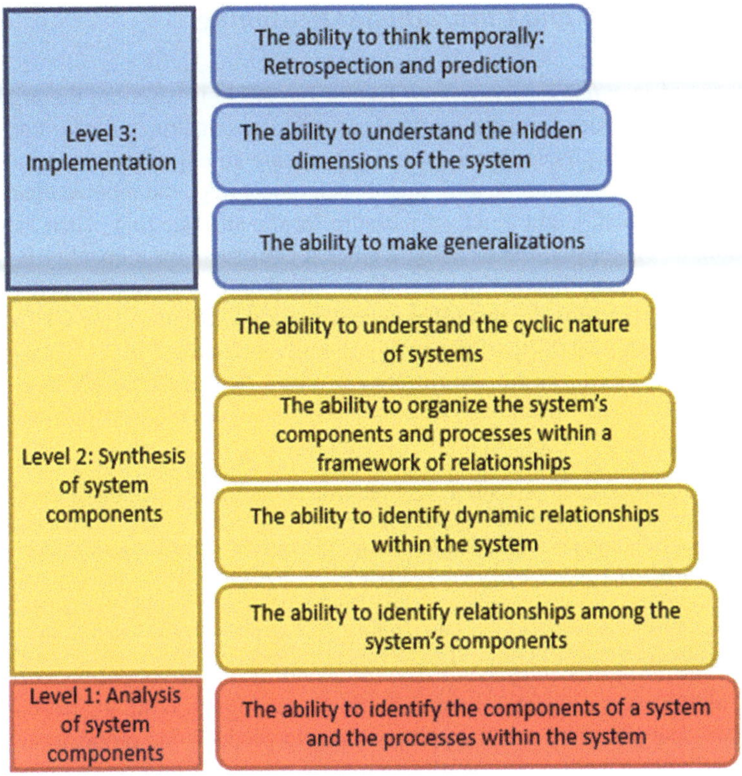

Fig. 2 The Systems Thinking Hierarchical (STH) model

The ability to identify dynamic relationships within the system and understand the cyclic nature of systems requires envisioning multiple pairs of quantities changing together over time, which requires the more specific quantitative skill of covariational reasoning because it encompasses "the very [mental] operations that enable one to see the invariant relationship between quantities in dynamic situations" (Thompson, 2011, p. 46). For instance, one needs to conceptualize the energy, *A*, emitted by the atmosphere, and the energy, *R*, emitted by the surface as changing simultaneously over time and in a continuous loop that enhances the surface temperature. Finally, quantitative reasoning is required to bring quantities, relationships, and dynamic relationships together to organize the system's components and processes within a coherent framework of relationships. Thus, quantitative and covariational reasoning skills play a role in understanding the energy budget as a system.

3.2 Quantitative and Covariational Reasoning

The theory of *quantitative reasoning* (Thompson, 2011) is based on the argument that students construct a quantity through an effortful cognitive process known as *quantification*. This process involves conceptualizing an object in a situation (e.g., the atmosphere), a measurable attribute of such object (e.g., the relative abundance of atmospheric GHG), and a unit of measure for the attribute (e.g., such as ppm). According to Thompson, the meaning a student constructs for a quantity is inseparable from the quantification process. Understanding the energy budget requires developing meaning for the quantities representing the abundance of GHG in the atmosphere and the intensity of the energy flows between the Sun, the surface, and the atmosphere, as well as the relationships that exist between such quantities.

The construct of *covariational reasoning* finds one of its earliest definitions in the work of Saldanha and Thompson (1998), who characterized it as:

> Someone holding in mind a sustained image of two quantities' values (magnitudes) simultaneously. It entails coupling the two quantities, so that, in one's understanding, a multiplicative object is formed of the two. As a multiplicative object, one tracks either quantity's value with the immediate, explicit, and persistent realization that, at every moment, the other quantity also has a value … An operative image of covariation is one in which a person imagines both quantities having been tracked for some duration, with the entailing correspondence being an emergent property of the image. (pp. 298–299)

The multiplicative object in their definition is analogous to the logical conjunction "and" that joins or unites two propositions to produce one proposition that is true if and only if both constituent propositions are true. Concerning covariation, the multiplicative object joins the corresponding values of two covarying quantities so that the student "mentally unites their attributes to make a new attribute that is, simultaneously, one and the other" (Thompson et al., 2017, p. 96). Saldanha and Thompson hypothesized that covariational reasoning may involve developmental levels. This hypothesis was further developed by Carlson et al. (2002), who proposed the *Covariation Framework* to examine and assess the covariational reasoning abilities of students. The framework describes, in increasing order of sophistication, five mental actions involved in reasoning about quantities that vary together (Table 1).

4 The Context of the Study

This chapter makes use of the results of a larger study (González, 2017) to illustrate how quantitative and covariational reasoning are involved in understanding climate change. The larger study investigated how three PSTs made sense of the mathematics of climate change and consisted of two phases and a mini lesson (Fig. 3). The PSTs were three female students—Kris, Pam, and Jodi (pseudonyms)—enrolled in a mathematics education program at a large public university in the Southeast of the United States. By the time the study took place, they had completed two calculus courses,

Table 1 Carlson et al.'s (2002) Covariation Framework

Mental action	Description of mental action	Behaviors
MA1	Coordinating the value of one variable with changes in the other	• Labeling the axes with verbal indications of coordinating the two variables (e.g., y changes with changes in x)
MA2	Coordinating the direction of change of one variable with changes in the other variable	• Constructing an increasing straight line • Verbalizing an awareness of the direction of change of output while considering changes in the input
MA3	Coordinating the amounts of change of one variable with changes in the other	• Plotting points/constructing secant lines • Verbalizing an awareness of the amount of change of the output while considering changes in the input
MA4	Coordinating the average rate-of-change of the function with uniform increments of change in the input variable	• Constructing contiguous secant lines for the domain • Verbalizing an awareness of the rate of change of the output (with respect to the input) while considering uniform increments of the input
MA5	Coordinating the instantaneous rate of change of the function with continuous changes in the independent variable for the entire domain of the function	• Constructing a smooth curve with clear indications of concavity changes • Verbalizing an awareness of the instantaneous changes in the rate of change for the entire domain of the function (direction of concavities and inflexion points are correct)

an introduction to higher mathematics course, and a mathematics content course for secondary teachers.

During phase 1 of the larger study, González explored the PSTs' conceptions of two quantities, concentration and irradiance, within contexts that did not involve the complexity of climate change. After phase 1 and before phase 2, each PST participated in a 30-min-long mini-lesson to prepare them for working on those tasks. That mini lesson included: (i) watching a video introducing the energy budget and the greenhouse effect and (ii) a Q&A session during which González answered questions and further elaborated on the concepts seen in the video. In phase 2, the study explored

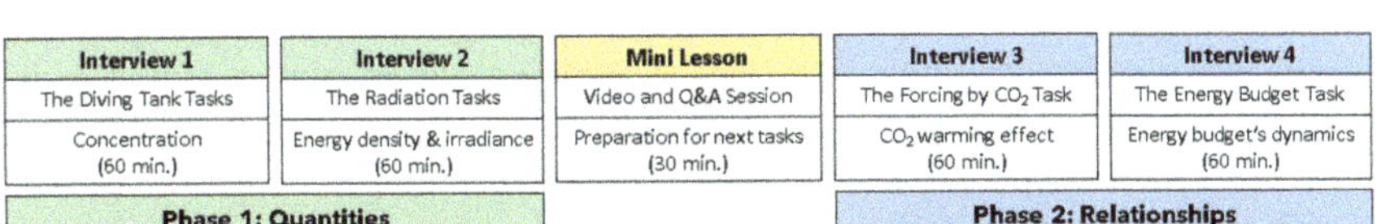

Fig. 3 The design of González's (2017) study

the PSTs' ability to make sense of relationships within the Earth's energy budget. Throughout the two phases, each PST completed a total of six mathematical tasks over the course of four, 60-min-long, individual, task-based interviews (Goldin, 2000). The interviews were video recorded and transcribed for the analysis.

This chapter contrasts the quantitative and covariational reasoning of Kris and Pam concerning their work on some of the tasks involved in phases 1 and 2 of González's study. These two PSTs were selected because their quantitative and covariational reasoning showed interesting differences that illustrate the discussion in the following section.

5 Quantitative Reasoning and Understanding Climate Change

The discussion of results is organized into three subsections. First, there is the discussion on Kris and Pam's preliminary work on making sense of concentration and irradiance. The second subsection discusses how Kris and Pam's quantitative reasoning is involved in making sense of the energy budget's interacting components in terms of quantities and relationships between them. Finally, the last subsection discusses how Kris and Pam's covariational reasoning is involved in conceptualizing dynamic relationships and cyclic processes within the energy budget.

5.1 Preliminary Work: Making Sense of (Unfamiliar) Quantities

Before making sense of the energy budget as a system, one needs to develop meaning for the quantities associated with the energy budget's interacting components and the units in which those quantities are measured. González (2017) developed a sequence of four mathematical tasks for the phase 1 of his study where the quantities concentration and irradiance emerge as solutions to particular situations. In what follows, I contrast the quantitative reasoning of Kris and Pam relative to their work in tasks 2 and 4 of the sequence (Table 2). I make use of Thompson's (1994) distinction between *ratio* and *rate* to discuss the ways Kris and Pam understood concentration measured in ppm (henceforth referred to as *ppm concentration*) and irradiance measured in $Js^{-1}m^{-2}$.

For Thompson, *ratio* or *internalized ratio* is a multiplicative comparison of two specific, non-varying quantities. For instance, the ppm concentration of tank A in the Diving Tank Task 2 is 362 ppm; this quantity can be conceptualized as a *ratio* in three ways: (i) as the multiplicative comparison of two specific measured volumes such as comparing 362 cm^3 of CO_2 with 1,000,000 cm^3 of air, (ii) as the multiplicative comparison of a specific measured volume and one unit of another volume such as

Table 2 The tasks involving ppm concentration and irradiance

Diving tank task 2

The *volume concentration* of gas *X*, denoted as Q_x, in an air mixture is the ratio

$$Q_x = \frac{\text{volume of gas X}}{\text{volume of air}}$$

Diving tanks also contain a small volume of carbon dioxide (CO_2). The table below shows the volumes of air and CO_2 of two diving tanks

Tank	Air (cm^3)	CO_2 (cm^3)
A	4,000,000	1448
B	800,000	316

(a) Calculate each tank's volume concentration of CO_2. Interpret your result in the context of this situation.

(b)When concentrations are small, they are often measured in ppm (parts per million), or the number of parts corresponding to a particular gas in 1,000,000 parts of air. Calculate each tank's *ppm concentration* of CO_2.

Radiation task 2

When the *energy density* of a metallic sheet of *daridium* increases by 2500 J/m^2, the sheet's temperature rises by 4 °C. In an experiment, two sheets were positioned at the same distance from two devices that produce radiation (see Figure)

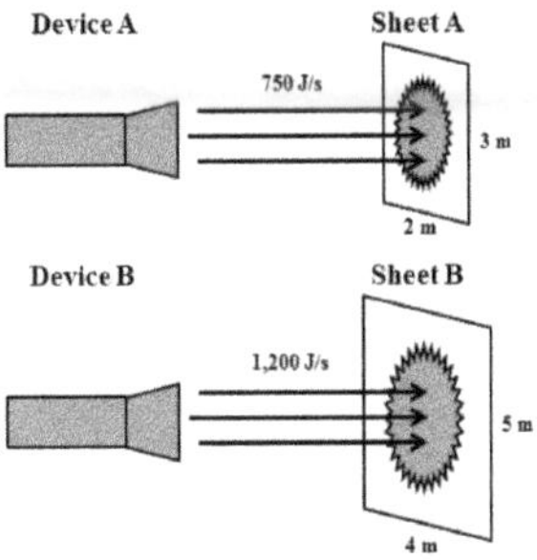

Device A radiates 750 J/s (Joules per second) toward sheet A and device B radiates 1200 J/s toward sheet B. If both sheets were at room temperature (around 15 °C) at the beginning of the experiment and both devices started radiating energy at the same time, then which sheet will first reach a temperature of 25 °C?

comparing 362 cm^3 of CO_2 with 1 m^3 of air, and (iii) as the quotient 362 resulting from multiplicatively comparing two specific but undetermined volumes (e.g., the ppm concentration of a specific tank for which we do not know the volumes of each gas it contains). Thompson included the last comparison within the category of ratio because the student conceptualizes 362 as the particular case of dividing the volumes of CO_2 and air for a specific tank so that those volumes, albeit undetermined or unknown, are conceived as fixed. In other words, the student does not necessarily conceptualize 362 as a constant multiplicative relationship between two volumes that can vary.

Thompson defined *rate* as a *reflectively abstracted constant ratio* because it symbolizes the structure of a ratio—dividend, divisor, and quotient—as a whole and emphasizes the constant multiplicative relationship between the constituent quantities as their values vary. Rate, therefore, is a ratio conceived as independent of particular measured magnitudes of the constituent quantities and represents a proportional relationship between two quantities that can vary. He also argued that a mature concept of rate requires an understanding of the relationship between the simultaneous *accumulation* of change in two quantities that covary and the *accrual* by which the accumulation is built. For instance, the irradiance incident to sheet A in the Radiation Task 2 is 125 $Js^{-1}m^{-2}$, which means that the increase in the average surface energy of sheet A after device A has been working for 3 s is the simultaneous accumulation of three accruals of 125 J/m^2 and three accruals of 1 s. Also, the student must realize that the relationship between the accumulated surface energy density and accumulated time, at any time during the experiment, is constant and equal to 125 $Js^{-1}m^{-2}$.

5.1.1 Concentration Measured in ppm

Before working on the Diving Tank Task 2, the PSTs worked on the Diving Tank Task 1 where they made sense of *volume concentration.* This quantity is a dimensionless ratio between the volume of a gas and the volume of air in which that gas is mixed. This task help PSTs become familiar with what concentration is used for. Thus, the Diving Tank Task 2 started with a reminder of how to calculate volume concentration and asked the PSTs to determine the volume concentration of CO_2 for two diving tanks. The second part of the task then asked the PSTs to determine the ppm concentration of CO_2 for those same tanks (Table 2).

Kris's Conceptualization of Concentration. Kris conceptualized ppm concentration as a ratio comparing a measured volume of CO_2 with 1,000,000 cm^3 of air. First, she divided the CO_2 volume by the air volume for each tank, obtaining the volume concentration of 0.000362 for tank A and 0.000395 for tank B. Next, she multiplied each one of those values by 1,000,000, to obtain the values 362 for tank A and 395 for tank B, which she interpreted as the ppm concentration of CO_2 for each tank. Kris explained her procedure in the following way:

I Why did you multiply 0.000362 by 1,000,000 to get the concentration in ppm?

K This [*points at 0.000362*] is the proportion of carbon dioxide out of the whole tank ... if we take 1,000,000 cm^3 of air from tank A and kind of transfer it to a new tank ... then the proportion would be the same in the second tank. So, we can take this proportion [*points at 0.000362*] and multiply it by the volume of the second tank to get the volume of carbon dioxide in the second tank, which is that [*points at 362*].

I Why did you transfer 1,000,000 cm^3 of air from tank A to a new tank?

K In order to measure the ppm level or units, whatever, we have to see the amount of CO_2 contained in 1,000,000 cm^3 of air. So, given that it was uniformly distributed kind of like the first problem, putting 1,000,000 of this [*points at tank A*] into a new tank would maintain the concentration, but we would be able to more easily see the ppm.

When Kris "took" 1,000,000 cm^3 of air from tank A and "transferred" it to a second tank, she stated that "the proportion would be the same in the second tank". The proportion that Kris was referring to was the value 0.000362. This suggests that she conceptualized the value 0.000362 as a *rate* describing a constant multiplicative relationship between two volumes that can vary. In contrast, Kris's conceptualization of ppm concentration appeared more consistent with Thompson's *ratio* because it involved a multiplicative comparison of a specific measured volume of CO_2 and 1,000,000 cm^3 of air. Kris utilized the product $0.000362 \times 1{,}000{,}000$ cm^3 to obtain the specific volume of CO_2 (362 cm^3) in her made-up tank, which suggests that she

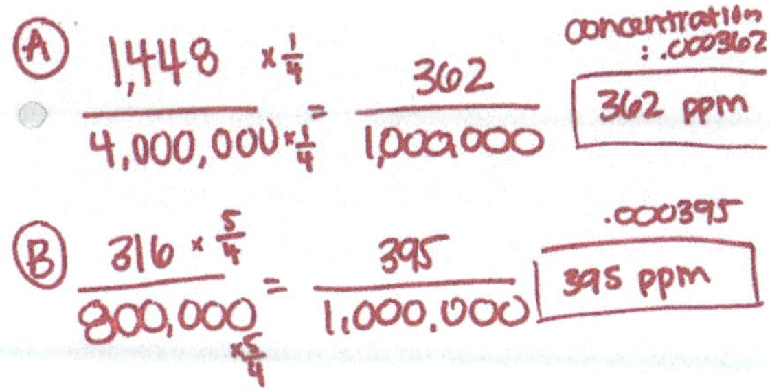

Fig. 4 Pam made use of equivalent fractions to find the ppm concentration

conceptualized 362 ppm as a multiplicative comparison of two specific measured volumes, 362 cm^3 of CO_2 and 1,000,000 cm^3 of air.

Pam's Conceptualization of Concentration. Pam's conceptualization of ppm concentration was neither strictly a *ratio* nor a fully realized *rate* yet. First, Pam set up a fraction formed by the CO_2 volume over the air volume for each tank, obtaining 1448/4,000,000 for tank A and 316/800,000 for tank B. Next, she found a new pair of fractions, each equivalent to one of the initial fractions, but with a denominator of 1,000,000 cm^3 so that their numerators indicated the ppm concentration (Fig. 4). She thus obtained 362/1,000,000 for tank A, representing 362 ppm, and 395/1,000,000 for tank B, representing 395 ppm. When I asked Pam to interpret the value 362 ppm, she said the following:

> It is kind of in that [*points at the unit "ppm"*]. Like, this [*points at "362"*] is parts per million … So, it's kind of like five miles per hour. So, if you go two hours, you're going 10 miles, because it's five miles every hour. So, every million you have is 362 parts. So, if you have 2,000,000, it's 724 parts, because every million increase, your 362 will increase. That's why I was able to do this, multiply by four. Because every million, like, every 4,000,000 you have, you have four 362's.

Let us consider her interpretation of the fraction 362/1,000,000; such fraction was not a quotient between two specific volumes (362 cm^3 of CO_2 and 1,000,000 cm^3 of air) but rather a "rule" for the simultaneous *accumulation* of an integer number n of *accruals* of volume of CO_2 and volume of air—CO_2 accumulates in accruals of 362 cm^3 while air accumulates in accruals of 1,000,000 cm^3. Although this may resemble the notion of *rate*, a closer inspection suggests that Pam conceptualized the accumulation as a repeated addition of accruals rather than a multiplicative relationship ("every 4,000,000 you have, you have four 362's"). That is to say, the fraction 362/1,000,000 represented a "rule" to simultaneously *add* an integer number n of accruals of CO_2 and air, hence it could not be considered a Thompson's *rate*.

5.1.2 Irradiance

Before working on the Radiation Task 2, the PSTs worked on the Radiation Task 1 where they were expected to find the radiation (energy) incident to a metallic sheet averaged over its surface, which is known as *energy density* and is measured in Joules per square meter (Jm^{-2}). They were asked to compare the energy density of two metallic sheets and conclude which one of them had the highest temperature. They conceptualized energy density as a quotient associating a fixed amount of energy with a unit of area. For instance, Pam claimed that sheet A was hotter than sheet B because the former's energy density (3125 Jm^{-2}) was higher than the latter's energy density (2812.5 Jm^{-2}). She then interpreted those magnitudes as follows:

> This is the same amount of space [points at "3,125" and "2,812.5"], they're both one square meter. So, this one [*points at sheet A*] has way more Joules in that one square meter than this one does [*points at sheet B*]. So, sheet A has got more Joules going to the same amount of space than [*sheet B*].

After they worked with energy density, the PSTs were presented with the Radiation Task 2, which required them to conceive another pair of metallic sheets receiving radiation at *different rates*. That time, they were tasked with finding out which sheet's temperature was increasing *the fastest*, which would involve making use of the quantity irradiance (Table 2).

Kris's Conceptualization of Irradiance. Kris conceptualized irradiance as a *time rate* indicating the amount of increase in energy density per second. For instance, this is how she found the irradiance of sheet A and interpreted its magnitude:

> We know that for every second device A is running, 750 Joules of radiation get put into sheet A. ... Because kind of like the other problem where we were talking about energy density, and we know to calculate energy density we need to divide the amount of energy by the area to get ... the energy per unit area, area unit [*sic.*], and so we divided 750 by 6, which is the area of sheet A because it's 2 meters by 3 meters. So, we got 125 Jm^{-2} [*sic.*] increase in energy density per second that device A is running.

The excerpt suggests a conception of irradiance that I would represent in terms of its units as follows: 125 $Js^{-1}m^{-2}$ was conceptualized as (125 Jm^{-2}) s^{-1}, meaning 125 Jm^{-2} increase in energy density per second. This conception resembles Thompson's *rate* since it represents a constant multiplicative relationship between two quantities (energy density and time) that can vary. It also describes the accumulation of energy density, in accruals of 125 Jm^{-2}, and time, in accruals of 1 s. Finally, Kris also related irradiance with a measure of *how fast* the sheet's temperature was increasing: "As long as we know the increase of energy density per second, then we can tell immediately ... which [sheet] is going to reach [25 °C] faster". To represent her reasoning, she drew two increasing lines in a coordinated plane: one with slope 125 $Js^{-1}m^{-2}$ representing the change in the energy density of sheet A with respect to time and another with slope 60 $Js^{-1}m^{-2}$ representing the change in the energy density of sheet B with respect to time (Fig. 5). Finally, she explained, "I know that more Joules are being added per second per area unit to sheet A than sheet B ... energy density is going to increase faster for sheet A than it is for sheet B".

Pam's Conceptualization of Irradiance. Pam conceptualized irradiance as neither a *ratio* nor a fully realized *rate* yet. The quotient 125 $Js^{-1}m^{-2}$ was not conceptualized as a multiplicative comparison but rather as an *additive relationship* (association) between an amount of energy per second and one square meter. For instance, when I

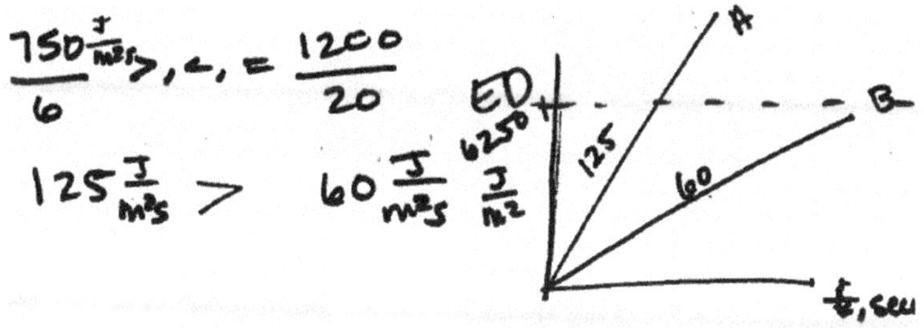

Fig. 5 Kris represented irradiance as the slope of a line

asked her to interpret the magnitude 125 $Js^{-1}m^{-2}$, she replied, "It's like every second, how many Joules are to one meter [*sic.*]. So, after one second has been, [*draws a rectangle formed by six squares and writes '125' in each square*]" (Fig. 6). Pam's response and drawing suggest that irradiance was conceptualized as an additive rule for the accumulation of accruals of radiation rate (125 Js^{-1}) and accruals of area (1 m^2). Pam's additive thinking was more prevalent when talking about irradiance without connecting it to temperature. When she made claims about the temperature, her approach was different; Pam compared the radiation rate 750 Js^{-1} to the radiation rate 1200 Js^{-1} *and separately* compared the area 6 m^2 to the area 20 m^2 in order to determine which sheets was heating up the fastest:

> Twelve hundred [Js^{-1}] is not too far over 750 [Js^{-1}], but there is a big difference between 6 square meters and 20 square meters … So, if I was just looking at them, I'd be like sheet A because there is less surface area … Smaller things heat up faster, but if they are heated up at a lower rate, then it would kind of depend.

This *between-state ratio* comparison—Pam compared two ratios by comparing those constituent quantities from the same measure space (Karagoz Akar, 2010)—suggests a multiplicative comparison between magnitudes of radiation rate and a multiplicative comparison between magnitudes of area. This is suggested when she stated that 1200 Js^{-1} "is not too far over" 750 Js^{-1} but that there was "a big difference" between 6 and 20 m^2. I interpret those comparisons as multiplicative because $1200/750 = 1.6 < 20/6 \approx 3.3$ but $1200 - 750 = 450 > 20 - 6 = 14$, which suggests that Pam was comparing 1200 and 750 (and 20 and 6) multiplicatively.

Although she made connections with temperature later, her conception of irradiance remained at the level of Thompson's *ratio* because she did not conceptualize it as a constant multiplicatively relationship between quantities that can vary. For example, González (2017) presented Pam with a hypothetical case in which only the irradiance of each sheet was known (she neither knew the dimensions nor the radiation rates). When González asked Pam if irradiance provided sufficient information to determine which sheet's surface temperature would first reach 25 °C, she said the following:

> That (irradiance) is not enough to know … because this can be one twenty-five (125 $Js^{-1}m^{-2}$), but the area could be 20 [m^2] and the rate could be 2,500 [Js^{-1}]. You don't

Fig. 6 Pam's representation of the value 125 $Js^{-1}m^{-2}$

> know anything, so I think this [*draws a circle enclosing "125 $Js^{-1}m^{-2}$"*] doesn't help you unless you know how big [the sheet] is or you know the rate [of radiation].

Pam's argument suggests that she did not recognize irradiance as a multiplicative relationship between radiation rate and area that remains constant. For her, irradiance represented an association between specific, non-varying quantities, thus radiation rates and areas must be compared for every situation before concluding what sheet is heating faster, even when the irradiance is known.

The conceptions that Kris and Pam developed for the quantities ppm concentration and irradiance would have an impact on their ability to understand the Earth's energy budget as a system. Some of these conceptions would be productive, while others not so much. For instance, their conceptions of ppm concentration supported the conceptualization and quantification of changes in the abundance of CO_2 in the atmosphere, and Kris's conception of irradiance helped her relate changes in irradiance within the climate system with changes in the planet's surface temperature. In contrast, Pam's conception of irradiance would prove to be a cognitive obstacle for understanding that relationship, as it will be discussed in the following sections.

5.2 *Making Sense of the Energy Budget as a System Quantitatively*

After working on simplified scenarios, the PSTs moved to working on tasks framed in the context of the Earth's energy budget. This would entail making sense of it as a system, which required three competencies from the STH model: identify components and processes within the system, identify relationships among the system's components, and organize the system's components and processes within a framework of relationships (Fig. 2). In this section, I discuss the role of Kris and Pam's quantitative reasoning in activating those competencies by helping them make sense of the energy budget's components and processes in terms of quantities and relationships. More specifically, I discuss how Kris and Pam's conceptualized measurable attributes of the system's components as quantities, such as the *concentration* of CO_2 in the atmosphere (C) or the *irradiance* of the energy flows between the Sun, the surface, and the atmosphere (S, R, and A). I also discuss how they made sense of the system's processes by establishing key relationships between those quantities, such as *radiative equilibrium* ($S + A = R$), *energy imbalance* ($N = (S + A) - R$), and *forcing by CO_2* ($F = f(C)$). This frame of relationships represents a quantitative structure that supported the PSTs' understanding of the energy budget as a system.

5.2.1 Radiative Equilibrium and Energy Imbalance

Before working on the phase 2 tasks (Fig. 3), the PSTs watched a video introducing the energy budget and had a Q&A session with the researcher to clarify questions about it. Then, they were asked to think about and model two important relationships, *radiative equilibrium* and the *planetary energy imbalance.* In Fig. 1, S and A represent the energy flowing into the surface from the Sun and from the atmosphere, respectively, and R represents the energy flowing out of the surface or the surface radiation. If $S + A = R$, then there is *radiative equilibrium* since the energy is flowing into the surface at the same rate it is flowing out of it. This results in a constant surface energy and surface temperature. In contrast, if for example we had $S + A > R$, then we say that there is a *planetary energy imbalance* because the energy is flowing into the surface at a higher rate than it is flowing out it. This results in an increase in the surface energy and the surface temperature. If we use N to represent the magnitude of that energy imbalance, then this quantity can be modeled it by $N = (S + A) - R$.

Kris Conceptualizes Radiative Equilibrium and Energy Imbalance. Kris conceptualized the planetary energy imbalance, N, as a difference that indicates the magnitude by which the amount of energy absorbed by the surface exceeds the amount of energy released by it, as seen in the following exchange that Kris and I had regarding her conceptualization of N (refer to Fig. 1 while reading the excerpt):

I How do we know if the budget is at equilibrium at the surface?
K Well, S plus A has to equal R [*writes* "$S + A = R$"], maybe?
I What does that mean?
K So, that means that the energy. Well, you said that this is continuously happening [*traces a circle with her finger over the diagram of the energy budget*]. So, that means all these quantities [*places her hand over the diagram*], the inputs and outputs, have to be equal at any certain time. So, the inputs have to equal the outputs at any given time. So, that means that the energy that the Earth's surface is absorbing, so that would be from the sun [*points at S*] and also from the greenhouse effect [*points at A*], and that has to equal the amount of energy that is being released [*points at R*]. So, I just abbreviate the greenhouse as A, and I added it to S, and that has to equal the amount that's being let out [*points at R*].
I So, if these two expressions, like S plus A and R, aren't equal, do we say that the budget is, is what?
K Imbalanced.
I So, if N measures the imbalance, would you be able to define N mathematically?

K So, if this is the case, S plus A minus R has to equal zero [*writes "$S + A - R = 0$"*], if that is the case. So, N is going to [*writes "$N = (S + A) - R$"*]. I guess you could consider zero a type of imbalance, but it is a zero imbalance, so that would imply that it is balanced, so I guess this should work [*draws a box around "$N = (S + A) - R$"*].

I So, N is a difference between inputs and the outputs. So, if N is zero?

K Then, they are equal, which means it's in equilibrium, radiative equilibrium.

I Awesome. And if N isn't zero?

K Then, it's imbalanced, they are not in equilibrium.

As the excerpt shows, Kris conceptualized radiative equilibrium and N in terms of relationships between energy flows S, R, and A, specifically $S + A = R$ and $N = (S + A) - R$, respectively. It is also important to point out that, for the case of radiative equilibrium, Kris argued that the inputs and outputs must be equal "at any given time", which suggests that she understood radiative equilibrium as a dynamic relationship. This hypothesis will gain more support later in the chapter and will hold true not only for radiative equilibrium but also for N. For now, let us further examine Kris's conceptualization of N.

The quantity N is in function of the energy flows S, A, and R, which are quantified in terms of irradiance with $Js^{-1}m^{-2}$ units. Therefore, it is important to examine Kris's conception of N in relation to her conceptions of irradiance, which she conceptualized as a time rate indicating how fast the energy density and the surface temperature increase. The following excerpt suggests that Kris conceptualized N as a difference between irradiances and, consequently, she thought of it as a measure of how fast the planet's surface energy and surface temperature were increasing:

I What does that ($N > 0$) mean in terms of temperature?

K That means temperature is increasing because, as we gain energy, as we saw at the beginning, the temperature increases.

I Great. So, equivalently, if N is negative, less than zero.

K That means temperature is decreasing when N is less than zero [*writes "$T\downarrow$ when $N < 0$"*] because this [*place her hand over the energy budget*] is losing energy.

It seems that for Kris N not only was a quantity defined by a difference but also a time rate so that $N > 0$ indicated an increase in the surface energy and the surface temperature, while $N < 0$ indicated a decrease in those two quantities. Moreover, she referred to the Radiation Task 2 when she said, "as we saw at the beginning, the temperature increases", thus directly connecting N with her understanding of irradiance. Her conception of N, along with the relationships she constructed for radiative equilibrium and planetary energy imbalance ($S + A = R$ and $N = (S + A) - R$, respectively), would later help Kris understand how CO_2 pollution contributes to global warming.

Pam Conceptualizes Radiative Equilibrium and Energy Imbalance. Pam also conceptualized N as a difference between the amount of energy that goes into the surface and the amount of energy that leaves it. When asked to represent N mathematically, she responded in the following way:

> So, it got S in it [*points at S*], and then minus R [*points at R*]. Ok, ok, alright, alright, I got this. So, R gets split up into L and B [*writes "R", draws two arrows emanating from it, and writes "L" and "B" at the end of each arrow, respectively*], and then B gets down back to Earth [*draws a downward arrow emanating from B and writes "A" at the end of it*]. So, S minus R plus half of B [*writes "*$S - R + \frac{1}{2} B$*"*] because this is B and it gets split up [*points at both A's*].

After it was pointed out that $\frac{1}{2}B = A$, Pam decided to write the formula $N = S - R + A$, establishing this important relationship. Although this difference is mathematically equivalent to Kris's formula for N, Pam's conceptualization of it was different in an important way. This became apparent when Pam talked about her idea of radiative equilibrium:

I What does balance mean for you in terms of energy? I noticed that you wrote down here S, R, and A. So, what would balance mean between those three letters?

P S plus A equals R.

I Ok, so you see S and A as inputs.

P Yes, and it is like here is something [*points at the surface*] and two people are putting things into it [*points at S and A*], and one person is taking that R out of it (the surface) … I don't know if S and A are equal, but I think that R is equal to S plus A. That is what I think of balance.

I We'll see that they are related in that way, but you are right: S and A are not necessarily equal. So, how did you realize when this [*points at the diagram of the energy budget*] is at equilibrium again?

P When the same amount [*points at S and A*], when no matter what goes in [*points at S and A*], the same amount goes out [*points at R*].

First, the excerpt shows that Pam arrived at a productive (and important) definition of radiative equilibrium as a relationship between energy flows: $R = S + A$. It is important to construct that relationship before one can think about how the budget can be thrown out of equilibrium. Second, Pam continuously referred to the energy flows as "amounts of energy" in the excerpt. This suggests that Pam conceptualized N as a difference between the total amount of energy absorbed by the surface and the total amount of energy released by it. In other words, N is a difference between an increment of energy ("two people are putting things into it") and a decrement of energy ("one person is taking that R out of it") so that N is the actual surface energy. This appears consistent with her conception of irradiance from the Radiation Task 2, where she conceptualized it as an amount of radiation rate (energy per second) associated with 1 m^2. Associations of this type tend to emphasize the quantity in direct proportion to the ratio (energy) and silence the other quantities in inverse propotion to it (time and area) (Howe et al., 2010). Thus, irradiance was an amount of energy in Pam's mind and N, as a difference between irradiances, was also an amount of energy: the surface energy to be precise.

Similar to her work on the Radiation Task 2, Pam related N to the planet's surface temperature, but the way she conceived such a relationship was not sufficiently accurate. When asked to interpret $N > 0$ in terms of the energy budget and the surface temperature, Pam said, "if it is positive, that means that S plus A, that means there is more [energy] coming in than coming out" and then added, "it means that it is getting hotter". Although her claim is correct, the connection she established between N and temperature must be interpreted with caution because there was little evidence that Pam understood such connection adequately. For instance, when asked to interpret $N < 0$, she said that it represented "colder temperature", but she also considered it impossible for N to be negative, as illustrated by the following excerpt where Pam compared the energy flows to amounts of water:

> Yeah, I guess I was thinking like, if we are talking about flows of water, you can't take more water out [*points at R*] than you put in [*points at S and A*]. So, if I put in a cup of water [*points at S and A*], you can't take a cup and a half out [*points at R*], and then have a negative amount of water [*points at the surface*]. So, I didn't think N could be negative.

The excerpt further supports the hypothesis that Pam conceptualized N as the amount of surface energy, hence it cannot be negative. This suggests that Pam may have interpreted $N > 0$ as "hotter temperature" and $N < 0$ as "colder temperature" to provide answers that she thought the interviewer wanted to hear; that is, Pam's reasoning had more to do with the perceived desirability of responses than with an understanding of the connection between N and temperature change.

5.2.2 The Forcing by CO_2

Kris and Pam examined other relationships while working on the Forcing by CO_2 Task (Table 3). In climate science, the term *forcing* refers to factors that have an impact on the planet's energy budget and tend to disrupt radiative equilibrium. Then, forcing by CO_2 refers to the impact that a change in the atmospheric CO_2 concentration, C, would have over the energy budget. Let us consider the energy budget in Fig. 1 ignoring, for the moment, variation over time. Let us also assume that C is such that there is radiative equilibrium; that is, $S + A = R$ or $(S + A) - R = 0$. If C were to increase, then so would the atmosphere's capacity to absorb and radiate energy (B and A would increase, respectively). We can thus think of the atmospheric energy flows B and A as functions of C. On the other hand, the solar energy flow, S, and the surface energy flow, R, do not depend (at least directly) on the value of C and remain constant. This causes an energy imbalance or *forcing by* CO_2 with magnitude $F = (S + A) - R > 0$.[2] Kris and Pam were tasked with drawing the graph depicting the relationship between C and F since we can think of F as ultimately depending on C or $F = f(C)$. Although Kris and Pam's approaches followed a path similar to the one described above, there were important differences in relation to their quantitative understanding of the situation.

Kris Relates CO_2 Concentration to the Forcing. To establish the relationship between F and C, Kris first noticed that the energy flows B, L, and A changed as

[2] The forcing, F, is a change in N **caused only** by the change in C so that $F = (S + A) - R$ if variation in N as time elapses is not considered. For the tasks in González's (2017) study, F can be interpreted as the value of N at $t = 0$.

Table 3 The Forcing by CO_2 Task

Let F and C be the *forcing by carbon dioxide* (CO_2), in $Js^{-1}\ m^{-2}$, and the *atmospheric CO_2 concentration*, in ppm, respectively. Use the diagram of the Earth's energy budget, what you learned about the greenhouse effect, and the formula $F = (S + A) – R$ to determine how F changes with respect to C. Then, draw the graph of $F = f(C)$

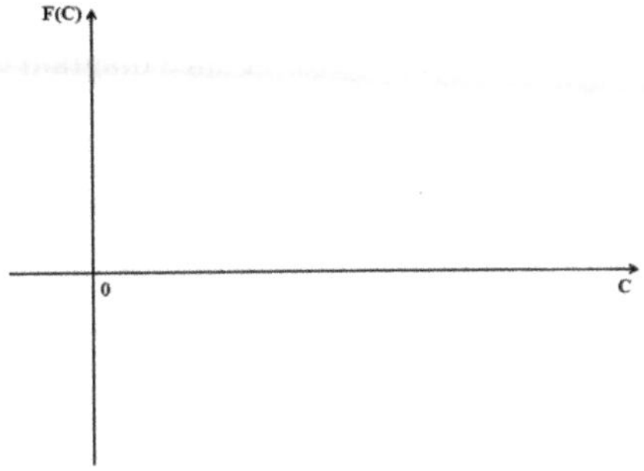

C increased *and* that the energy flows S and R remained constant as C increased. More specifically, she noticed that A increased as C increased while S and R did not change. Since $F = (S + A) – R$, Kris realized that F must increase as C increases, establishing the relationship $F = f(C)$ and connecting it to changes in the surface temperature.

> If you increase the concentration of CO_2, B is going to increase because there are more CO_2 molecules to absorb the heat, so less it is going to be leaked [*points at L*] … OK, so S stays the same [*pauses*]. Wait, hold on [*writes "A = B/2"*]. If B increases, then A is going to increase, and S and R stay the same [*pauses*]. So, [F] is going to be positive … More [energy] is being absorbed [*points at B*], so less it's leaking [*points at L*], so that means more gets put back in [*points at A*] … So, that means the temperature is increasing because it is gaining energy.

Kris made use of quantities and relationships to describe the processes relating the different energy budget's components (e.g., "B is going to increase because there are more CO_2 molecules" → relationship between C and B; "More [energy] is being absorbed [*points at B*], so less it's leaking [*points at L*], so that means more gets put back in [*points at A*]" → relationship between B, L, and A; "So, [F] is going to be positive" → $F = (S + A) – R > 0$). In doing so, Kris created a quantitative structure that gave coherence to the energy budget as a whole and "showed" how it works. Such structure also helped her realize that there was a link between CO_2 pollution and global warming ("So, that means the temperature is increasing because it is gaining energy"). This highlights the role of quantitative reasoning in developing an understanding of the energy budget as a system and how that system responded to CO_2 pollution.

Pam Relates CO_2 Concentration to the Forcing. Pam encountered more difficulties than Kris while working on the Forcing by CO_2 Task. When Pam's reasoning appeared to be stalled, González (the interviewer) saw the need to intervene to help Pam progress on the task. The intervention was not meant to mentor Pam but to suggest processes within the energy budget she could consider; this allowed him to explore how she made sense of those processes quantitatively. These interventions during clinical interviews are methodological strategies that Moore (2010) has termed *exploratory teaching interview*, an adaptation of the teaching experiment methodology described by Steffe and Thompson (2000).

One intervention consisted of giving Pam a diagram of the energy budget, similar to the one shown in Fig. 1, with the following values: $S = 240$ Js^{-1}m^{-2}, $R = 390$ J s^{-1}m^{-2}, and $B = 300$ J s^{-1}m^{-2}. These values represented the energy budget's initial conditions (for $t = 0$) and were meant to show radiative equilibrium numerically since $F = (240 + 150) - 390 = 0$. Then, the value $B = 300$ was changed to $B = 310$ to simulate the effects of an instantaneous increase in C at $t = 0$. The following dialogue illustrates how Pam reacted to the exploratory teaching intervention:

I Let's imagine we increase the concentration of CO_2 by a certain amount. This results in B growing from 300 to 310. So, this flow changes [*point at B*], this flow changes [*point at L*], and these two change [*point at both A's*].
P They've just got bigger.
I Exactly, could you now calculate the value of F corresponding to the new values of energy flows?
P But I don't know what S is now.
I It is still 240 because we are just making changes in the atmosphere, and S does not depend on the atmosphere's composition. So, S is 240 and R remains at 390 as well.
P Except B [*writes "F = (S + A) – R"*] … So, B is 310; that means A is now 155. So, we have 240 plus 155 minus 390 [*writes "F = (240 + 155) – 390 = 5"*].
I This value [*point at 5*] is a change in the energy imbalance caused by a change in the concentration of CO_2. That is a forcing by CO_2, that is the value of F.
P Ah! So, when the CO_2 increases, F increases.

Working with particular values for the energy flows helped Pam establish relationships between quantities. For instance, she noticed that B and A "just got bigger" at the moment CO_2 concentration increased (B and A depend on C), or that the forcing, F, increases "when the CO_2 increases" (F is in function of C). As the interview continued, Pam would make additional references to these relationships, suggesting she conceptualized a quantitative structure that helped her see how the energy budget works and the link between CO_2 pollution and global warming:

> When the forcing by CO_2 is positive, so that means there is more going in than coming out … Because as we go up [*moves her index finger over the graph*], the B gets bigger, and bigger, and bigger, and the L gets smaller, and smaller, and smaller. So, there is more [energy] going into the Earth, so it is hotter, the temperature of the Earth is hotter.

The previous excerpt shows that Pam understood that an increase of CO_2 enhanced the atmosphere's capacity to absorb energy ("B gets bigger, and bigger") and reduced the amount of energy escaping to space ("L gets smaller, and smaller"). Since more energy is trapped into the climate system, the planet's surface temperature increases.

5.3 Conceptualizing Dynamic Relationships and Cyclical Processes

The previous sections discussed how quantitative reasoning is involved in making sense of the components and processes of the energy budget (system) in terms of quantities and relationships. After that quantitative structure was developed, the next step is understanding how it "behaves" over time (dynamics relationships) as a response to being thrown out of radiative equilibrium by an increase in the CO_2 in the atmosphere. The response also includes the process by which radiative equilibrium is restored through a continuous energy exchange between the atmosphere and the surface (feedback loop). To accomplish these, two more competencies from the STH model are required: identify dynamic relationships within the system and understand the cyclic nature of systems (Fig. 2).

5.3.1 Conceptualizing the System's Dynamic Relationships

Kris and Pam explored dynamic relationships in the energy budget by working on the Energy Budget Task (Table 4), which involved making sense of a simplified scenario where a unique and instantaneous pulse of CO_2 is released toward the atmosphere at time $t = 0$ years. The pulse makes the atmospheric CO_2 concentration, C, increase from $C = C_0$ to $C = C_0 + \Delta C$. Let us imagine that the atmospheric energy flows B, L, and A (Fig. 1) change instantaneously in response to the instantaneous increase in C so that an energy imbalance $N = (S + A) - R > 0$ is produced at time $t = 0$ years (forcing by CO_2). In the formula $N = (S + A) - R$, the variable F was replaced by the variable N to indicate that the difference now changes over time t. Therefore, the first part of the task was to visualize and graph the way N changes as t increases. Here is when covariational reasoning is required to visualize that and others dynamic relationships.

Let $N = g(t)$ be the way of denoting the covariation between N and t, then $N = g(0) > 0$ indicates that the energy inflow into the surface, $S + A$, is larger than the energy outflow, R, at $t = 0$. As a result, the surface energy and, consequently, the surface temperature will increase as t increases. As the surface warms up, it radiates more infrared energy toward the atmosphere (R increases as t increases), enhancing the atmosphere's temperature. As the atmosphere warms up, it radiates more infrared energy in both directions, toward space and back to the surface (the A's increase as t increases). This further enhances the surface temperature so that R keeps increasing as t increases. This energy feedback loop between the surface and the atmosphere, known as the *greenhouse effect*, causes R to continue to increase so that the difference $N = (S + A) - R$ decreases (asymptotically) toward zero as t elapses; this process is responsible for the transition toward a new radiative equilibrium. To help Kris and Pam develop a sense for such dynamic relationships in the energy budget, the following three recursive rules were given to them: $B_t = (31/39) \cdot R_t$, $A_t = ½ \cdot B_t$, and $R_{t+1} = S + A_t$.

Table 4 The Energy Budget Task

A change in the concentration of CO_2 in the atmosphere can result in an imbalance in the Earth's energy budget. This initial imbalance is known as forcing by CO_2. We want to examine how the net planetary energy imbalance, N, and the planet's average surface temperature, T, vary as time increases after the forcing. Use what you learned about the energy budget, the greenhouse effect, and the formula $N = (S + A) - R$ to infer how
(a) N varies over time and sketch its graph.
(b) T varies over time and sketch its graph.

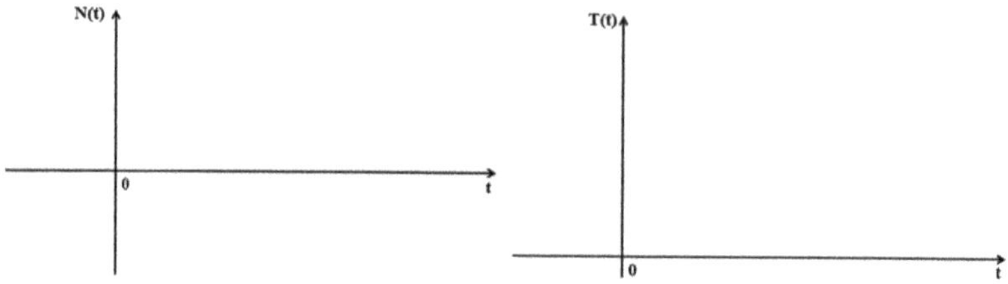

Kris Conceptualizes and Graphs $N = g(t)$. To construct the graph of $N = g(t)$, Kris used the recursive rules to determine four consecutive values for B, A, and R, which she recorded on a diagram of the energy budget (Fig. 7). Then, she paid close attention to how B was increasing by decreasing amounts as t increased by 1-year increments and interpreted that variation in the following way:

> Ok, so what I am seeing is that we are reaching a new equilibrium because, after this initial increase of absorption of 320 (B increasing from 300 to 320), initially. The next cycle [*with a capped marker, traces a circle connecting B, A, and R*] is going to absorb 328, which is an increase of eight … The next cycle [*with a capped marker, traces another circle connecting B, A, and R*] causes the absorption to increase by an amount of 3.5, which is less than eight. And then, the next cycle [*with a capped marker, traces another circle connecting B, A, and R*] causes the absorption to increase by 1.417. Well, this difference right here, between 320 and 328, is greater than the difference between these two values, the 328 and the 331.5, and the difference between these two [*points at the B-values "332.917" and "331.5"*] is less than those [*points at the B-values "331.5" and "328"*], which is less than that [*points at the B-values "328" and "320"*]. That tells me that there is eventually going to be a limit … Yeah, it's going to reach a new equilibrium point somewhere. Hey, it's going to look like our sensitivity function.

Following her analysis, Kris freehandedly drew an accurate graph of $N = g(t)$ for $t > 0$, without the need for N-values (Fig. 8). First, Kris coordinated and compared amounts of change in B for 1-year increments in t (MA3 covariational reasoning), to anticipate (and justify) the concavity of the graph and its asymptotic decrease towards zero. Her anticipation is clear when she said, "it's going to look like our sensitivity function", referring to a graph she drew while working on the Forcing by CO_2 Task which resembled an exponential-decay type of curve. She also interpreted the asymptotic decrease toward zero as the energy budget transitioning to radiative equilibrium after the forcing ("it's going to reach a new equilibrium point somewhere"), which is a key realization about the energy budget's response to an increase in CO_2.

$$B_t = \frac{32}{39} \cdot R_t$$

$$A_t = \frac{1}{2} \cdot B_t$$

$$R_{t+1} = S + A_t$$

t	R_t	B_t	A_t
0	390.0	320.0	160.0
1	400.0	328.2	164.1
2	404.1	331.6	165.8
3	405.8	332.9	166.5

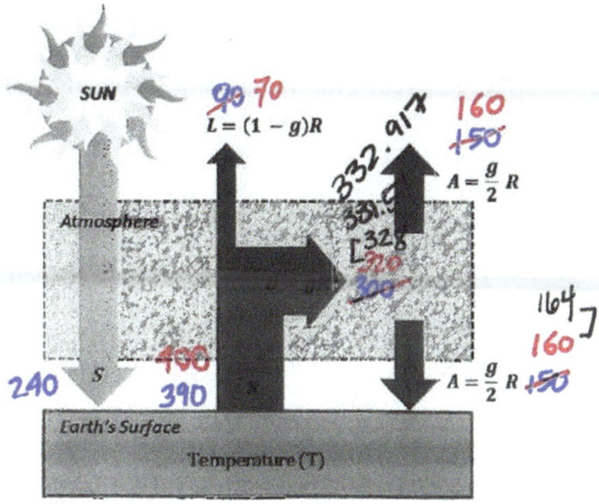

Fig. 7 Kris used recursive rules to compute consecutive values for the energy flows

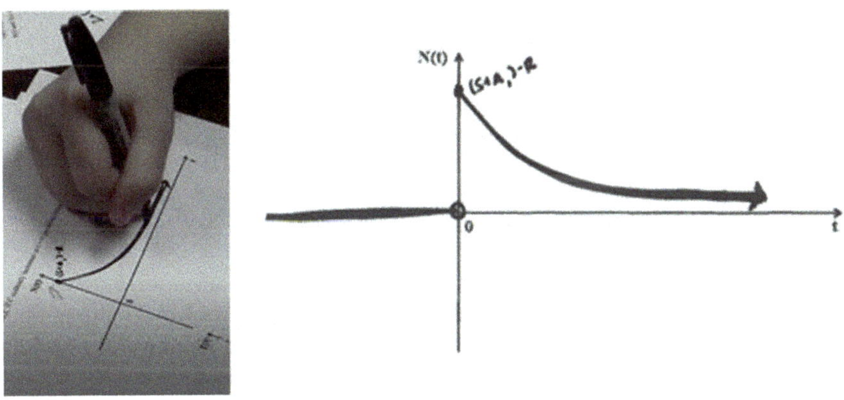

Fig. 8 Kris's graph of $N = g(t)$

Second, although Kris only talked about B increasing by *decreasing amounts* for uniform increments in t, she might have perceived that all other energy flows in the budget were changing by decreasing amounts. She drew an accurate graph of $N = g(t)$, a quantity that depends on quantities such as A and R. Since her graph showed N decreasing by decreasing amounts, it was likely that Kris not only envisioned B increasing by decreasing amounts but also A and R. This suggests that Kris developed a quantitative structure that allowed her to visualize all energy flows simultaneously varying (decreasing or increasing) by *decreasing amounts of change* as t increased by 1-year increments. This reveal how quantitative and covariational reasoning can be a powerful combination to develop a coherent view of a system, such as the energy budget, while one can visualize it *evolving dynamically as a whole.*

Pam Conceptualizes and Graphs $N = g(t)$. Pam utilized the recursive rules to determine three consecutive values for B, A, and R. However, Pam experienced difficulties with interpreting the variation of values to describe the way N was changing with respect to t. As a result, her account of the relationship $N = g(t)$ was vague and limited to describing its direction of change alone (MA2 covariational reasoning). While reading the following excerpt, please keep in mind that Pam worked with $N = (S + A) - R = (240 + 155) - 390 = 5$ for $t = 0$:

I OK, so you said N was five at $t = 0$. Would N increase or decrease after that?

P It was five, and then it would decrease to be zero again [*pauses*]. Let me think about this one more time. So, this was the original [*points at R*], it was 390, like 240 plus 150. That was when [N] was zero, and then we increased the concentration, just here [*points at B*], to affect everything else in the atmosphere ... So, at $t = 0$, N was five because at that time is when we increased [the CO_2 concentration], and that is when N turned into five, right? So, $t = 0$, N was five [*writes "t_0: $N = 5$"*]. But then, 395 got absorbed, wait [*pauses*] ... OK! This is what I am thinking. We increased time zero (the CO_2 concentration increased at $t = 0$), but I am pretty sure [N] would balance itself back out.

I OK, did you learn that by watching the video?

P That is what I heard. That is what I am thinking I got it from.

I So, we know the energy imbalance should decrease over time.

P Yeah, from five back down to zero. So, I think [the graph] could go from like here, like this was five at $t = 0$ [*points at the top of the vertical axis*]. I think it is going to go [*with a capped marker, traces a concave-upward, decreasing graph*]. But I think it is going to hit zero pretty quickly [*with a capped marker, traces a decreasing line*].

The excerpt shows that Pam envisioned N decreasing as t increased ("[N] would decrease to be zero again"), which represents evidence of MA2 covariational reasoning, and related that pattern with the transition towards radiative equilibrium ("I am pretty sure [N] would balance itself back out"). However, she did not adventure a graph for $N = g(t)$, which suggested that Pam needed additional support to visualize in *what way* N was decreasing with respect to time.

González (2017) recommended to Pam that she used the formula $N = (S + A) - R$ to find values that would help her visualize how N was changing with respect to t. She thus determined a series of N-values and then compared amounts of change in N for 1-year increments in t (MA3 covariational reasoning). This helped her conceptualize $N = g(t)$ as a function that decreases by decreasing amounts.

> [N] decreased pretty quickly at first from five to two [*points at $N = 5$, and then at $N = 1.98$*] ... like three units of that [*points at $Js^{-1}m^{-2}$*], Joules per second per meter square [*sic.*]. And then, it decreased by about one [*points at $N = 1.98$, and then at $N = 0.82$*] Joules per second per meter square [*sic.*]. So, it went from about three to about one. I am assuming [N] is going to decrease by a little bit, and a little bit, and a little bit, until it reaches zero again.

After saying that, Pam freehandedly drew an accurate graph to represent $N = g(t)$ (Fig. 9). With this, Pam appeared to have conceptualized the dynamic relationships

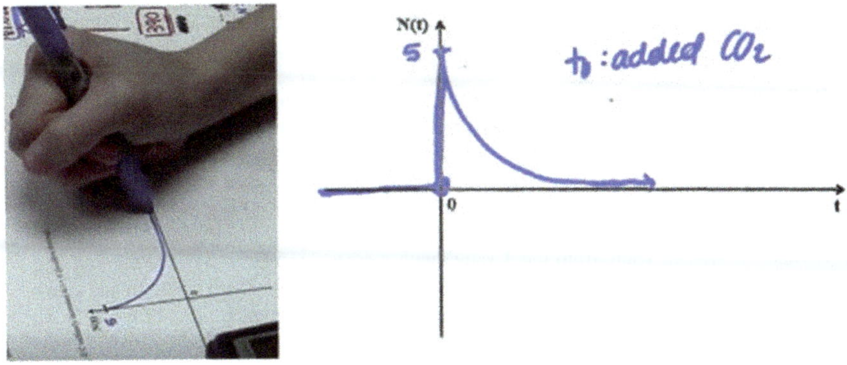

Fig. 9 Pam's graph of $N = g(t)$

involved in the energy budget's response to the increase in CO_2. This highlights the role of covariational reasoning in activating the systems thinking competence of identifying a system's dynamic relationships.

5.3.2 Conceptualizing the System's Cyclical Processes (Feedback Loops)

For the second part of the Energy Budget Task (Table 4), Kris and Pam were asked to interpret $N = g(t)$ in terms of changes in the planet's average surface temperature, T, with respect to time, t. They were expected to draw a graph of that relationship, which I hereafter represent as $T = h(t)$. To correctly interpret N, it is important to conceptualize it as the rate of change of the surface energy changes with respect to time. It is also important to visualize the *energy feedback loop* between the surface and the atmosphere as a special type of covariation between the energy flows R and A (Fig. 1). Let us remember that an increase in the CO_2 concentration, C, results in a warming effect over the planet's surface (A increases as C increases), which in turn results in an increase in the emission of infrared energy towards the atmosphere (R increases as t increases). This warms the atmosphere, producing an increase in the emission of infrared energy back to the surface (A increases as t increases). A higher A further warms the surface, enhancing T, which enhances R, which in turn enhances A again, and the cycle repeats. Therefore, that "circular" covariation between R and A represents a *balancing feedback loop* that enhances T over time *and* is responsible for the energy budget transitioning towards a new radiative equilibrium.

Kris's Conceptualization of the Feedback Loop. When Kris was asked to describe how T changes as t increases, she attended to the relationships between the energy flows R, B, and A as they change simultaneously and dynamically. One of such relationships was the covariation between R and A and the feedback loop that it represented. Notice how Kris became aware that such a feedback loop was responsible for $T = h(t)$ being an increasing function and for the energy budget's transitioning towards a new radiative equilibrium:

> Well, [the surface] keeps in taking. I think it is warming up because once we added more CO_2, that is less of the emitted energy that is getting just like shut out passed the atmosphere, leaked from it. So then, more of it is going to be absorbed by the atmosphere ... Whatever is absorbed by the atmosphere [*points at B*] is going to be absorbed back into the [*points at the surface*], well half of that plus the sun's energy [*points at S*] is going to be absorbed by the Earth, which is going to keep increasing, as we saw with like the 400 [*points at the R-value of "400"*]. Then, from the *A* value [*with a capped marker, traces the top half of a circle, going from R to A*], just with the *A*, [the surface] absorbs 160, and then we add a new *R*-value [*with the capped marker, traces the bottom half of the circle, going from A to R*], whatever that was, and then [the surface] absorbs 164 [*with the capped marker, re-traces the top half of the circle, going from R to A*]. So, I think it is going to keep increasing [*draws an increasing, concave-downward graph for T = g(t) that appears to have an upper limit or a horizontal asymptote that she labels as "new equilibrium temperature"*].

Kris first realized that a considerable portion of the surface radiation, *R*, remains "trapped" in the energy budget through a continuous energy exchange between the atmosphere and the surface; this trapped energy enhances *T* as *t* increases. Kris also conceptualized the energy feedback loop between the surface and the atmosphere in terms of covariation between *R* and *A*. She conceptualized *R* and *A* as simultaneously increasing as *t* elapsed (MA2) and as obeying a circular causality relationship (i.e., an increase in *A causes R* to increase, which in turn *causes A* to increase again and so on). Kris communicated that circular causality between *R* and *A* with gestures such as tracing the top half of the circle, going from *R* to *A*, and the bottom half of the circle, going from *A* to *R*. These two realizations, the simultaneity of change and circular causality, appeared central to be able to conceptualize a feedback loop in terms of covariation between two quantities.

Kris also drew an accurate graph of $T = h(t)$, which showed an asymptotic increase towards a new higher value of temperature (Fig. 10). The shape of her graph and her labeling of the asymptote as "new equilibrium temperature" indicate that Kris realized that the energy feedback loop between the surface and the atmosphere was responsible for the transition toward a new radiative equilibrium. When she was asked about the asymptote, Kris compared amounts of change in *A* for 1-year increments in *t* (MA3) to justify it:

> New equilibrium temperature, approximately that [*draws a mark on the vertical axis next to the label "new equilibrium temperature"*] ... Because the difference of the temperature, or whatever, was the four initially. Well, we changed it to ten (*A* "jumps" from 150 to 160 at $t = 0$), then it went to four (*A* increases from 160 to 164 when *t* increases from 0 to 1). So, it is increasing at a decreasing rate.

Kris's MA3 covariational reasoning supported the conceptualization of the covariation between *R* and *A* as a *balancing feedback loop* that not only enhances $T = h(t)$ but also "moves" the energy budget toward a new radiative equilibrium. I am not claiming that Kris was aware that the energy exchange between the surface and the atmosphere represents a balancing feedback loop, but rather I am claiming that Kris inferred a *balancing quality* of the energy exchange between the surface and

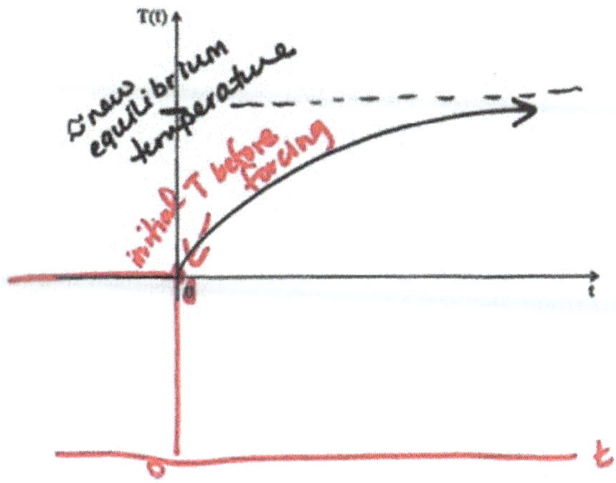

Fig. 10 Kris's graph of $T = h(t)$

the atmosphere by noticing that A increased by decreasing amounts of change as t increased.

Pam's Conceptualization of the Feedback Loop. Pam encountered difficulties with interpreting $N = g(t)$ in terms of changes in the surface energy and the surface temperature. On the one hand, Pam interpreted the increasing direction of change of R, B, and A with respect to t as an indication that the surface energy and the surface temperature were increasing as t increased. On the other hand, Pam also interpreted the decreasing direction of N with respect to t as an indication that the surface energy and the surface temperature were decreasing as t increased. The following excerpt illustrates Pam's difficulties:

I Is the energy budget gaining energy or losing energy?

P I think it's losing [*long pause*]. Oh, it's losing. No, it's gaining, this is gaining … If we look, it keeps increasing [*points at the values of R, B, and A*], my numbers are getting bigger, and bigger. So, I think the energy budget increases.

I Let's see these two [values] right here [*draws a box around* $N_0 = 5$ *and another around* $N_2 = 0.82$]. For which one of those values is the energy budget gaining energy?

P Well, from here to here, it is losing energy [*points at* $N_0 = 5$*, and then at* $N_2 = 0.82$]. But, I thought [*pauses*]. I am thinking of energy balance as … these numbers are increasing [*points at the values of R, B, and A*], but the N is the planetary energy imbalance. Oh! So, it is decreasing. I don't know why I said increasing. I mean, I know why I said increasing, but that was because I was reading it incorrectly.

I Wait, what is increasing or decreasing?

P I see these numbers increasing [*points at the values of R, B, and A*], but this N is the energy imbalance, the budget. I've forgotten these are pretty much the same thing … So, it is decreasing [*draws an arrow from* $N_0 = 5$ *to* $N_2 = 0.82$].

I The planetary energy imbalance is decreasing, so it means the imbalance is getting smaller.
P So, it's coming to balance.
I Alright. So, is this [energy] budget losing or gaining energy?
P I want to say gaining because these numbers are getting so big [*points at the values of R, B, and A*].
I So, are you going with gaining, or are you going with losing?
P The planetary energy imbalance is decreasing [*pauses*].
I Right, the imbalance is getting smaller ... So, we are getting closer and closer to equilibrium.
P Right, yes.
I So, are we gaining or losing energy?
P Aaah! [*Laughs nervously*]. Losing, we are losing because if we have gained energy [the surface] would get hotter, but it is not getting hotter because N is smaller so it's cooling off. So, it was too hot because it was increased, like we added so much [CO_2], but now it's trying to cool itself off. So, the energy is decreasing.

As the excerpt shows, Pam was having difficulties with deciding whether the surface energy and the surface temperature were increasing or decreasing as t increased. A possible explanation for this conflict can be traced back to Pam's conceptions of the quantities S, R, A, and N. On the one hand, she conceptualized S, A, and R as amounts of energy entering or leaving the surface. Since R and A were increasing as t increased, she might have interpreted that as more energy "circulating" at the surface level, and thus the surface temperature must have been increasing. On the other hand, she conceptualized N as the magnitude of the surface energy. Therefore, she might have interpreted a decreasing $N = g(t)$ as the surface energy decreasing as t increased.

Pam ultimately decided that the surface energy and the surface temperature were decreasing because $N = g(t)$ was a decreasing function of time. She justified her decision by indicating that the surface was releasing more energy than the energy it was absorbing. This argument revealed that Pam did not quite conceptualize the energy feedback loop between the surface and the atmosphere:

> Because a lot [of energy] is going in, but more is coming out, like R increases as A increases. So, as our A increases, R increases, but our S is staying the same, but A is always less than R, so more is coming out [*pauses*]. So, the Earth is trying to cool itself off, so the temperature is decreasing from here to here [*sequentially points at "N = 5" and "N = 0.82"*]. Yes [*her tone expresses contentment*] ... We are getting more hotter here [*sic.*] [*points at "N = 5" and pauses*] because more is going into the Earth and less is coming out. But here [*points at "N = 0.82"*] more is going in, but more is coming out than here [*points at "N = 5"*], so it's cooler [*sequentially points at "N = 5", "N = 1.98", and "N = 0.82"*]. It is getting cooler because if N is high, then the temperature is high. So, as time goes on, it's normal. It's like, we have a temperature of zero before this time starts [*points at the "t" labeling the horizontal axis and pauses*]. No, we don't have a temperature of zero, we have a temperature like balance temperature, and then as t hits zero, it increases, so it's about here [*points at the top of the vertical axis (T-axis)*], and then it cools off [*with a capped marker, traces a decreasing, concave-downward curve as the graph of $T = g(t)$*] until it is back in its balancing point [*points at the right end of the horizontal axis (t-axis)*].

The excerpt suggests that Pam conceptualized R and A as increasing in tandem (MA2). However, she did not quite conceptualize a circular causality between those quantities. She appeared to be *aware* that the increase in A caused an increase in R but *less aware* that the increase in R also caused an increase in A. Pam's response assumed R to be the amount of energy *leaving* the surface, ignoring that an important fraction of it, represented by A, is reabsorbed by the surface, thus overlooking the causality relationship from R back to A ("So, as our A increases, R increases, but our S is staying the same, but A is always less than R, so more is coming out. So, the Earth is trying to cool itself off"). In summary, Pam could conceptualize R and A as increasing in tandem but not in terms of circular causality, which represented an obstacle to understand the energy feedback loop between the surface and the atmosphere.

The combined effect of Pam's conception of N as the surface energy and Pam's inability to conceptualize a circular causality relationship between R and A led her to conclude that $T = h(t)$ was a decreasing function of time. This was illustrated in the previous excerpt where she said that T decreases as N decreases from 5 to 0.82 $Js^{-1}m^{-2}$ ("so it's cooler [*sequentially points at "$N = 5$", "$N = 1.98$", and "$N = 0.82$"*]. It is getting cooler because if N is high, then the temperature is high"). Then, she freehandedly drew a decreasing, concave-downward graph intercepting the t-axis at some $t > 0$. She did not justify her choice of concavity and, when faced with the contradiction that her graph showed that T could be zero, she scratched a section of the graph and replaced it with a concave-upward section (Fig. 11). This suggests that Pam did not interpret the graph's shape as describing the way T was decreasing with respect to t. Therefore, Pam was only aware of the decreasing direction of the relationship $T = h(t)$ (MA2 covariational reasoning). This indicates that Pam concluded that the planet was "cooling off" after an increase in CO_2, which represents an inaccurate (and contradictory) view of the real impact of CO_2 pollution on global warming. Pam's conclusion may open the door for misconceptions regarding global warming and climate change (e.g., "as soon as we stop CO_2 emission, global warming will stop" or "as long as we keep current emission, global warming would not get any worse").

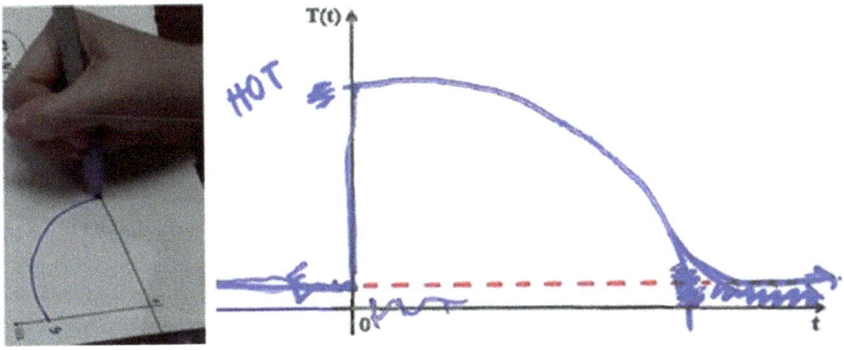

Fig. 11 Pam's graph of $T = h(t)$

6 Conclusions and Implications

Making sense of the climate system and climate change requires what are known as systems thinking competencies, or the ability to understand and model real phenomena as complex systems (Ghosh, 2017; Orgill et al., 2019; Renert, 2011; Roychoudhury et al., 2017). This chapter discussed the role of quantitative and covariational reasoning in activating the systems thinking competencies involved in understanding two key concepts of climate change: the *Earth's energy budget* and the *link between CO_2 pollution and global warming*. Making sense of the energy budget involves an *analysis* of its components and processes followed by a *synthesis* thereof into a coherent framework of relationships. These two levels of the Systems Thinking Hierarchical (STH) model for competencies (Assaraf & Orion, 2005) can be activated through the development of a quantitative structure formed by: (i) quantities associated with the energy budget's components (*concentration*, C, to quantity abundance of CO_2 in the atmosphere and *irradiance* to quantify the energy flows S, R, L, B, and A that circulate between the Sun, the surface and the atmosphere); (ii) relationships between those quantities to represent the energy budget's processes (radiative equilibrium, $S + A = R$; the planetary energy imbalance, $N = (S + A) - R$; the radiative forcing by CO_2, $F = f(C)$); and (iii) dynamic and cyclical relationships describing the energy budget's response to increasing CO_2 in the atmosphere ($N = g(t)$ decreases less rapidly as t increases; the circular causality relationship between R and A; $T = h(t)$ increases less rapidly as t increases). The development of such dynamic quantitative structure required a combination of Kris and Pam's quantitative and covariational reasoning abilities.

A possible sequence to develop that quantitative structure can start with helping students and teacher conceptualize two unfamiliar quantities (and their units of measure) that commonly appear in mathematical descriptions of climate change: ppm concentration and irradiance. Once students and teachers have constructed meaning for them, they can move to conduct the *analysis* and *synthesis* of components and processes involved in the Earth's energy budget. These two levels of the STH model include five competencies (Fig. 2), three of which predominately require the application of quantitative reasoning: identify the system components and process, identify relationships among the system components, and organize the system's components and processes within a framework of relationships. This framework, along with the meanings Kris and Pam developed for concentration and irradiance, constitutes the quantitative structure that shows how the energy budget works. Then, students and teachers can continue with the other two competencies in the *synthesis* level of the STH model (Fig. 2): identify dynamic relationships within the system and understand the cyclical nature of systems. These competencies are necessary to fully establish the link between CO_2 pollution and global warming and require the application of covariational reasoning.

The ideas discussed above have implications for teaching and research related to both mathematics education and science education. Concerning teaching, the chapter shows that it is possible to attend to school mathematics curriculum requirements while promoting climate change education. For instance, Kris and Pam had opportunities to activate, apply, and further develop their understanding of ratios and rates, functions and covariation, properties of graphs, rate of change, modeling, among other topics in mathematics. The tasks Kris and Pam worked on also address important topics and competencies described in the Common Core State Standard for Mathematics[3] (CCSSI, 2010). This suggests that mathematics educators can be protagonists in promoting awareness of climate change and encouraging students to take action to address it. Also, emerging empirical evidence suggests that developing the PSTs' quantitative and covariational reasoning while modeling climate change represents a starting point to prepare PSTs to incorporate this socio-scientific issue into their future mathematics classroom (González, 2021).

The chapter also provides empirical evidence that quantitative and covariational reasoning play a significant role in supporting the understanding of at least two important topics in science education: the climate system and climate change. This, in turn, points to the importance of strengthening students' quantitative and covariational reasoning in the context of learning this concept in science classes. I would extend the suggestion to include preservice science teachers, who do not feel prepared to include climate change into their future classroom for different reasons, one of which is that they consider the mathematics to be too advance (Boon, 2010; Namdar, 2018). Considering the development of preservice science teachers' quantitative and covariational reasoning abilities during their preparation programs can support efforts to prepare them to teach climate change in their future classrooms. This is crucial considering that the Next Generation Science Standards (NRC, 2013) positioned weather, climate, and global climate change as core ideas to be taught in middle grades and high school science classes.

In general, the ideas discussed in this chapter indicate that quantitative and covariational reasoning can be a bridge between mathematics and science education that allows us to explore topics that require both bodies of knowledge to be applied. The chapter shows that quantitative and covariational reasoning can activate or mediate the development of systems thinking competencies. Authors from disciplines as diverse as biology, chemistry, engineering, economy, and mathematics have proposed that current global, complex, politically charged, socio-scientific issues (universal income, evolution, pandemic and vaccines, climate change, etc.) require STEM professionals to understand them as complex systems (Ghosh, 2017; Orgill et al., 2019; Renert, 2011; Richmond, 1997; Roychoudhury et al., 2017; Schuler et al., 2018). Therefore, it is important to further research the role that quantitative and covariational reasoning may have in preparing STEM professionals to deal with current issues that require systems thinking competencies. Furthermore, exploring current, global, complex, politically charged, socio-scientific issues through quantitative and covariational reasoning can contribute to preparing STEM professionals, and citizens in general, to make informed decisions about how these issues may affect

[3] These topics and competences include HS-NQ 1 through 3, HS-F-IF 2, HS-F-BF 1, and modeling with mathematics in general.

their lives and navigate diverse claims about them in the media. In this sense, quantitative and covariational reasoning may have an important role to play in supporting the socially critical thinking of citizens, which suggests another focus of future research.

References

Assaraf, O.B.-Z., & Orion, N. (2005). Development of system thinking skills in the context of earth system education. *Journal of Research in Science Teaching, 42*, 518–560.

Ayene, M., Krick, J., Damitie, B., Ingerman, A., & Thacker, B. (2019). A holistic picture of physics student conceptions of energy quantization, the photon concept, and light quanta interference. *International Journal of Science and Mathematics Education, 17*, 1049–1070.

Barwell, R. (2013a). Climate change. In A. Coles, R. Barwell, T. Cotton, J. Winter, & L. Brown (Eds.), *Teaching mathematics as if the planet matters* (pp. 31–49). Routledge.

Barwell, R. (2013b). The mathematical formatting of climate change: Critical mathematics education and post-normal science. *Research in Mathematics Education, 15*(1), 1–16.

Boon, H. J. (2010). Climate change? Who knows? A comparison of secondary students and preservice teachers. *Australian Journal of Teacher Education, 35*(1), 104–120.

Carlson, M., Jacobs, S., Coe, E., Larsen, S., & Hsu, E. (2002). Applying covariational reasoning while modeling dynamic events: A framework and a study. *Journal for Research in Mathematics Education, 33*(5), 352–378.

Common Core State Standards Initiative (CCSSI) (2010). *Common core state standards for mathematics.* Retrieved from http://www.corestandards.org/assets/CCSSi_Math%20Standards.pdf

de Berg, K. C. (2012). A study of first-year chemistry students' understanding of solution concentration at the tertiary level. *Chemistry Education Research and Practice, 13*(1), 8–16.

Ghosh, A. (2017). *Dynamic systems for everyone.* Springer International Publishing.

Goldin, G. A. (2000). A scientific perspective on structured, task-based interviews in mathematics education research. In A. Kelly & R. A. Lesh (Eds.), *Handbook of research design in mathematics and science education* (pp. 517–545). Lawrence Erlbaum Associates.

González, D. (2017). *Preservice mathematics teachers' understandings of intensive quantities and functions involved in introductory mathematical models for global warming* (Doctoral dissertation, University of Georgia). Retrieved from https://getd.libs.uga.edu/pdfs/gonzalez-martinez_dario_a_201708_phd.pdf

González, D. A. (2021). The progression of preservice teachers' covariational reasoning as they model global warming. *Journal of Mathematical Behavior, 62*, 1–22. https://doi.org/10.1016/j.jmathb.2021.100859

Howe, C., Nunes, T., & Bryant, P. (2010). Intensive quantities: Why they matter to developmental research. *British Journal of Developmental Psychology, 28*, 307–329.

Intergovernmental Panel on Climate Change (IPCC). (2018). *Global warming of 1.5 °C. An IPCC special report on the impacts of global warming of 1.5 °C above pre-industrial levels and related global greenhouse gas emission pathways, in the context of strengthening the global response to the threat of climate change, sustainable development, and efforts to eradicate poverty* (in press).

Karagoz Akar, G. (2010). Different levels of reasoning in within state ratio conception and the conceptualization of rate: A possible example. In P. Brosnan, D. B. Erchick, & L. Flevares (Eds.), *Proceedings of the 32nd annual meeting of the North American chapter of the international group for the psychology of mathematics education* (Vol. 6, pp. 711–719).

Lambert, J. L., & Bleicher, R. E. (2013). Climate change in the preservice teacher's mind. *Journal of Science Teacher Education, 24*, 999–1022.

Liu, X., & McKeough, A. (2005). Developmental growth in students' concept of energy: Analysis of selected items from the TIMSS database. *Journal of Research in Science Teaching, 42*(5), 493–517.

Moore, K. C. (2010). Relationships between quantitative reasoning and students' problem solving behaviors. In *Proceedings of the fourteenth annual conference on research in undergraduate mathematics education* (pp. 298–313). Portland State University.

Namdar, B. (2018). Teaching global climate change to pre-service middle school teachers through inquiry activities. *Research in Science and Technology Education, 36*(4), 440–462.

National Research Council (NRC) (2013). *The next generation science standards.* The National Academies Press.

Orgill, M., York, S., & MacKellar, J. (2019). Introduction to systems thinking for the chemistry education community. *Journal of Chemical Education, 96*, 2720–2729.

Raviolo, A., Farré, A. S., & Schroh, N. T. (2021). Students' understanding of molar concentration. *Chemistry Education Research and Practice, 22*, 486–497.

Renert, M. (2011). Mathematics for life: Sustainable mathematics education. *For the Learning of Mathematics, 31*(1), 20–25.

Richmond, B. (1997). The "thinking" in systems thinking: How can we make it easier to master? *The Systems Thinker, 8*(2), 1–5.

Riess, W., & Mischo, C. (2010). Promoting systems thinking through biology lessons. *International Journal for Science Education, 32*, 705–725.

Roychoudhury, A., Shepardson, D. P., Hirsch, A., Niyogi, D., Mehta, J., & Top, S. (2017). The need to introduce systems thinking in teaching climate change. *Science Educator, 25*(2), 73–81.

Saldanha, L. A., & Thompson, P. W. (1998). *Re-thinking covariation from a quantitative perspective: Simultaneous continuous variation.* Paper presented at the 20th annual meeting North American chapter of the international group for the psychology of mathematics education, Raleigh, NC, United States.

Schuler, S., Fanta, D., Rosenkraenzer, F., & Riess, W. (2018). Systems thinking within the scope of education for sustainable development (ESD)—A heuristic competence model as a basis for (science) teacher education. *Journal of Geography in Higher Education, 42*(2), 192–204.

Sewell, D., Rayner, P. J., Shank, D. B., Guy, S., Lilburn, S. D., Saber, S., & Kashima, Y. (2017). Causal knowledge promotes behavioral self-regulation: An example using climate change dynamics. *PLoS ONE, 12*(9), 1–19.

Steffe, L. P., & Thompson, P. W. (2000). Teaching experiment methodology: Underlying principles and essential elements. In R. Lesh & A. E. Kelly (Eds.), *Research design in mathematics and science education* (pp. 267–307). Erlbaum.

Thompson, P. (2011). Quantitative reasoning and mathematical modeling. In L. L. Hatfield, S. Chamberlain, & S. Belbase (Eds.), *New perspectives and directions for collaborative research in mathematics education* (Vol. 1, pp. 33–57). University of Wyoming.

Thompson, P. (1994). The development of the concept of speed and its relationship to concepts of rate. In G. Harel & J. Confrey (Eds.), *The development of multiplicative reasoning in the learning of mathematics* (pp. 181–234). SUNY Press.

Thompson, P. W., Hatfield, N. J., Yoon, H., Joshua, S., & Byerley, C. (2017). Covariational reasoning among U.S. and South Korean secondary mathematics teachers. *Journal of Mathematical Behavior, 48*, 95–111.

Weber, E., Ellis, A., Kulow, T., & Ozgur, Z. (2014). 6 principles for quantitative reasoning and modeling. *Mathematics Teacher, 108*(1), 24–30.

Operationalizing and Assessing Quantitative Reasoning in Introductory Physics

Suzanne White Brahmia and Alexis Olsho

1 Introduction

Learning to symbolize and reason about the covarying relationships between abstract quantities, while being introduced to over 100 new physical quantities, characterizes a typical student's experience in an introductory algebra-based or calculus-based physics course. Students who are enrolled in a physics course at the upper high school or early university levels are typically also enrolled in a course in algebra 2 (functions, equations and inequalities, logarithmic and exponential relationships, and polynomial equations), precalculus, or calculus. There is an opportunity for mathematics instruction to help enrich students' experiences mathematizing in physics contexts, and for physics instruction to help students develop better conceptual understanding of the mathematics that they use. This chapter seeks to make connections between the mathematics and physics worlds, inspiring instruction that can result in a deeper understanding and appreciation of the mathematical nuances of the symbolic models that describe the physical world. What follows is written to help bridge these two instructional worlds.

Quantitative literacy (QL) is the ability to adequately use elementary mathematical tools to interpret and manipulate quantitative data and ideas that arise in individuals' private, civic, and work lives (Gillman, 2005). We also note that quantitative literacy requires an *inclination* to describe real-world phenomena mathematically. Quantitatively literate individuals recognize the value in considering mathematics as a way to understand and reason about real-life situations. In this chapter we consider Physics

S. White Brahmia (✉)
University of Washington, Seattle, WA, USA
e-mail: brahmia@uw.edu

A. Olsho
United States Air Force Academy, USAF Academy, CO, USA
e-mail: alexis.olsho.ctr@afacademy.af.edu

G. Karagöz Akar et al. (eds.), *Quantitative Reasoning in Mathematics and Science Education*, Mathematics Education in the Digital Era 21,
https://doi.org/10.1007/978-3-031-14553-7_12

Quantitative Literacy (PQL), i.e., quantitative literacy in the context of (introductory) physics, and argue that it is a pedagogically and intellectually fertile actualization of QL.

Introductory physics uses familiar mathematics in distinct ways to describe the world and make meaning. To an expert, a physics equation "tells the story" of an interaction or process. Quantitative modeling, in which patterns are expressed using mathematical functions that relate physical quantities to each other, is the backbone of PQL. For example, when reading the equation,

$$x(t) = +20\,\mathrm{m} + (-3\,\mathrm{m/s})t + \frac{1}{2}\left(-9.8\,\mathrm{m/s^2}\right)t^2$$

an expert may quickly construct a mental story of how the position of a projectile varies with time, starting 20 m above the ground and launched with a speed of 3 m/s vertically downward. The one-dimensional coordinate system is determined in this case by the physical fact that the acceleration due to gravity points downward, toward the earth. Part of the challenge of learning physics is developing the ability to decode symbolic representations in this manner.

While the ability to describe the physical world quantitatively as exemplified above is a goal of introductory physics courses, little has been done to determine specific, assessable learning objectives related to PQL. This may be, in part, due to a lack of self-awareness on the part of instructors about what PQL entails, and how they, as experts, reason quantitatively in contexts of introductory physics. There is a growing body of literature that seeks to better clarify what PQL entails in introductory physics (Bajracharya et al., 2012; Boudreaux et al., 2020; Eichenlaub & Redish, 2019; Eriksson et al., 2018; Hayes & Wittmann, 2010; Huynh & Sayre, 2018; Redish, 2021; Torigoe & Gladding, 2011; White Brahmia et al., 2020, 2021). This section builds on that prior work. In order to frame improving quantitative literacy in a physics instructional context, we first operationalize *physics quantitative literacy* (*PQL*) in Sect. 2. Next, in Sect. 3, we outline introductory physics learning objectives that can help instructors meet the broad goal of developing students' PQL, and suggest areas of overlap with concurrent mathematics courses. Lastly, in Sect. 4, we describe an assessment instrument we've developed to help instructors determine whether or not their instructional methods are helping students meet the PQL learning objectives.

2 Operationalizing Physics Quantitative Literacy

PQL relies on a blend of conceptual and procedural mathematics and physics content to formulate and apply quantitative models to describe the physical world. Figure 1 shows a visual representation of the process of quantitative modeling in physics, beginning with observations that can lead to creation of *base quantities*. We define base quantities, such as time, position, and change in position, as those that can be created from observations and a single type of measurement. Quantitative modeling

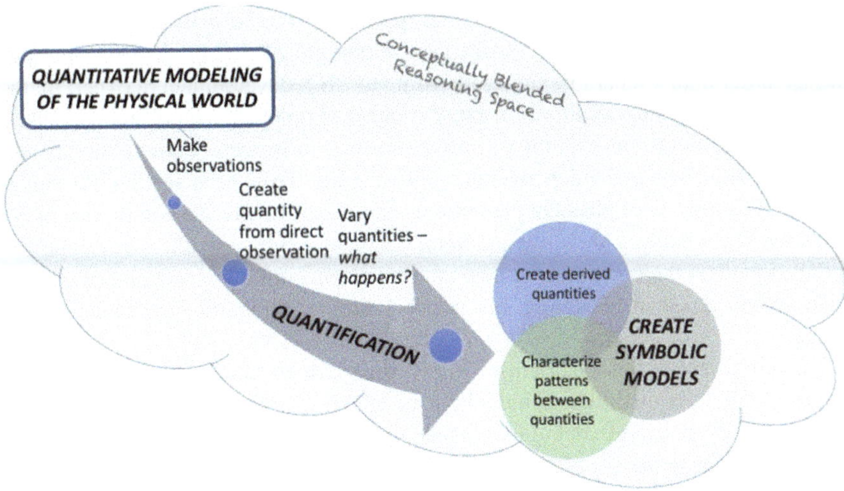

Fig. 1 Quantitative modeling in physics

continues with the exploration of these quantities and their relationships to each other, resulting in derived or composite quantities (such as velocity or speed), established relationships between quantities, and more formal symbolic models. In this section, we operationalize PQL by describing quantitative modeling as outlined in Fig. 1—the process of mathematizing the physical world.

The equation $x(t) = +20\,\text{m} + (-3\,\text{m/s})t + \frac{1}{2}(-9.8\,\text{m/s}^2)t^2$ in the introductory vignette above is an instantiation of the kinematics equation $x(t) = x_o + v_o t + \frac{1}{2}at^2$, which describes the time-dependent position of an object moving with constant acceleration a and initial velocity v_o from initial position x_o. It is introduced in the first week of almost all college-level introductory physics courses. This equation is a result of the quantitative modeling process in Fig. 1. The first step in the process is observations leading to the creation of base quantities position and time. These quantities appear in the kinematics equation as variables (x and t) and a parameter (initial position x_o). Consideration of how the base quantities vary and covary leads to the derived quantities of velocity and acceleration, which appear as parameters v_o and a. It also leads to characterizations of patterns between the quantities: position can be described as a function of time (x can be expressed as $x(t)$) and depends on the "accumulation" of displacement due to motion characterized by initial velocity v_o and acceleration a. The result is a symbolic model, the general kinematics equation $x(t) = x_o + v_o t + \frac{1}{2}at^2$.

As Fig. 1 depicts, quantitative modeling occurs in a conceptually blended mental space. Quantitative modeling in physics is not simply "doing mathematics with physics quantities." It requires a novel combination of mathematical and physical reasoning. Conceptual blending theory (CBT) (Fauconnier & Turner, 2002) provides a framework for characterizing this combination. Fauconnier and Turner describe a cognitive process in which a unique mental space is formed from two (or more) separate mental spaces. The blended space can be thought of as a product of the input spaces, rather than a separable sum. According to CBT, development of expert mathematization in physics would occur not through a simple addition of new elements (physics quantities) to an existing cognitive structure (arithmetic), but rather through the creation of a new and independent cognitive space. This space, in which creative, quantitative analysis of physical phenomena can occur, involves a continuous interdependence of thinking about the mathematical and physical worlds. Development of PQL involves the creation of a new cognitive space that depends on both mathematical and physical reasoning, but is not a simple, separable sum of these two spaces.

The remainder of this section uses Fig. 1 as a guide to fully operationalize PQL. Section 2.1 details quantitative modeling, of which quantification is a foundation. In Sect. 2.2, we discuss in detail two facets of quantitative modeling that are particularly important in the contexts of introductory-level physics: reasoning about sign and signed quantities; and covariational reasoning with quantities. While reasoning about sign and covariational reasoning have been well-researched by the mathematics education community, recent work by the authors and their collaborators suggest these modes of reasoning as used in physics contexts by physics experts are distinct from the analogous modes in mathematical contexts (White Brahmia et al., 2020). Characterization of these types of reasoning with physics quantities is necessary to understand quantification and quantitative modeling in physics courses, especially for developing assessable learning objectives.

2.1 Quantitative Modeling in Physics

Quantification is a facet of quantitative modeling, and generates the building blocks for the mathematical descriptions involved in quantitative modeling. Thompson defines quantification as "the process of conceptualizing an object and an attribute of it so that the attribute has a unit of measure, and the attribute's measure entails a proportional relationship... with its unit" (Thompson, 2011, p. 37). For example, a bus's motion can be quantified by a velocity (combining the mathematical objects of ratio and vector) relative to the ground. Thompson considers quantification to be "a root of mathematical thinking," and argues that learners develop their mathematics from reasoning about quantities. In work involving middle school algebra students, Ellis (2007) claims that modes of mathematical structural reasoning are more likely to develop when students practice with quantities that are composed of other quantities through multiplication or division, rather than the strictly numerical patterns and algorithms common to school mathematics. Ellis claims it is precisely these kinds of quantities that help develop students' abilities to create powerful generalizations. White Brahmia (2019) argues that quantification is the overlooked first step in the modeling process in physics instruction.

Quantification in introductory-level physics courses is typically not generative. Students are rarely asked to create new quantities to describe attributes. Instead, quantification in introductory-level physics courses is focused on the understanding and use of introduced quantities to describe processes and physical phenomena. Students are asked to participate in quantitative modeling with already-defined physical quantities.

Just as conceptual understanding of mathematical operations enriches cognition, so too does understanding the meaning and calculation of introduced quantities. Consider the two common framings of division as the process of *sharing* or *segmenting,* as described by Thompson and Saldanha (2003). Sharing is the partitioning of a number into some number of equal-sized portions (e.g., $\frac{12}{3} = 4$ shares in each of 3 portions). Segmenting is portioning out a number in groups of a given size (e.g., $\frac{12}{4} = 3$ portions of size 4). Thompson and Saldanha (2003) demonstrate that "operational understanding of division entails a conceptual isomorphism between" sharing and segmenting. These framings are productive in the context of numbers and can help new learners to visualize the meaning of division. Moreover, they are productive for students in many "real-life" scenarios. Contrast, however, this conceptual understanding of ratio and division with the construction of velocity as a vector quantity. Velocity can be understood by framing division as an operation which relates (Thompson et al., 2014) a change in position, which is a vector, to a time interval, which is a scalar, and produces a quotient entirely different from the dividend and the divisor. Velocity as the vector rate of change of position has its own physical meaning. Thompson et al. (2014) argue that understanding of a ratio quantity created by comparing two quantities of different natures is equivalent to understanding "relative magnitude" and note "high-level scientific reasoning that involves physical quantities typically involves conceiving of relative magnitudes." In our experience, many students coming out of mathematics courses lack this understanding. We also find that it is uncommon for physics instructors to make explicit this difference when introducing velocity—that division is now performed for a different reason than it was when calculating, for example, the duration of a process that takes one-fourth as long as another, $\frac{22\,\mathrm{s}}{4} = 5.5\,\mathrm{s}$.

We note that PQL includes an *inclination* or habit-of-mind to quantify or create quantitative models, hereafter referred to as "models." The modeling shown in Fig. 1 begins with observations of the world, which may lead to quantification for individuals with high QL. Ability to think mathematically is not enough; it must be accompanied by a recognition that the physical world can be described quantitatively, and an inclination to develop and understand the model.

Observation and quantification are crucial first steps in developing models in physics. Modeling can also result in novel composite quantities. Acceleration is one such composite physical quantity: $\vec{a}$ is the ratio of a change in velocity $\Delta\vec{v}$ and an interval of timeΔt. The creation of acceleration as a quantity is a result of a quantitative model: Galileo famously wrestled with the mathematical decision of whether to describe accelerated motion with a ratio of change in velocity to distance traveled or change in velocity to elapsed time. His choice of the latter led to the formal concept of acceleration, a foundation for the subsequent Newtonian synthesis. The quantitative modeling demonstrated in the introductory vignette involves both a procedural and conceptual mastery of the prerequisite mathematics (Redish & Kuo, 2015; Thompson, 2011). Gray and Tall (1994) describe this combination of procedural and conceptual mastery in mathematical contexts as *proceptual* understanding and name it as a target learning goal for mathematics courses. Gray and Tall (1994) highlight the distinction between procedural efficiency and conceptual understanding, explaining that "the symbol $\frac{3}{4}$ stands for both the process of division and the concept of fraction." In the terms of Thompson and Saldanha (2003), proceptual understanding involves both the conceptualization of fraction, and the conceptualization and action of division.

We argue that quantitative modeling also requires proceptual understanding of physics quantities themselves. Consider the quantity *average velocity*, $\vec{v}_{av} = \frac{\Delta \vec{x}}{\Delta t}$. A physics student with a proceptual understanding of velocity would be procedurally proficient at determining an object's average velocity by dividing its displacement by the elapsed time, as well as understand conceptually that the ratio is a quantity unto itself, $\vec{v}_{av}$, with its own properties and meaning.

We also argue that to succeed in physics courses, it may not be enough to understand the mathematics as taught in mathematics courses. In introductory physics, "flexibility" with mathematics is expected of students—they are expected to understand and apply mathematics in ways that are different than they may have been taught in prior mathematics courses. This flexibility is a hallmark of expert-like reasoning in physics (Sherin, 2001; Vlassis, 2004). A physics expert is able to distinguish between a negative sign used to indicate the type of electric charge in surplus in a given system (Olsho et al., 2021), and one used to indicate the direction of a component of an electric field relative to an assigned coordinate system (White Brahmia et al., 2020); a product may indicate an increase or accumulation of a quantity, or the creation of a new quantity. Physics experts readily interpret these aspects of the mathematization of physical systems (Fig. 2).

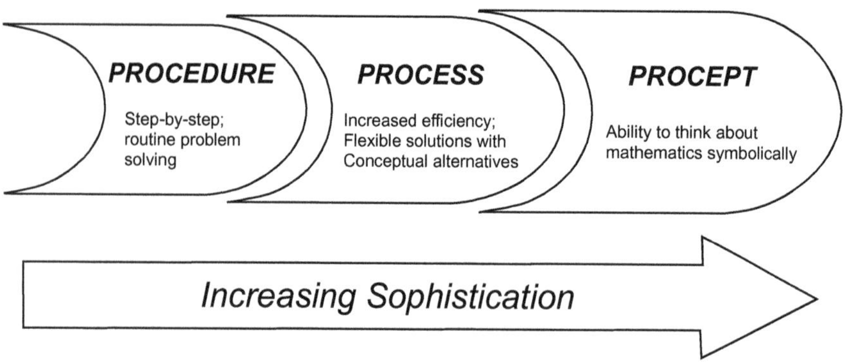

Fig. 2 Proceptual development, adapted from Tall (2008)

Familiarity with multiple representations is a foundation for modeling in physics (Brewe, 2008). This familiarity facilitates the expert-like habit of seeking coherence between varied representations of quantities and relationships. Because quantitative modeling requires proceptual understanding of the mathematics used to relate physics quantities, as well as familiarity with the physics quantities themselves, we suggest that students must have some experience with the multiple representations taught in mathematics and physics courses (e.g., symbolic, graphical, and diagrammatic). The ability to think abstractly about physics quantities allows for greater understanding of the physical phenomena or qualities that the quantities represent; for example, students are able to consider the meaning of the quantity "electric potential" beyond its algebraic representation. Familiarity with multiple representations and physics quantities also allows students to make useful generalizations about quantities—they are able to consider similarities and differences between disparate vector quantities such as electric field and acceleration, or scalar quantities such as mass and charge. A proceptual understanding of the mathematics may help develop a deeper understanding of the physics quantities, which can, in turn, deepen understanding of the mathematics (Sealey & Thompson, 2016).

Success in a physics course requires conceptualizing models that were generated by someone else; moreover, students are expected to understand the symbolizing of quantity and covariational relationships between quantities as if they created the models themselves. This depth of understanding involves recognizing patterns and decoding symbolic models. In the following section, we explicate these cognitive activities by focusing on two areas of reasoning central to the quantitative models featured in introductory-level physics.

2.2 *Facets of Quantitative Reasoning in Introductory Physics*

In this section, we discuss two facets of quantitative modeling that are of particular importance in introductory physics: reasoning about sign and signed quantities; and covariational reasoning, including reasoning about compound quantities. As discussed earlier, reasoning about sign is of particular importance to quantification of base quantities in physics, while covariational reasoning plays a substantial role in development of quantitative models and quantification of composite or derived quantities.

2.2.1 Reasoning About Sign and Signed Quantities in Physics

Negative integers represent a more cognitively difficult mathematical object than positive integers do for pre-college mathematics students (Bishop et al., 2014). Mathematics education researchers have isolated a variety of "natures of negativity" fundamental to algebraic reasoning in the context of high school algebra—the many meanings of the negative sign that must be distinguished and understood for students to develop understanding (Gallardo & Rojano, 1994; Nunes, 1993; Thompson & Dreyfus, 1988; Vlassis, 2004). These various meanings of the negative sign, which will be discussed in greater detail below, form the foundation for scientific quantification, where the mathematical properties of negative numbers are well-suited to represent natural processes and quantities. Recognition that the negative sign has different meanings in different contexts, and correct interpretation of the meaning of a negative sign in a given context—called "flexibility" with negativity by mathematics education researcher Vlassis (2004)—is a known challenge in mathematics education. There is mounting evidence that reasoning about negative quantity poses a significant hurdle for physics students at the introductory level and beyond (Bajracharya et al., 2012; Ceuppens et al., 2019; Eriksson et al., 2018; Hayes & Wittmann, 2010; Huynh & Sayre, 2018; White Brahmia et al., 2020).

In physics, as in mathematics, it is convention that an unsigned quantity is a positive quantity (e.g., "5 μC" is taken to mean a charge of $+ 5$ μC). While research indicates that students are not facile at interpreting the meaning of negative signs specifically, we suggest that it is the presence of an explicit sign associated with a quantity that results in the difficulty. Indeed, physics education researchers report that a majority of students enrolled in a calculus-based physics course struggled to make meaning of negative *and* positive quantities in spite of completing Calculus I and more advanced courses in mathematics (White Brahmia & Boudreaux, 2016, 2017). In our discussion below, we focus on negativity and use of the negative sign (as by convention, that is the context in which use of an explicit sign is necessary), but suggest the applicability to sign and signed quantities more generally.

Flexibility with negativity and interpretation of the negative sign in different physics contexts plays an important role in both quantification specifically and quantitative modeling more generally. Sherin's (2001) "symbolic forms" were developed to explain how successful physics students interpret and create equations. Sherin suggested that students associate symbolic patterns with physical and mathematical meaning. Work by mathematics and science education researchers has expanded Sherin's original list of symbolic forms (Dorko & Speer, 2015; Rodrigues et al., 2019; White Brahmia, 2019). While mathematics education researchers identified a "measurement" symbolic form as consisting of magnitude, units, and exponent (Dorko & Speer, 2015), research in physics contexts suggests a "quantity" symbolic form consisting of *sign,* value, and units, where the sign carries physical meaning related to the specific quantity (White Brahmia, 2019). These two symbolic forms are shown in Fig. 3.

The difference between the symbolic forms speaks to the importance of sign when considering physics quantities. Quantities representing change, such as $\Delta v = v_{\text{final}} - v_{\text{inital}}$ (i.e., change in speed), are fundamental to introductory-level physics but are discussed less in mathematics course. The "quantity" symbolic form includes the expectation of a sign associated with each quantity, which in the case of Δv informs whether the speed is increasing or decreasing. Expressed using the "measurement" symbolic form, Δv would only consist of magnitude and units, and omits important information about the nature of the change. The inclusion of sign allows for a more complete description of an object's motion.

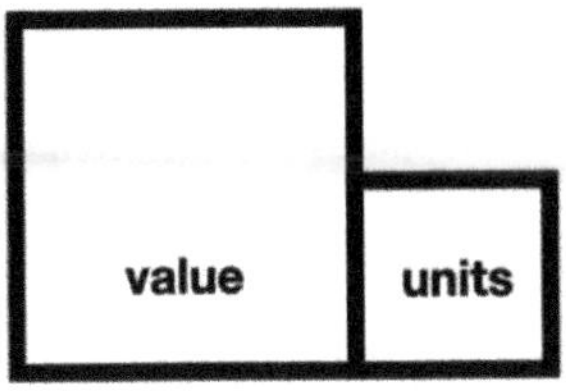

(a) Creating Quantity: Measurement (Dorko & Speer, 2015)

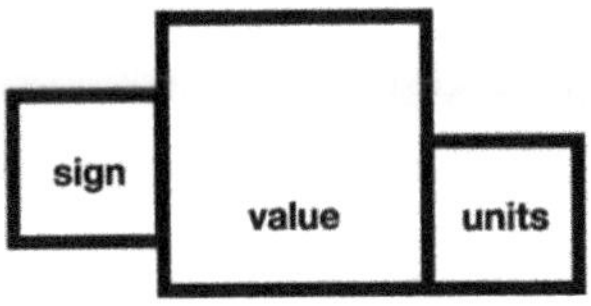

(b) Quantity (White Brahmia, 2019)

Fig. 3 Symbolic forms relevant to physics quantities

The meaning of the sign is of particular significance for scalar quantities, where the meaning may be consistent with mathematical conventions (comparison to a reference or zero, such as for temperature) or could be part of a model or physics convention (for example, heat Q is negative for a system when thermal energy is transferred out of that system). Introductory-level physics students also see a use of sign that is idiosyncratic to physics: sign as an indication of type, as with electric charge, where the sign of the net charge on an object indicates the *type* of charge (positive or negative) in surplus on the object (Olsho et al., 2021).

Sign plays an important role for vector quantities as well. Vector quantities are always interpreted geometrically (i.e., having a magnitude and a direction) in college-level physics courses; a negative sign associated with a vector or vector component thus indicates its direction, either relative to a defined coordinate system ($F_x = -3\text{N}$) or to another vector, as in the quantitative statement of Newton's Third Law ($\vec{F}_{12} = -\vec{F}_{21}$).

For quantitative modeling more generally, students must consider the meaning of negative (and positive) signs when they are used to model physical relationships or processes, or to compare or combine quantities. In these cases, positive and negative signs can be used to describe how quantities relate to each other, or as part of the operations of addition and subtraction—divergent uses of the same symbols. Students are introduced to expressions that relate quantities that oppose or are opposite to each other. Even when used to indicate the operation of subtraction, the negative sign has varied meanings in physics contexts. To describe the many meanings of the negative signs in the contexts of introductory-level physics, White Brahmia, et al. (2020) developed a framework of the natures of negativity in introductory physics, based on an analogous framework in the context of algebra (Vlassis, 2004). An abbreviated version of the physics framework is shown in Table 1. The framework outlines three uses or facets of the negative sign in physics: as associated with a single *quantity*; as used describe a *relationship* between multiple quantities; and as used to denote the *operation* of subtraction. As seen in Table 1, each of these facets is itself multifaceted, which is an indication of the many nuances of negativity in physics contexts.

Table 1 Abbreviated version of the framework of natures of negativity in introductory physics (White Brahmia et al., 2020)

(Q) Quantity	(R) Relationship	(O) Operation
1. Scalar	**1. Opposes**	**1. Removal (physical)**
a. Type (charge only)	**2. Opposite**	**2. Difference (temporal)**
b. Change, rate of change	**3. Relative Orientation**	**4. Removal (modeling)**
c. Comparison to reference	**4. Negative exponents**	**3. Difference (other)**
d. Models, convention		
2. Vector component		

Boldface indicates a facet of a main nature of negativity

Use of the negative sign to convey physical meaning is a basis of quantitative modeling. Even at the college introductory level, combinations of positive and negative signs are necessary to model processes and relationships. Further, the negative sign associated with a given quantity can have multiple correct interpretations. For example, when a force does negative work on a system, it can be interpreted as an indication that the force acts to decrease the mechanical energy of the system. The negative sign also indicates that the force is applied in a direction opposite to the direction of the displacement of the system. White Brahmia and Boudreaux (2017) found that students who understood that a force does negative work on a system when applied in a direction opposite to the system's displacement were more likely to understand that a net negative work is associated with a decrease in the system's energy. The researchers interpreted this result as an indication that a mathematical understanding about the scalar product catalyzed a more robust understanding about the change in system energy. This is an example of how understanding positive and negative signs is associated with more complete understanding of physics quantities, and the quantities' meanings within physics models (White Brahmia, 2019).

2.2.2 Covariational Reasoning in Physics

Covariational reasoning, "the cognitive activities involved in coordinating two varying quantities while attending to the ways in which they change in relation to each other" (Carlson et al., 2002, p. 354) has been shown to be strongly associated with student success in calculus by mathematics education researchers (Carlson et al., 2002; Saldanha & Thompson, 1998; Thompson, 1994). Physics covariational reasoning plays a substantial role in physics quantitative modeling. It involves finding the relationship between quantities, and representing that relationship symbolically. These are both key facets of quantitative modeling as depicted in Fig. 1. In college-level introductory physics courses, students are routinely asked to describe how quantities relate to each other, and how a change in one quantity affects another quantity.

However, few studies by physics education researchers have explored how covariational reasoning is used in introductory physics contexts. A study reported on by Zimmerman et al. (2020a, 2020b) suggests that covariational reasoning in physics graduate students ("experts" in introductory physics contexts) differs in some ways from that in mathematics graduate students, as reported by Hobson and Moore (2017). In particular, physics experts display a number of specific behaviors—one of which will be described in the paragraphs below—that allow them to consider the relationship between two variables while reducing or even eliminating the formal, novel covariational reasoning seen in mathematics experts in similar contexts. These behaviors allow for physicists to engage in reasoning about the quantities themselves, as well as the relationship between the quantities, in a way that is not typically necessary in mathematical contexts. For this reason, we call the covariational reasoning done by physics experts "covariational reasoning with quantities" or simply "physics covariational reasoning."

Zimmerman et al. (2020a, 2020b) have identified a number of behavior in physics experts that seem to facilitate covariational reasoning. In what follows we focus on a particular instance of the overarching physics expert behavior which Zimmerman et al. (2020a, 2020b) call "compiled relationships": the use and creation of defined relationships between two quantities that may or may not be in the problem statement in order to help address the relationship between two quantities in the specified task. We suggest that this is a cognitive activity that is distinct in physics covariational reasoning, and that it allows for greater focus on the meaning of physics quantities. The use of compiled relationships as an expert behavior relies on the fact that there are relatively few functions that make up the models encountered in a college-level physics course—most involve linear or inverse relationships, basic trigonometric functions, simple quadratics, or exponential decay. Most physical contexts at this level can be mathematized with just this handful of functions, with which expert physicists become very familiar. Therefore, physics experts come to expect one of these common functions, and readily mathematize tasks that involve novel covariational reasoning for mathematics experts—for whom any function is possible.

The behavior encompassed by the compiled relationships category has several facets. Here, we define a facet which we call "automatic mathematization" (Zimmerman et al., 2020a, 2020b) which illustrates a key difference between the way physics experts and mathematics experts approach quantitative modeling tasks that involve covariational reasoning. *Automatic mathematization* is the almost-immediate, automatic assignment of a known functional relationship between quantities. This mathematization is typically guided by the physics and draws on well-tested models of nature. It may be as simple as a learned rule such as "force decreases as $\frac{1}{r^2}$" or more complex, requiring identification of a physical phenomenon in a particular context and then mathematizing. An example of the latter was seen in interviews with physics graduate students who were asked to draw a graph relating intensity of light in liquid as a function of the depth of the water. Several of the interviewees recognized that a decrease in light intensity with increasing distance from the light source was due to the physical phenomenon of scattering. These graduate students then assumed that the intensity would therefore decrease exponentially with increasing distance from the source, connecting the physical phenomenon of scattering to the function $f(x) = e^{-x}$. Zimmerman et al. (2020a, 2020b) report on several other physics-specific expert behaviors that were not reported on in the studies of mathematics graduate students by Hobson and Moore (2017), Moore (2014). They conclude that physics covariational reasoning is built on a proceptual understanding of quantities themselves, and a handful of functions. The physics expert behaviors described above—and others—allow physicists to make sense of the quantities, through their physical interpretation, and the mathematical relationships between quantities simultaneously. We believe that this blended sensemaking is characteristic of physics covariational reasoning, and therefore, of quantitative modeling in physics.

In this section, we have described our work exploring how experts reason quantitatively. In Fig. 1 we outline the reasoning that goes into generating and interpreting symbolic models in physics. The quantitative modeling demonstrated in the vignette in the introduction exemplifies this reasoning process, where the position and time are quantities that emerge from direct observation and the velocity and acceleration are *derived* quantities that characterize the motion. Unlike "measures" in mathematics, physics quantities typically include a sign that carries its own important meaning. The covariational relationship between quantities is symbolized in the kinematics equation shown.

By identifying these sophisticated reasoning patterns, we create targets for assessable PQL-related learning objectives—discussed in the next section—for students enrolled in introductory physics courses.

3 Assessable PQL Learning Objectives

Having operationalized PQL in the previous section through frameworks that characterize expert reasoning, in this section we describe the development of assessable PQL learning objectives for the college-level introductory physics course, using expert PQL as a target. We note that explicit PQL learning objectives in introductory physics are uncommon, largely because the kind of reasoning outlined in the previous section is assumed by most physics instructors to be developed in the prerequisite mathematics courses. There is a gap between physics and mathematics instruction that this work seeks to help close.

Developing learning objectives (LOs) that can help guide instructional efforts toward effective development of PQL builds on the sustained and productive department-wide efforts developing undergraduate physics course learning objectives (Chasteen et al., 2011). In this section we discuss evidence-based PQL learning objectives, and in the next, an example of an assessment instrument that can be used to assess the effectiveness of instruction at meeting some of these objectives.

3.1 Methodology

The methodology we describe here for developing LOs discusses the overall development process, and also includes LOs that are not associated with PQL. The remainder of the chapter focuses specifically on the subset of LOs associated with developing PQL.

At the outset, we recognized that effective LOs articulate values shared by a broad group of instructors. Our first step in creating a succinct set of assessable learning objectives for the introductory physics sequence involved consolidating the outcomes of prior systematic efforts by the physics education research community, representing hundreds of the researcher's hours spent collaborating with departmental colleagues. Past department-level efforts in the United States have focused mainly on courses beyond the introductory level, which rely on a proceptual understanding of calculus. PQL at the introductory level helps build the foundation for the calculus-thinking that underpins modeling in physics; we approached this project through the lens of conceptually understanding the mathematical foundations of algebraic physics models.

In order to develop a set of LOs that are broadly appealing and recognizable to most instructors, we started with the existing LOs from a variety of widely respected sources.[1] We conducted a card-sorting task with those LOs, and supplemented the results where appropriate. Learning scientists have used card-sorting tasks to investigate mental organization of disciplinary knowledge (Chi et al., 1981; Schoenfeld & Herrmann, 1982). Experts are given cards showing various content with no pre-established groupings. They are then asked to sort the cards into groups that they feel make the most sense, and describe each group. The first author (SWB), a physics education research postdoc (whose dissertation specialization was surface science), and a senior astrophysics graduate student with extensive teaching and curriculum development experience, employed a card-sorting task with learning objectives that span the introductory physics course. On each card was a single objective. The researchers independently sorted the objectives into groups, then discussed their groups, and modified their sortings until they reached agreement.

The overall structure of the resulting learning objectives is hierarchical (see Fig. 4) and includes a novel level not seen in other efforts—*sequence-level* objectives that span the entire introductory physics sequence. The sequence level includes a limited number of LOs that blend the professional science practices and physics habits-of-mind characteristics of high-functioning STEM professionals. We recognize that this level of learning takes a long time and may not be measurable over the course of one term. It is mainly at the sequence level that we include objectives designed to develop skills that are strongly associated with PQL. Developing a proceptual understanding

---

[1] We looked to the high-quality practices of NGSS and the College Board, which have been carefully crafted over several years, for guidance in developing our own learning objectives at the sequence level (NSTA; College Board, 2020). They created the individual sequence-level LOs used in the sorting task by gathering the LOs from the multiple sources (Beichner, 2011; Etkina et al., 2006; Kozminski et al., 2014; LGBT + Physicists, 2013; SEI).

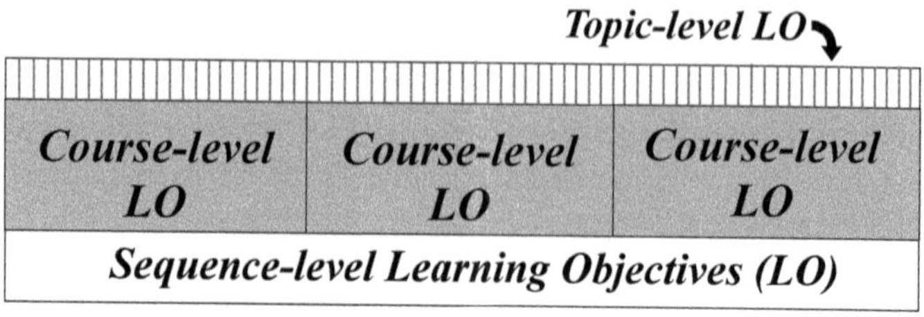

Fig. 4 Hierarchical structure of LOs

of models in physics also happens slowly, over the period of multiple sequential courses.

The course-level objectives include 10–15 overarching content themes that are specific to that course. Lastly, unit-level objectives, often thought of as the specific content, correspond in duration to a typical chapter in a college course.

The introductory physics sequence-level learning objectives resulting from the card-sorting task are listed in Table 2. The resulting consensus includes three themes around which similar LOs clustered: physics habits of mind, understanding models and limits, and professionalism and workplace skills.

Note that there is no separation between lab and lecture course objectives. While some objectives lend themselves better to the lab, there is considerable cognitive overlap. There is compelling evidence that it takes both laboratory and lecture/recitation experiences for these learning objectives to be met. Labs which emphasize following instructions and the development of technical laboratory skills miss an opportunity to help students develop the ability to design ways to answer scientific questions (Canright & White Brahmia, 2021; Etkina, 2015).

We share an example that is ubiquitous in physics: the inverse-square covariational relationship, which is central to many physics models (e.g., Coulomb's law, Newton's Law of Gravitation, light and sound intensity). The following example shows the learning objectives that are part of developing reasoning associated with Coulomb's law, a $\frac{1}{r^2}$ force, and the associated field in an electromagnetism course (typically the second term in an introductory sequence), and demonstrates how the levels shown in Fig. 4 differ:

- The relevant unit-level LOs include:
  - **Analogy to Gravitation**: Use Newton's 3rd law to reason about the force vector direction along a line connecting the two interacting objects; Use the $\frac{1}{r^2}$ structure of the gravitational and electrical forces to reason covariationally about similarities in the interactions between massive objects and between charged objects.
  - **Coulomb's Law**: State Coulomb's Law in equation form and explain the covariational relationship between the electrostatic force and (1) the magnitude of the charges, and (2) the separation of the charges
- The relevant course-level LOs include:

Table 2 Sequence-level learning objectives

HM: physics habits of mind
HM-1. Translation between physical and symbolic world: develop the inclination and ability to translate between the physical and symbolic worlds in an effort to quantitatively reason about how nature works
HM-2. Reasoning with physical quantities: reason abstractly and quantitatively with new scalar and vector quantities: make physical sense of the quantities and mastering their mathematical structures
HM-3. Multiple representations: create and translate between multiple representations of the same concept (e.g., text, equations, graphs, diagrams)
HM-4. Problem articulation: articulate what it is that needs to be solved in a particular problem, what is known and represent them using a non-verbal representation
HM-5. Perseverance: recognize that wrong turns are valuable in learning the material, recover from mistakes, and persisting in working to the solution even when there is no clear path to the endpoint
HM-6. Sensemaking with quantity: effectively use unit reasoning, vector and scalar natures and limiting cases to make sense of answers
HM-7. Order of magnitude and reasonableness: anticipate the order of magnitude to judge the reasonableness of measurements and calculations
HM-8. Reasoning based on mathematical structure: look for and make use of patterns associated with mathematical structure to reason across contexts and scale
HM-9. Recognizing uncertainties: be able to recognize that all measured quantities have inherent uncertainties
ML: understanding models and their limits in physics
ML-1. Making observations: form a scientific question, design and carry out experiments to look for patterns
ML-2. Developing a model: analyze and interpret data while attending to uncertainty in measurement and construct explanations based on patterns in the data
ML-3. Reasoning with mathematical models: develop and use mathematical models and explanations, construct viable arguments, engage in argumentation from evidence and critique reasoning of others
ML-4. Model limitation: articulate assumptions made when applying a model, and the range over which a particular model is a valid description of nature
ML-5. Model testing: design an experiment to test the model and make a prediction of the outcome based on it
ML-6. Scientific judgment: analyze and interpret data from a testing experiment while attending to uncertainty, and make a scientific judgment about the outcome
PW: professionalism and workplace
PW-1. Collective intelligence: recognizing the two features of high collective intelligence, and monitoring social climate to optimize these features (equitable speaking turns, social sensitivity)
PW-2. Collaboration: able to articulate affordances that a group brings to arriving at a creative solution, knowing what the roles are that members of effective groups t
PW-3. Inclusion: demonstrate effective communication skills in the context of a recitation or lab group that results in whole-group meaningful participation

(continued)

Table 2 (continued)

PW: professionalism and workplace
PW-4. Communicating physics: be able to communicate physics in written and oral forms
PW-5. Independent Learning: recognizing and acting on confusion: be able to articulate specifically the source of confusion and taking action to move beyond that difficulty (e.g., office hours, group study)
PW-6. Skepticism toward conclusions: recognize that scientific conclusions—whether from an outside source or from your own calculations—may be incorrect, and develop the ability to check these conclusions with simple calculations, 3rd party information, and/or common sense

The PQL-specific LOs appear in bold italic

- **Electric Force and Field**: Apply Coulomb's Law and the superposition principle to find the net force and field due to a distribution of charges
- **Sophisticated Quantities in E&M**: Distinguish between the vector and scalar nature of EM quantities and the role of $\pm$ signs

• The relevant sequence-level LOs include:

- **ML-3: Reasoning with mathematical models**: Develop and use mathematical models and explanations, construct viable arguments, engage in argumentation from evidence.
- **HM-8: Reasoning based on mathematical structure**: Look for and make use of patterns associated with mathematical structure to reason across contexts and scale.

In the remainder of this chapter, we focus on sequence-level objectives because PQL develops over repeated exposure, at a different rate for all students. The expectation is that by the time students have completed the introductory sequence of physics, these objectives will have been met. Sequence-level objectives in turn strongly influenced the course-level objectives, and the streamlining of the unit goals. We next look closely at the specific PQL sequence-level LOs.

3.2 *Sequence-Level Learning Objectives*

A subset of the sequence-level learning objectives that target PQL specifically is indicated by bold italics in Table 2. We suspect that mathematics instructors will find these familiar, and likely see overlap with their own learning objectives. We see great potential to embolden student learning, both in mathematics and in physics, if both disciplines can emphasize mathematical reasoning that is highly valued in physics. We focus here on three of the learning objectives from Table 2 to better clarify why they matter, and how they might overlap with mathematics instruction: HM-1, HM-3, and HM-6 (see Table 3).

Table 3 Sample LOs with examples from physics

HM-1	HM-3	HM-6
Translation between physical and symbolic world	Multiple representations	Sensemaking with quantity
• *Positive and negative signs* (e.g., electric charge, one dimensional velocity, displacement, acceleration) • *Summation* of conserved quantities (e.g., energy, momentum) • *Unit vectors* to represent direction of vector quantities (e.g., force, displacement, electric field)	• *Interpretation of slope and area under curve in graphs* (e.g., position vs. time, pressure vs. volume, force vs. displacement) • *Verbal interpretation of equations* (e.g., example in introduction) • *Force diagrams* to represent direction and magnitude of vector quantities (e.g., Newton's laws, statics)	• *Limiting cases* What happens in a given model for very large/small and zero values of a quantity? • *Dimensional analysis* are the units of an answer consistent? Does a model make sense in the physical world? • *Vector versus scalar reasoning* does "direction" carry meaning for a given quantity? (e.g., force, energy, momentum, time)

HM-1, **Translation between the physical and symbolic world**, is a continuous mental action of experts in physics, relying heavily on mathematical symbols to convey deep meaning. Addition and subtraction can be performed only with like quantities, and the operations carry different meaning than the integers that carry the same symbols, as was demonstrated in the introduction of this chapter.

HM-3, **Multiple Representations**, is brought to life in the vignette at the opening of this chapter. The reliance on particular representations and the inclination to seek coherence between them is a hallmark of expert behavior around making sense of models. Equations are ubiquitous in all physics contexts. Graphical representations of position, velocity, and acceleration as a function of time are an instructional platform kinematics, bar charts are commonly used to keep track of conserved quantities, and vector diagrams are foundational in the studies of solid and fluid statics and dynamics.

HM-6, **Sensemaking with quantity**, encompasses exploring the limiting cases of single and multivariable models, using the units in a calculation both to guide and to check for sensemaking, and exploring physical-world implication of the vector or scalar nature of a quantity. As an example of the latter, multiplication and division create entirely new quantities with unique properties. Work is a scalar product of two vectors (force and displacement); it is neither force nor displacement, and not a vector. Nonetheless, students routinely conflate work and force, not differentiating between the product and a factor, or a scalar and a vector.

We have gathered a substantial amount of evidence for face validity of the sequence-level LOs. The language has been modified iteratively based on a series of interviews with faculty until the LOs reached a steady state in which they are both understood as intended and valued by instructors. Much work remains before it becomes standard practice across most institutions that undergraduate physics instruction is designed to meet evidence-based objectives, and measures of the effectiveness of instruction are based on them. In their current form, the LOs described in

Table 2 are used at our institutions with a broad set of instructors, with an associated outcome of facilitating consensus about course content, assessments, professional development, and modifications to courses.

In this section we've demonstrated the ubiquity and importance of quantification, symbolizing and modeling to physics reasoning, and provided PQL learning objectives that reflect their value to instruction. In what follows we describe an instrument that can be used to assess whether or not instruction is meeting these objectives.

4 The Physics Inventory of Quantitative Literacy

Despite the importance of physics quantitative literacy as a learning outcome in introductory physics courses, there is a dearth of instruments to assess its development. To address this need, we developed, with collaborators Smith, Boudreaux, Eaton, and Zimmerman, the *Physics Inventory of Quantitative Literacy* (PIQL), a multiple-choice reasoning inventory (White Brahmia et al., 2021). Various concept inventories, such as the *Force Concept Inventory* (Hestenes et al., 1992) and the *Force and Motion Conceptual Evaluation* (Thornton & Sokoloff, 1998) in physics, and the *Precalculus Concept Assessment* (Carlson et al., 2010) and *Calculus Concept Inventory* (Epstein, 2006) in mathematics, have raised awareness of student difficulties, leading to directed instructional interventions and improvements in curricula, and we believe that the PIQL can have an analogous impact on physics instruction. There are, however, several aspects of the PIQL that set it apart from concept inventories:

1. Instead of focusing on a single physics concept or level of mathematics, the PIQL was developed to assess facets of mathematical *reasoning* (i.e., PQL) that are important in introductory physics, and foundational to subsequent physics courses.
2. The PIQL has several "multiple-choice multiple-response" items (i.e., multiple choice questions for which there may be more than one correct answer, and for which students are asked to choose all responses that they believe are correct), which allow us to probe both conceptual mathematics and conceptual physics features of student reasoning in a given context.
3. The PIQL is designed to assess development of PQL throughout an entire introductory physics course sequence, rather than providing a measurement of concept mastery for a single course.

As the PIQL is intended to assess PQL and its development with instruction in physics, the items focus on the types of quantification and quantitative modeling that are important in introductory physics: reasoning about sign and signed quantities, and covariational reasoning. Covariational reasoning in particular is foundational to the mathematics course that is prerequisite to introductory physics courses (precalculus), and several PIQL items are adapted from items from the Precalculus Concept Assessment (Carlson et al., 2010). In addition, some PIQL items assess student reasoning about ratios and proportions; while this type of reasoning is related to covariational reasoning, we treat it as a distinct category for PIQL items. Proportional reasoning represents a domain of quantification that is particularly relevant for introductory physics, where many models involve linear relationships and many quantities are ratios of other quantities (Boudreaux et al., 2020). Reasoning about sign and signed quantities and covariational reasoning are key to quantification and quantitative modeling, as described in Sect. 2. The PIQL's focus on these facets of mathematical reasoning in physics contexts makes it an important metric for assessing whether PQL-related learning objectives are being met, particularly those in the *HM: Physics Habits of Mind* and *ML: Understanding models and their limits in physics* categories. PIQL items are not focused on procedural mathematics or calculations, which are also important in introductory physics and are well-served by meeting the mathematics prerequisites for physics. The conceptual mathematics and quantitative reasoning embodied in the PIQL are a foundation for the mathematics used in introductory physics courses at the college level, and are not typically an outcome of the prerequisite mathematics courses.

Expert-like PQL is firmly rooted in a well-formed conceptual blend of physics concepts and proceptual understanding of precalculus and algebra, as discussed in Sect. 2; therefore, novel PIQL items were developed on the theoretical foundation of Conceptual Blending Theory (Fauconnier & Turner, 2002), as well as Sherin's (2001) symbolic forms. Readers interested in the process of item develop- ment based on these theoretical frameworks should see the journal article describing the PIQL's development and validation (White Brahmia et al., 2021).

Here, we describe three items from the PIQL and relate them to the learning objectives described in Table 2, chosen to exemplify the LOs highlighted in Table 3. The first item was written to probe student understanding of sign and signed quantities, and assesses learning objective HM-1 primarily, in addition to HM- 2, HM-6, and HM-8. The second involves covariational reasoning, and assesses learning objective HM-3 primarily, along with HM-1, HM-2, HM-6, HM-8, and ML-3. The third also involves covariational reasoning, focusing on evaluation of an algebraic limit. It primarily assesses learning objective HM-6, as well as HM-1, HM-8, and ML-3.

The *Electric field* question (see Fig. 5) asks students to determine the meaning of a negative sign associated with a component of a vector quantity. In introductory physics contexts, the most useful and intuitive interpretation of a vector quantity is a geometric interpretation. Students learn that a vector is a quantity with a magnitude and a direction. Therefore, the sign associated with a vector component indicates its *direction* relative to a defined coordinate system. We find, however, that students struggle to make meaning of the sign of vector components that represent unfamiliar quantities (White Brahmia & Boudreaux, 2017). This is especially true for quantities such as electric field, and others related to electromagnetism. We believe that, for many students enrolled in college-level introductory physics courses, a lack of intuition and experience with quantities of electromagnetism, as well as unfamiliarity with the mathematical abstraction of vector fields obscures the meaning of the sign. This is despite the fact that the meaning of the sign of a vector component is understood by students in the more familiar context of mechanics. This question serves

Recall that the electric field is a vector quantity. At a location on the x-axis, the x-component of the electric field is measured to be $E_x = -10$ N/C, and the y and z-components of the field are measured to be zero.

Consider the following statements about this situation. Select the statement(s) that **must be true.** ***Choose all that apply.***

a. The source of the field is a negative charge.
b. At that location of the x-axis, the field is in the negative x-direction.
c. In this field, the motion of any charged particle is opposite to the direction of the field.
d. The magnitude of the electric field is decreasing.
e. The magnitude of the electric field at that location is 10 N/C less than it is at the origin.

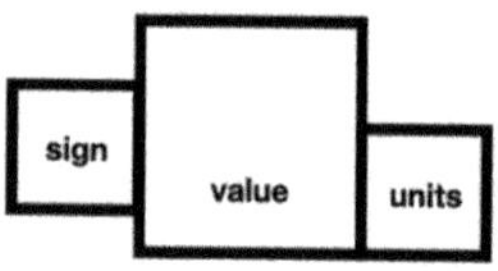

Fig. 5 *Electric field*, PIQL multiple-choice multiple-response item (top) that exemplifies a proceptual understanding of the "sign" aspect of White Brahmia's "quantity" symbolic form (bottom). The correct response is b

as an assessment of PQL-related learning objective HM-1 in particular: students are expected to translate the physical attribute of direction into a symbolic representation using sign.

The *Plant Growth* question, shown in Fig. 6, is an item that assesses students' graphical interpretation and covariational reasoning, and is based on an item from the Precalculus Concept Assessment (Carlson et al., 2010). The item features a graph with time as the independent variable, and growth rate as the dependent variable. Students are asked to compare not the growth rates of the two plants, but the amount of growth of the two plants over the period depicted in the graph. To do this, students could recognize that plant A grows at a faster rate for the entire time shown in the graph, and therefore grows more; or, a student could recognize that the area under each curve represents the accumulated growth of the associated plant. Both of these strategies require students to interpret the quantities depicted in the graph, and how they can use those quantities to compare a third, related quantity. This item is particularly relevant to sequence-level learning objectives HM-3. Students are expected to make sense of graphical representations of quantities: when the independent variable is time, and the dependent variable is a time rate of change of a given quantity, the area under the curve represents the accumulation of that quantity.

The *Fish* item, shown in Fig. 7, is also adapted from the Precalculus Concept Assessment (Carlson et al., 2010), and assesses covariational reasoning in an algebraic (rather than graphical) context. To answer, students need to determine that the expression given for $N(t)$ increases with increasing t, by recognizing that the numerator grows more quickly than the denominator. They must also recognize that answering the item requires a determination of a limit, and determine the value of the limit of the given algebraic expression. One way to determine the value of the

The graph at right represents the *growth rate* vs. *time* for two plants. Which of the following statements best describes the growth of the two plants from $t = 0$ to $t = 1$ month?

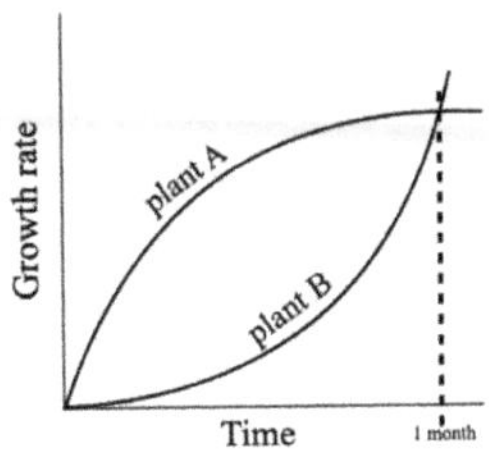

a. Plants A and B have the same amount of growth.
b. Plant A has experienced more growth than plant B.
c. Plant B has experienced more growth than plant A.
d. The graph does not provide enough information to compare the growth of the two plants.

Fig. 6 PIQL item that exemplifies the covariational reasoning used in introductory physics contexts, and understanding of graphical representations of quantitative models. The correct answer is b

The wildlife game commission released 500 fish into a lake. The function $N(t)$ defined by

$$N(t) = \frac{600t + 500}{0.5t + 1}$$

represents the approximate number of fish in the lake as a function of time (in years). Which one of the following best describes how the number of fish in the lake changes over time?

a. The number of fish gets larger each year, but does not exceed 500.
b. The number of fish gets larger each year, but does not exceed 1200.
c. The number of fish gets smaller each year, but does not get smaller than 500.
d. The number of fish gets larger each year, but does not exceed 600.
e. The number of fish gets smaller each year, but does not get smaller than 1200.

Fig. 7 *Fish*, PIQL multiple-choice single-response item, that assesses students' covariational reasoning in an algebraic context. The correct answer is b

limit of the expression is to rewrite the expression as

$$N(t) = \frac{600(t + 5/6)}{0.5(t + 2)} = 1200\frac{t + 5/6}{t + 2}.$$

As t gets large, the fraction approaches 1 from below; thus, as t increases, $N(t)$ approaches 1200 from below. We note that while the wording of the answer choices is such that less rigorous reasoning can be employed to find the correct answer, recognition of the necessity of taking a limit is central to this item. This item is well-aligned with learning objective HM-6: students must recognize the need to consider the limit of an expression for large values of t.

Interestingly, though experts categorize the items on the PIQL as primarily using proportional reasoning, reasoning about sign and signed quantities, or covariational reasoning, both exploratory and confirmatory factor analyses of student responses on the steady-state version indicated that the items on the inventory were not separable into these constructs from the students' perspective. This indicates that, from the students' perspective, the PIQL may assess a single construct (i.e., physics quantitative literacy) and that the three facets of reasoning are deeply interconnected in physics contexts for students. Student interviews as well as targeted psychometric analyses are consistent with this interpretation (White Brahmia et al., 2021). This supports our classification of the PIQL as a *reasoning* inventory, rather than a concept inventory, and that it is an appropriate metric for assessing PQL-related learning objectives, which are focused on reasoning rather than specific mathematical or physical concepts.

5 Conclusion

In this paper we define Physics Quantitative Literacy (PQL) and describe its central role in physics thinking. We operationalize PQL in the context of quantification and modeling, with a focus on covariational reasoning and reasoning about sign and signed quantities. We also demonstrate that PQL is not only central to physics learning, but has a strong overlap with concepts in algebra, and precalculus as well. We then describe our process for developing sequence-level assessable PQL learning objectives for the introductory physics sequence, and present the current version of those objectives. We note that mathematics educators are likely to see overlap with their own learning objectives for algebra and precalculus courses. It is with optimism for this synergistic potential between the disciplines that we include physics learning objectives in this chapter. Lastly, we describe an assessment instrument designed to assess some of these learning objectives, the Physics Inventory of Quantitative Literacy (PIQL), a reliable and valid reasoning inventory that assesses students' physics quantitative literacy as it develops with instruction in introductory physics courses. Results from Classical Test Theory provide evidence for its validity and reliability, and both exploratory and confirmatory factor analyses suggest that it is a single-factor instrument. We interpret the factor analysis results as an indication that the PIQL tests a single construct that we call Physics Quantitative Literacy (PQL). We presented the PIQL as a useful metric for assessing PQL-related learning objectives, and as a step toward establishing metrics for learning objectives for calculus-based introductory physics courses.

Students come to their STEM courses having succeeded in their prerequisite mathematics courses, yet they typically encounter an unfamiliar experience with the mathematics they "know." Many fail to make effective connections with their prior learning experience in order to function in the new one. There is a strong need for a proceptual facility with some of the mathematics from prerequisite courses, as relied on in introductory mathematics-based STEM courses (like physics), to be part of the students' learning progression through these courses.

We conclude by encouraging education researchers and curriculum developers from mathematics and mathematics-based disciplines, like physics, to explore the overlap between our disciplines in the work that we do. We are teaching the same students. Exploring the interface of their course-taking experiences and mutually supporting our collective learning objectives holds potential for symbiotic learning. In addition, collaboration at the interface opens possibilities of realizing new learning outcomes that may even include a more creative and generative approach in both disciplines. We consider the work in this chapter to be one step in that direction.

Acknowledgements We thank Trevor Smith and Philip Eaton for their contributions to the various statistical analyses described in this chapter, Charlotte Zimmerman and Andrew Boudreaux for their contributions to the reasoning frameworks that guided to the discussion of quantification and quantitative modeling, and we thank Chaz Ruggieri and Sean Brennan for their contribution to the development of the physics learning objectives. This work is supported by the National Science Foundation under Grants No. DUE-1832836, DUE-1832880, DUE-1833050, DGE-1762114, and an NRC Research Associateship award at Air Force Research Laboratory.

References

Bajracharya, R. R., Wemyss, T. M., & Thompson, J. R. (2012). Student interpretation of the signs of definite integrals using graphical representations. *AIP Conference Proceedings, 1413*(1), 111–114. Retrieved from https://doi.org/10.1063/1.3680006

Beichner, R. J. (2011, April). The student-centered active learning environment for undergraduate programs (SCALE-UP) project. In *APS New England section spring meeting abstracts* (p. G1-001).

Bishop, J. P., Lamb, L. L., Philipp, R. A., Whitacre, I., Schappelle, B. P., & Lewis, M. L. (2014). Obstacles and affordances for integer reasoning: An analysis of children's thinking and the history of mathematics. *Journal for Research in Mathematics Education, 45*(1), 19–61.

Boudreaux, A., Kanim, S. E., Olsho, A., Brahmia, S. W., Zimmerman, C., & Smith, T. I. (2020). Toward a framework for the natures of proportional reasoning in introductory physics. In *Proceedings of the annual physics education research conference* (pp. 45–50).

Brewe, E. (2008). Modeling theory applied: Modeling instruction in introductory physics. *American Journal of Physics, 76*(12), 1155–1160.

Canright, J. P., & White Brahmia, S. (2021, August 4–5). Developing expertlike epistemologies about physics empirical discovery using virtual reality. In *Physics education research conference 2021* (pp. 75–80). Virtual conference.

Carlson, M., Jacobs, S., Coe, E., Larsen, S., & Hsu, E. (2002). Applying co-variational reasoning while modeling dynamic events: A framework and a study. *Journal for Research in Mathematics Education, 33*(5), 352–378. https://doi.org/10.2307/4149958

Carlson, M., Oehrtman, M., & Engelke, N. (2010). The precalculus concept assessment: A tool for assessing students' reasoning abilities and understandings. *Cognition and Instruction, 28*(2), 113–145.

Ceuppens, S., Bollen, L., Deprez, J., Dehaene, W., & De Cock, M. (2019). 9th grade students' understanding and strategies when solving x (t) problems in 1D kinematics and y (x) problems in mathematics. *Physical Review Physics Education Research, 15*(1), 10101.

Chasteen, S., Perkins, K., Beale, P., Pollock, S., & Wieman, C. (2011). A thoughtful approach to instruction: Course transformation for the rest of us. *Journal of College Science Teaching, 40*(4), 24–30.

Chi, M. T., Feltovich, P. J., & Glaser, R. (1981). Categorization and representation of physics problems by experts and novices. *Cognitive Science, 5*(2), 121–152.

College Board (2020). *AP physics 1: Course and exam description.* https://apcentral.collegeboard.org/. Accessed 2021-01-13.

Dorko, A., & Speer, N. (2015). Calculus students' understanding of area and volume units. *Investigations in Mathematics Learning, 8*(1), 23–46. https://doi.org/10.1080/24727466.2015.11790346

Eichenlaub, M., & Redish, E. (2019). Blending physical knowledge with mathematical form in physics problem solving. In *Mathematics in physics education* (pp. 127–151). Springer International Publishing. https://doi.org/10.1007/978-3-030-04627-9_6

Ellis, A. B. (2007). The influence of reasoning with emergent quantities on students' generalizations. *Cognition and Instruction, 25*(4), 439–478. https://doi.org/10.1080/07370000701632397

Epstein, J. (2006). *The calculus concept inventory* (pp. 60–67). National STEM Assessment.

Eriksson, M., Linder, C., & Eriksson, U. (2018). Towards understanding learning challenges involving sign convention in introductory level kinematics. *Physics education research conference (PERC)*.

Etkina, E. (2015). Millikan award lecture: Students of physics—Listeners, observers, or collaborative participants in physics scientific practices? *American Journal of Physics, 83*(8), 669–679. https://doi.org/10.1119/1.4923432

Etkina, E., Van Heuvelen, A., White-Brahmia, S., Brookes, D. T., Gentile, M., Murthy, S., & Warren, A. (2006). Scientific abilities and their assessment. *Physical Review Special Topics-Physics Education Research, 2*(2), 020103.

Fauconnier, G., & Turner, M. (2002). *The way we think: Conceptual blending and the mind's hidden complexities*. Basic Books.

Gallardo, A., & Rojano, T. (1994). School algebra—Syntactic difficulties in the operativity. In *Proceedings of the XVI international group for the psychology of mathematics education, North American chapter* (Vol. 1, pp. 265–272).

Gillman, R. (2005). *Sigmaa on quantitative literacy introduction.* http://sigmaa.maa.org/ql/about.php

Gray, E. M. & Tall, D.O. (1994). *Duality, ambiguity, and flexibility: A "proceptual" view of simple arithmetic* (Vol. 25, Tech. Rep. No. 2). Retrieved from https://www.jstor.org/stable/749505

Hayes, K., & Wittmann, M. C. (2010). The role of sign in students' modeling of scalar equations. *The Physics Teacher, 28*(4), 246–249.

Hestenes, D., Wells, M., & Swackhamer, G. (1992). Force concept inventory. *The Physics Teacher, 30*(3), 141–158.

Hobson, N. L. F., & Moore, K. C. (2017). Exploring experts' covariational reasoning. In K. Moore & P. Thompson (Eds.), *20th annual conference on research in undergraduate mathematics education* (pp. 664–672).

Huynh, T., & Sayre, E. C. (2018). *Blending mathematical and physical negativeness.* arXiv:1803.01447

Kozminski, J., Lewandowski, H., Beverly, N., Lindaas, S., Deardorff, D., Reagan, A., Dietz, R., Tagg, R., EblenZayas, M., Williams, J., & Hobbs, R. (2014). *AAPT recommendations for the undergraduate physics laboratory curriculum.* AAPT.

LGBT+ Physicists (2013). *Supporting LGBT+ physicists: A best practices guide for departments.* https://lgbtphysicists.org/files/BestPracticesGuide.pdf. Accessed January 13, 2021.

Moore, K. C. (2014). Quantitative reasoning and the sine function: The case of Zac. *Journal for Research in Mathematics Education, 45*(1), 102–138.

NSTA (n.d.). *Access the next generation science standards by topic.* https://ngss.nsta.org/AccessStandardsByTopic.aspx. Accessed January 13, 2021.

Nunes, T. (1993). Learning mathematics: Perspectives from everyday life. In *Schools, mathematics, and the world of reality* (pp. 61–78).

Olsho, A., White Brahmia, S., Smith, T., & Boudreaux, A. (2021). When negative is not "less than zero": Electric charge as a signed quantity. *The Physics Teacher, 59*(4), 253–256.

Redish, E. (2021). Using math in physics: 4. Toy models. *The Physics Teacher*, *59*(9), 683–688.

Redish, E., & Kuo, E. (2015). Language of physics, language of math: Disciplinary culture and dynamic epistemology. *Science & Education, 24*(5–6), 561–590.

Rodriguez, J.-M. G., Bain, K., & Towns, M. H. (2019). Graphical forms: The adaptation of Sherin's symbolic forms for the analysis of graphical reasoning across disciplines. *International Journal of Science and Mathematics Education*, 1–17.

Saldanha, L. A., & Thompson, P. (1998). Re-thinking co-variation from a quantitative perspective: Simultaneous continuous variation. In S. B. Berenson & W. N. Coulombe (Eds.), *Annual meeting of the psychology of mathematics education—North America* (Vol. 1, pp. 298–304). North Carolina State University.

Schoenfeld, A. H., & Herrmann, D. J. (1982). Problem perception and knowledge structure in expert and novice mathematical problem solvers. *Journal of Experimental Psychology: Learning, Memory, and Cognition, 8*(5), 484.

Sealey, V., & Thompson, J. (2016). Students' interpretation and justification of "backward" definite integrals. In *Proceedings of the 19th annual conference on RUME.*

SEI. *Science education initiative at University of Colorado, Boulder*. http://www.colorado.edu/sei/fac-resources/learning-goals.html

SEI. *Carl Wieman science education initiative at the University of British Columbia.* https://cwsei.ubc.ca/resources/instructor/learninggoals

Sherin, B. L. (2001). How students understand physics equations. *Cognition and Instruction, 19*(4), 479–541.

Tall, D. (2008). The transition to formal thinking in mathematics. *Mathematics Education Research Journal, 20*(2), 5–24.

Thompson, P., & Dreyfus, T. (1988). Integers as transformations. *Journal for Research in Mathematics Education*, 115–133.

Thompson, P. (2011). Quantitative reasoning and mathematical modeling. In S. A. Chamberlain, L. L. Hatfield, & S. Belbase (Eds.), *New perspectives and directions for collaborative research in mathematics education: Papers from a planning conference for WISDOM^E* (p. 33).

Thompson, P. (1994). Images of rate and operational understanding of the fundamental theorem of calculus. *Educational Studies in Mathematics*. https://doi.org/10.1007/BF01273664

Thompson, P., Carlson, M. P., Byerley, C., & Hatfield, N. (2014). Schemes for thinking with magnitudes: A hypothesis about foundational reasoning abilities in algebra 1, 2, 3. *Epistemic Algebraic Students: Emerging Models of Students' Algebraic Knowing, 4*, 1–24.

Thompson, P., & Saldanha, L. (2003). Fractions and multiplicative reasoning. In J. Kilpatrick, W. G. Martin, & D. Schifter (Eds.), *Research companion to principles and standards for school mathematics* (pp. 95–113). National Council of Teachers of Mathematics.

Thornton, R. K., & Sokoloff, D. R. (1998). Assessing student learning of Newton's laws: The force and motion conceptual evaluation and the evaluation of active learning laboratory and lecture curricula. *American Journal of Physics, 66*(4), 338–352.

Torigoe, E. T., & Gladding, G. E. (2011). Connecting symbolic difficulties with failure in physics. *American Journal of Physics, 79*(1), 133–140.

Vlassis, J. (2004). Making sense of the minus sign or becoming flexible in "negativity." *Learning and Instruction, 14*(5), 469–484. https://doi.org/10.1016/j.learninstruc.2004.06.012

White Brahmia, S., & Boudreaux, A. (2016). Exploring student understanding of the negative sign in introductory physics contexts. In *Proceedings of the 19th annual conference on research in undergraduate mathematics education.*

White Brahmia, S., & Boudreaux, A. (2017). Signed quantities: Mathematics based majors struggle to make meaning. In *Proceedings of the 20th annual conference on research in undergraduate mathematics education.*

White Brahmia, S. (2019). Quantification and its importance to modeling in introductory physics. *European Journal of Physics, 40*(4), 044001.

White Brahmia, S., Olsho, A., Smith, T. I., & Boudreaux, A. (2020). Framework for the natures of negativity in introductory physics. *Physical Review Physics Education Research, 16*(1), 010120. https://doi.org/10.1103/PhysRevPhysEducRes.16.010120

White Brahmia, S., Olsho, A., Smith, T. I., Boudreaux, A., Eaton, P., & Zimmerman, C. (2021). Physics inventory of quantitative literacy: A tool for assessing mathematical reasoning in introductory physics. *Physical Review Physics Education Research, 17*(2), 020129.
Zimmerman, C., Olsho, A., Boudreaux, A., Smith, T. I., Eaton, P., & White Brahmia, S. (2020a). Exploring student facility with "goes like" reasoning in introductory physics. In S. Wolf, M. Bennett, & B. Frank (Eds.), *Physics education research conference 2020* (pp. 605–610). Virtual conference. https://doi.org/10.1119/perc.2020.pr.Zimmerman
Zimmerman, C., Olsho, A., Loverude, M. E., & White Brahmia, S. (2020b). Identifying covariational reasoning behaviors in expert physicists in graphing tasks. In *Research in undergraduate mathematics education conference proceedings 2020.* https://doi.org/10.5642/jhummath.201602.17

www.ingramcontent.com/pod-product-compliance
Ingram Content Group UK Ltd.
Pitfield, Milton Keynes, MK11 3LW, UK
UKHW020704041026
15590UKWH00001B/471
* 9 7 8 3 0 3 1 1 4 5 5 5 1 *